Evolution from a Thermodynamic Perspective

Carl F Jordan

Evolution from a Thermodynamic Perspective

Implications for Species Conservation and Agricultural Sustainability

Springer

Carl F Jordan
Odum School of Ecology
University of Georgia
Athens, GA, USA

ISBN 978-3-030-85185-9 ISBN 978-3-030-85186-6 (eBook)
https://doi.org/10.1007/978-3-030-85186-6

This Springer imprint is published by the registered company Springer Nature Switzerland AG
The registered company address is: Gewerbestrasse 11, 6330 Cham, Switzerland

For Sybille, inspired by a love of nature

Preface

The Thermodynamic Theater and an Evolutionary Play:

Species are the Players, the Theme is Power

In G. Evelyn Hutchinson's *The Ecological Theater and the Evolutionary Play*, individual species are the players and the theater ranges from a mountain top in Kenya to spruce forests in Maine. In the metaphor, the ecological theater is a matrix of geology, climate, and species, the environment within which the evolutionary play takes place. The plot is adaptation to the environment. The role of each species in the play is defined by its "niche," a particular subdivision of the matrix. Although Hutchinson recognized the role of energy flow in organizing trophic levels in ecosystems, the thermodynamic perspective developed by his student H.T. Odum brought energy flow to the forefront of ecological theory. Energy flow through natural systems has been a basic concept in the development of biophysical theory applied to the origin of life. It is central to models of ecosystems and to the thermodynamic perspective of ecology and evolution.

The perspective can be imagined as a drama with Darwin's finches as actors in an evolutionary play performed in the thermodynamic theater of the Galápagos Islands. Because of the isolation of these islands, species were few and trophic webs were simple compared to those of South America, the probable origin of the finches. When the ancestral birds first arrived, they encountered seeds that offered abundant and unexploited sources of energy. Following the original colonization, some subgroups emigrated to other islands due to an expanding population and resultant increased competition for food. In these other islands, they found seeds of different sizes. The finches' beaks may not have been well adapted to this new source of energy but through several generations of natural selection, those offspring with beaks best suited to this new energy source were those that survived and reproduced most successfully.

Recently, in the Galápagos, there has been an invasion of highly competitive alien plant species, blackberry, guava, and lantana. Seed dispersal by finches increased the range and seed germination of these species. What happens next depends on random improvisations in the play. One might be that Galápagos hawks switch their feeding preference to finches from iguanas that feast on grubs. A resultant increase in iguanas means there are fewer grubs that develop into beetles that consume the blackberries. As a result, the highly competitive blackberries thrive. That is only one of many possibilities. The evolution of populations and their interactions is unpredictable or stochastic because of the extremely high number of combinations that are possible, and the extreme sensitivity of any combination to small changes in the environment or in species behavior. In contrast, the thermodynamic fate of the undisturbed ecosystem is deterministic. Regardless of the species interactions involved, the power output of the Galápagos Islands' ecosystems will increase as food webs expand and energy flow increases.

Athens, GA, USA Carl F Jordan

Acknowledgments

As a graduate student in 1964, I read H.T. Odum's paper "Time's speed regulator: The optimum efficiency for maximum power output in physical and biological systems," but I did not understand it. Even today, I suspect that few ecologists understand the title and its implications. It took me a year working with Tom Odum in the Puerto Rican rain forest and 40 years of professional experience to fully realize its importance for evolutionary theory. Odum's personal energy and his enthusiasm has been the inspiration for me to write this interpretation not only for its implications for evolutionary theory but also for conservation of species diversity and agricultural sustainability.

E.P. "Gene" Odum and Frank Golley supported my efforts to use the systems perspective with my classes and graduate students at the Institute (now School) of Ecology, University of Georgia.

John Tedrow, my professor of soil morphology while I was in graduate school at Rutgers University, encouraged me to the pursue the concept of biogeochemical cycling that in 1963 was a leading-edge concept.

To these mentors, I am grateful.

Acknowledgments

Contents

Part III Conservation

Abbreviations

AID	Agency for International Development
ANOVA	Analysis of Variance
ATP	Adenosine Triphosphate
BDFF	Biological Dynamics of Forest Fragments
CATIE	Centro Agronómico Tropical de Investigación y Ensañanza
Cm	Centimeter
C/N	Carbon-Nitrogen ratio
C:N:P	Carbon:Nitrogen:Phosphorus
DNA	Deoxyribonucleic acid
Eco-Evo-Devo	Ecological Evolutionary Developmental Biology
EJ	Exajoule (10^{18}joules)
EROI	Energy Returned on (energy) Invested
FAO	Food and Agriculture Organization
FIV	Feline Immunodeficiency Virus
GDP	Gross Domestic Product
HIV	Human Immunodeficiency Virus
IPBES	The Intergovernmental Science-Policy Platform on Biodiversity and Ecosystem Services
IPCC	The Intergovernmental Panel on Climate Change
K-T	The Cretaceous–Tertiary (K–T) extinction,
LAI	Leaf Area Index
LMEP	Law of Maximum Entropy Production
LTER	Long Term Ecological Research
MPH	Miles per Hour (=1.6 Kilometers per Hour)
MRTS	Marginal Rate of Technical Substitution
NEA	Net Energy Analysis
NPP	Net Primary Productivity
Pg	Gross Photosynthesis
P/R	Photosynthesis-Respiration ratio

R	Respiration
SARE	Sustainable Agriculture Research and Education
UNESCO	United Nations Educational, Scientific and Cultural Organization
USDA	United States Dept. of Agriculture

Part I
Theory

To Understand Economics, Follow the Money; To Understand Ecosystems, Follow the Energy

In an economic system, information in the form of money flows in one direction and goods and services flow in the other. To understand why people do what they do, follow the money. The flow of money keeps the system organized, although organization of the economic system is not the reason people follow the money. Organization is just an inevitable consequence, the result of workings of the invisible hand of the economic marketplace as people pursue their quest for goods. In ecosystems, energy flow is in one direction, sensory information in the other. The flow of energy keeps the system organized, although organization is not the reason species follow energy. System organization is just an inevitable consequence, the result of workings of the invisible hand of the thermodynamic marketplace as species pursue their quest for energy.

Chapter 1
Two Views of Ecology, Evolution, and Conservation

1.1 Why I Wrote this Book

Theodore Roosevelt came to the Territory of Dakota in 1883 and "took hold of two cattle ranches". The objective it seems, was to test his own mettle. "In that land we led a free and hardy life, with horse and with rifle. We worked under the scorching midsummer sun, when the wide plains shimmered and wavered in the heat; and we knew the freezing misery of riding night guard round the cattle in the late fall round-up. In the winter we rode through blinding blizzards, when the driven snow-dust burned our face. We knew toil and hardship and hunger and thirst" (Roosevelt 1913). Roosevelt also sought a chance to hunt the big game of North America before they disappeared, but soon he realized that the decimation of bison, and the eradication of elk, bighorn sheep, deer and other game species was the loss of a magnificent heritage. He saw the effects of overgrazing, and suffered the loss of his ranches because of it. While many still considered natural resources inexhaustible, Roosevelt would write: "We have become great because of the lavish use of our resources. But the time has come to inquire seriously what will happen when our forests are gone, when the coal, the iron, the oil, and the gas are exhausted, when the soils have still further impoverished and washed into the streams, polluting the rivers, denuding the fields and obstructing navigation" (Qualtieri 2019). Conservation increasingly became one of Roosevelt's main concerns. After becoming president in 1901, Roosevelt used his authority to protect approximately 230 million acres of public land. (National Park Service 2019). He recognized that to save species, their ecosystem had to be protected.

Since Roosevelt's time there has been a series of laws aimed at protecting wildlife species, including the Lacey Act of 1900 which bans illegal trafficking of wildlife, the Migratory Bird Conservation act of 1929, a 1937 treaty prohibiting the hunting of right and gray whales, the Bald Eagle Protection Act of 1940, and the Endangered Species Preservation Act of 1966. None of these however, brought as

C. F. Jordan, *Evolution from a Thermodynamic Perspective*,
https://doi.org/10.1007/978-3-030-85186-6_1

much national attention to the loss of species as did "Silent Spring" by Rachel Carson (1962). The book prompted meetings of concerned scientists in the U.S. to provide recommendations to Congress that was formulating an act to deal with species loss. I was a participant at one of these meetings at the University of Michigan in 1971. The panel was dominated by evolutionary biologists, population ecologists and community ecologists and the focus was on species populations and their reaction to disturbance and pollution. As the only ecosystem ecologist present, I argued that we should also consider ecosystems, as species cannot exist outside of their place in an ecosystem's trophic structure. I was taken aback by comments indicating doubt that ecosystems even existed. The result of that panel, and many other similar discussions was that threatened species could best be protected by endangered species legislation. Consequently the "Endangered Species Act", was passed in 1973. (Environmental Protection Agency 2019).

Box 1.1
While the Endangered Species Act requires restoration of endangered species' habitat, "habitat" is not the same thing as "ecosystem". Habitat restoration means re-establishing the place or environment where a plant or animal naturally or normally lives and grows. That is not enough. Ecosystem restoration is necessary. Ecosystem restoration means re-establishing the feedback flows of energy and nutrients that ensure stability of the trophic-dynamic structure within which each species exists. It requires a dynamic equilibrium between autotrophs, herbivores, carnivores, and decomposers.

The first backlash against the Act occurred in 1973 when the snail darter (*Percina tanasi*) that inhabited the Little Tennessee River was recognized as an endangered species. As a result, construction of a proposed Tellico dam was delayed due to a lawsuit that eventually reached the Supreme Court, where protection for the darter was affirmed. The supposed economic benefits of the dam however, were deemed more important by Congress than the esthetic, ecological, educational, recreational, and scientific value of the darter. As a result, Congress passed an amendment to the Act that excluded the Tellico project from the Endangered Species Act (Department of Justice 2019). Plater (2013) has provided a documented history of this episode in environmental law. He criticized the national media because supporters of the darter were labeled by the press as "hypocrites as well as fools, environmental extremists manipulating the darter to misuse the law over a technicality". Once newspapers and television news defined the case as an insignificant fish versus an economically important dam project, the environmental case was doomed. The reaction against environmental advocates might not have been so vituperative had the law been focused on an endangered ecosystem, the Little Tennessee River that provided recreation and an unimpeded waterway for boaters and for trout, a cool-water species (Peeples 2015). Those reasons could have resonated with the public. Nevertheless, an Endangered Ecosystem (instead of Species) Act probably could not have stopped the Tellico Dam, but it might have prevented the invective heaped on the ecological community for defending what was viewed as an inconsequential organism.

Ecologists got more bashing 17 years later as a result of the northern spotted owl controversy. The owl's ecosystem is old-growth forests of the Pacific Northwest (Gruell et al. 1999) and the listing of the bird effectively stopped logging and reduced employment in parts of the region. Several federal agencies published an economic report that estimated that communities in the Pacific Northwest would lose 25,000 jobs by 1995 as a result of reduced harvest volumes. It claimed that there would be an increase in suicide, violence, divorce, social upheaval, and school closings (Forest History Society 2019), all because of the northern spotted owl that was merely a variety of a more southern species. The controversy re-emerged in 2020 when disastrous fires broke out in the old-growth forests of Oregon that are the owl's habitat. Thinning the forest is critical for reducing combustible fuel and lowering the risk of wildfires growing and spreading quickly (Carlton and Elinson 2020). Thinning is not a good option though because loggers like to take the big trees and leave the saplings that are preeminent in spreading the fires. Intentionally igniting low intensity fires is another way to clear the underbrush and slash so that fires cannot jump into the canopy of the larger trees. In pre-colonial times, lightning strikes periodically set fire to the forest and kept the understory open, but Smokey the Bear's Twentieth Century call to prevent fires of any size has led to the accumulation of slash that facilitates cataclysmic fires (Schiffman 2020). Carrying out periodic controlled burns and preserving the ecosystem would be a better way to conserve the owl.

The snail darter debacle compelled me to write the article "Do Ecosystems Exist? (Jordan 1981). I felt it would be easier to convince people that species could better be preserved if the ecosystem were to be the basis of conservation efforts. However the only defense I could muster for the ecosystem concept was a description of the *Heliconius-Passiflora-Anguria/Gurania* complex of tropical rain forests described by Gilbert (1973). It consists of complicated functional relationships which seem too improbable to evolve except in the context of energy flow and feedback through an ecosystem.

The *Heliconius* sp. Butterfly lays its eggs only on a species of *Passiflora* sp., a vine which has chemical defenses which are effective against most larvae but not that of *Heliconius*.

1. The chemical defense of the plant appears to be incorporated into the Heliconius larvae and to act as a chemical deterrent against predators of *Heliconius*.
2. The *Passiflora* vines have extrafloral nectar glands which attract ants and wasps that parasitize eggs laid on the Passiflora.
3. However, female *Heliconius* inspect *Passiflora* plants before ovipositing to ensure that no ants are present.
4. Females return to the same *Passiflora* vines regularly over a period of up to 6 months. Eggs are laid only when other eggs are not present, presumably to guarantee an adequate food supply for the eggs to be laid.

5. The extended period of egg laying is the result of the nutrition which the butterfly gets from the pollen of two other genera of vines, *Anguria* sp. and *Gurania* sp.
6. An extended period of nutrition is the result of the continuous male flowering of *Anguria* and *Gurania.*
7. Continuous male flowering of *Anguria* and *Gurania* is advantageous because of the sporadic appearance of female flowers of these genera.
8. The continuous male flowering of *Anguria* and *Gurania* combined with the long life and regular flower visits by *Heliconius* results in high probability of female pollination, despite the rarity of female flowering.

My defense was flawed because feedback develops at the population level as well as the ecosystem level. The feedback between species that resulted in co-evolution of the *Heliconius-Passiflora-Anguria/Gurania* complex is not evidence for the existence of ecosystems. Co-evolution of animals and plants as described in the first International Congress of Systematics and Evolutionary Biology (Gilbert and Raven 1973) could have evolved simply from feedback between species.

1.1.1 Dualities Still Impede Conservation Efforts

Since I wrote that article in 1981 there have been tremendous gains in understanding ecosystem function and evolution, particularly among ecosystem modelers who have used a thermodynamic perspective (Nielsen et al. 2020). Nevertheless, there still is a great divide between ecologists that take the "top-down" (ecosystem) view of control and those that favor the "bottom-up" (population) approach. Ecosystem ecologists love species as much as population ecologists do, but species cannot exist without an ecosystem (Di Marco et al. 2019). Focusing on endangered ecosystems may be a more effective way to save species because it also saves their ecosystem as well as important services of nature. Take for example the mangroves of the Indian Ocean. They are ecosystems that support a critically endangered species *Bruguiera hainesii* (Ono et al. 2016). Mangroves have important functions such as protecting shoreline and providing habitat for juveniles of many commercially important fish species. (Whitmore et al. 2005), but in Asia, mangroves are being cleared to establish shrimp farms. Protecting the mangrove ecosystem would protect the embedded services of nature as well as endangered species.

1.2 The Intergovernmental Science-Policy Platform of Biodiversity and Ecosystem Services (IPBES)

Species diversity, in the oceans and on land, is decreasing more rapidly than at any time in the history of the Earth (World Ocean Review 2010, United Nations 2019). There is no disagreement among scientists and conservationists that extraordinary action must be taken to halt these extinctions that undermine human well-being (Nature News 2018). An important step in this direction was the Intergovernmental Science-Policy Platform of Biodiversity and Ecosystem Services (IPBES) formed in 2012, patterned after the Nobel-winning Intergovernmental Panel on Climate Change (IPCC). The IPCC has been successful in bringing attention to threats to the world's economy resulting from a predicted higher average global temperature, a rising sea level, and increased severity of storms (Union of Concerned Scientists 2018a). While there is some disagreement as to whether these changes are happening or will happen, there is little dispute that if these changes do occur, it will have a huge impact on agriculture and other resource systems, and the global economy in general. It is important to take action to reduce atmospheric pollutants that could cause climate change (Union of Concerned Scientists 2018b). It is due to the unanimity of opinion of IPCC scientists as to the dangers of climate change that their warnings are being taken seriously. The problem for the IPBES was that little agreement has occurred among scientists and other representatives of the 129 member governments as to why biodiversity is important and why extinctions pose a threat to global welfare (Masood 2018). Deep misunderstandings and divisions still exist between population and ecosystem ecologists (Nature Editorial 2019).

The ecosystem view focuses on how interactions between species are functional links that stabilize ecosystems by controlling flows of energy and nutrients (Costanza et al. 2017). It is a view that is appealing to politicians and government officials because these interactions can be viewed as "services of nature" that benefit the economy (Jordan 1995; Daily 1997; Daily et al. 1997) and to which a monetary value can theoretically be assigned. This approach has been an anathema for population ecologists because it appears indifferent about which species provide the functional links. Their position is that the value of biodiversity lies in its cultural importance (Díaz et al. 2018). It depends on the existence of particular species that are important for health and nutrition, and for religious and ceremonial purposes. Their existence, like life, cannot be given an economic value.

Indigenous peoples use wild plants for religious, ceremonial, recreational, and medicinal purposes including birth control, and for stimulation and relaxation. There are over 1000 known plant species that produce hallucinogenic effects (Schultes et al. 1992). Curare, used to make poison arrows is made from a variety of plants from the Amazon region. Sour manioc (*Manihot esculenta*), an important source of carbohydrate for humans in tropical regions, produces hydrocyanic acid that protects the roots and leaves from attack by insects and bacteria (Montagnini and Jordan 1983). It allows the roots to be stored in the ground until it is convenient to harvest them. Biodiversity has values, not only to indigenous populations but also to dominant cultures. Many medicinal plants are still used in industrialized societies. Quinine is an anti-malarial drug derived from the bark of the tropical cinchona tree used in developed countries for the treatment of severe malaria when other drugs are ineffective (Achan et al. 2011).

1.2.1 *Targets for Conservation*

Since 2010, targets for conserving species have shaped policy and galvanized efforts to halt species loss worldwide. Specific targets such as doubling the number of tigers in the wild have been set by the World Wildlife Foundation (2020). Subtargets have been set by the Convention on Biological Diversity (2020) such as conservation of genetic diversity. Although the International Union for the Conservation of Nature have published criteria for a "Red List" of Endangered Ecosystems (Oskin 2013), no targets exist for ecosystem conservation, despite the wealth of evidence showing that their health and functions are essential to the processes that maintain all life (Watson et al. 2020). At present, major efforts for species conservation are based on the abundance of a few conspicuous and charismatic species such as tigers, pandas, and whales. The problem here is you have to save the ecosystem to save the species. Species cannot survive unless their support systems survive, a support system that includes the species they feed upon, and the energy and nutrients that support the fed-upon species. Jørgensen et al. (2016) have proposed ecosystem-based management called "resilience-based" management appropriate for fisheries in marine and coastal ecosystems. It shifts focus from the conservation of targeted species, frequently commercially important ones, to the active management of functional groups supporting critical processes and the maintenance of ecosystem services (Hughes et al. 2005). It emphasizes the importance of species interactions and ecological roles including that of humans in sustaining ecosystem resilience.

1.3 Evolving Objectives

I began this book by researching papers that would help defend the ecosystem view as a complementary approach for conserving species. It soon became evident that the twentieth century view of ecosystems was becoming outdated. Recent biophysical theory on the origin of life, and the development of cybernetic models of ecosystems have advanced what can be called the thermodynamic perspective of ecology and evolution. This is not a new idea. It goes back to the writings of Ludwig Boltzmann in the late nineteenth century, Schrödinger's 1944 volume *What is Life?*, and the theories of maximum power developed by Alfred Lotka and H.T. Odum in the twentieth century. The idea is that the complexity of life that we see today could not have developed based on selection of species only for their ability to propagate heritable characteristics which affect their reproductive success. Reproductive ability is the basis for selection in Neo-Darwinian theory, but there is nothing inherent in reproductive success alone that would necessarily result in higher forms of life. Reproductive success alone leads only to more of the same. Organisms need no more than one cell to survive, so there is no evolutionary advantage to multicellularity if survival is the only characteristic for which organisms are selected.

Higher forms of life could only have evolved through selection for a characteristic that gives rise to multi-cellularity. Thermodynamic theory posits that the characteristic needed for multi-cellularity is power output. Increased power output gives a competitive advantage to multi-cellular organisms for acquiring yet more energy in a deviation amplifying feedback cycle.

For the past several decades, a group of ecosystem modelers have been developing a series of "working hypotheses" that explain how ecosystems function from a thermodynamic systems perspective (Nielsen et al. 2020). The hypotheses have been influenced by theories of The Physics of Living Systems Group at the Massachusetts Institute of Technology whose mission is to decipher the physical principles of biological organization (MIT 2020). They have concluded that life must invariably be fueled by the "production of entropy" (England 2020) through an increase in complexity and organization of species and ecosystems. Life produces entropy by borrowing energy from the global entropic gradient in which solar energy is dissipated into heat as it re-radiates from the earth. As the energy borrowed from this gradient increases, power output of species and ecosystems increases. Over evolutionary time, species and ecosystems augment their stored entropy reciprocally and simultaneously by way of feedback that results in greater complexity of structure and function.

I found that looking at the origin of ecosystems and their subsequent evolutionary development leads to a more challenging and exciting view of conservation than a simple defense of the ecosystem concept. The challenge begins by first looking at the mechanisms that underlie the simultaneous and reciprocal control of species and ecosystems, and showing how that control affects both population and ecosystem dynamics. Then it looks back in evolutionary time to the very first ecosystems, shows how thermodynamic principles have guided their development over 2 billion years, and explains why "storage of entropy" is a powerful natural force that ensured the thermodynamic flow of energy through ecosystems remained intact despite five major global extinctions. The challenge concludes by showing how storage of entropy underlies conservation of species.

1.3.1 Literature Review

It is easy to write an academic book about an esoteric subject, because the audience already is familiar with the paradigm and its esoteric jargon. More challenging is to write a book for a general audience or a textbook on an esoteric subject, but one does it because the subject is important outside the narrow bounds of academia. There are many books about Neo-Darwinian evolution. Some are directed towards specialists, others toward a more general audience. Fewer books have been written about evolution from a thermodynamic perspective. "Entropy, Information, and Evolution" is a specialized series of papers edited by Weber et al. (1988) published because "with the advent of nonequilibrium thermodynamics, the time seemed ripe to explore the inter-relations between thermodynamics and evolutionary biology"

(Hull 1988). "Evolution as Entropy" (Brooks and Wiley 1986) was criticized by Collier (1986) who said that it confused biological information with thermodynamic entropy, but if life can be viewed as stored entropy (England 2013, Horowitz and England (2017)), Brooks and Wiley may have been correct. H.T. Odum's "Environment, Power and Society" (Odum 1971) touches on ecosystem evolution, but his energy circuit language is highly abstract and the symbols are difficult to interpret relative to what ecologists see in the field. Moore and deRuiter (2012) rely on mathematical formulations that take a picture of ecosystems at one instant in time but do not capture the dynamic nature of evolution. Morowitz (1968), and Kauffman (2000) focused primarily on how feedback systems operate to increase organization. "A New Ecology Systems Perspective" (Nielsen et al. 2020) is the current definitive publication for thermodynamic ecology, and while it does touch on applied aspects, it speaks only to academic theoreticians. There is only one book that interprets evolution from a thermodynamic perspective in a non-esoteric style, namely "Into the Cool" by Schneider and Sagan (2005). It has been criticized by Bahar (2008) for being hyperbolic, by Corning (2015) for being teleological and by Farmer (2005) who complained that no mechanisms were proposed for selection of traits that led to evolution of higher forms of life. However, these criticisms do not diminish the importance of the theory.

1.3.2 Updating Ecosystem Ecology

Eugene Odum's "Fundamentals of Ecology" (Odum 1959) introduced ecosystem ecology in a way that was understandable beyond the limits of academia. It has had a major impact on our approach to conservation and sustainability, but recent advances in thermodynamic understanding of ecosystem function means that ecosystem ecology needs an update. The update must include theory about the thermodynamic genesis of ecosystems and their evolution. The theory lays the foundation for understanding the role of feedback for conservation of species diversity and sustainability of natural resources.

References

J. Achan, A.O. Talisuna, A. Erhart, Quinine, an old anti-malarial drug in a modern world: Role in the treatment of malaria. Malar. J. **10**(144) (2011)

S. Bahar, Review: Into the Cool. Reports of the National Center for Science Education 25 (5–6), (2008). https://ncse.ngo/review-cool. Accessed 7 Dec 2020

D.R. Brooks, W.O. Wiley, *Evolution as Entropy: Toward a Unified Theory of Biology* (University of Chicago Press, Chicago, 1986)

J. Carlton, Z. Elinson, Owl protections questioned as blazes rage. Wall. Str. J. **2020**, A3 (2020)

R. Carson, *Silent Spring* (Houghton Mifflin, Boston, 1962)

J. Collier, Entropy in evolution. Biol. Philos. **1**, 5–24 (1986)

Convention on Biological Diversity, (2020). https://www.cbd.int/2010-target/goals-targets.shtml. Accessed 14 Dec 2020

P. Corning, Review of "Into the Cool". Institute for the Study of Complex Systems. Review withdrawn in 2019 (2015)

R. Costanza, R. de Groot, L. Braat, I. Kubiszewsk, L. Floramonti, P. Sutto, S. Farber, M. Grasso, Twenty years of ecosystem services: How far have we come and how far do we still need to go? Ecosyst. Serv. **28**(A), 1–16 (2017)

G.C. Daily, *Nature's Services: Societal Dependence on Natural Ecosystems* (Island Press, Washington, DC, 1997)

G.C. Daily, S. Alexander, P.R. Ehrlich, L. Goulder, J. Lubchenco, P.A. Matson, H.A. Mooney, S. Postel, S.H. Schneider, D. Tilman, G.M. Woodwell, *Ecosystem Services: Benefits Supplied to Human Societies by Natural Ecosystems. Issues in Ecology 2* (Ecological Society of America, 1997)

Department of Justice, Tennessee Valley Authority Vs. Hill. (2019), https://www.justice.gov/enrd/tennessee-valley-authority-v-hill. Accessed 17 Dec 2020

M. Di Marco, S. Ferrier, T.C. Harwood, A.J. Hoskins, J.E.M. Watson, Wilderness areas halve the extinction risk of terrestrial biodiversity. Nature **573**, 582–585 (2019)

Díaz S, Pascual U, Stenseke M and 30 others (2018) Assessing nature's contributions to people. Science 359: (6373) 270–272

J.L. England, Statistical physics of self-replication. J. Chem. Phys. **139**, 121923 (2013). https://doi.org/10.1063/1.4818538. Accessed 17 Dec 2020

J.L. England, *Every Life Is on Fire: How Thermodynamics Explains the Origins of Living Things* (Basic Books, New York, 2020)

Environmental Protection Agency, Summary of the Endangered Species Act. (2019), https://www.epa.gov/laws-regulations/summary-endangered-species-act. Accessed 17 Dec 2020

J.D. Farmer, Cool is not enough. Nature **436**, 627–628 (2005)

Forest History Society, The Northern Spotted Owl. (2019), https://foresthistory.org/research-explore/us-forest-service-history/policy-and-law/wildlife-management/the-northern-spotted-owl/. Accessed 17 Dec 2020

L.E. Gilbert, Ecological consequences of a co-evolved mutualism between butterflies and plants, in *Co-evolution of animals and plants. Symposium V. First International Congress of Systematic and Evolutionary Biology*, ed. by L. E. Gilbert, P. H. Raven, (University of Texas, Austin, 1973), pp. 210–240

L.E. Gilbert, P.H. Raven, Co-evolution of animals and plants. In Symposium V. First International Congress of Systematic and Evolutionary Biology. University of Texas, Austin (1973)

G.E. Gruell, W.C. Schmidt, S.F. Arno, W.J. Reich, Natural forest succession and fire history, in *Eighty-eight years of change in a managed ponderosa pine forest*, General Technical Reports RMRS-GTR-23, ed. by H. Y. Smith, S. F. Arno, (U.S. Department of Agriculture, Rocky Mountain Research Station, Ogden, UT, 1999), pp. 5–9

J.M. Horowitz, J.L. England, Spontaneous fine-tuning to environment in many-species chemical reaction networks. PNAS **114**(29), 7565–7570 (2017)

T.P. Hughes, D.R. Bellwood, C. Folke, R.S. Steneck, J. Wilson, New paradigms for supporting the resilience of marine ecosystems. Trends Ecol. Evol. **20**(7), 380–386 (2005)

D.L. Hull, Introduction, in *Entropy, Information, and Evolution*, ed. by B. H. Weber, D. J. Depew, J. D. Smith, (MIT Press, Cambridge, MA, 1988), pp. 1–8

C.F. Jordan, Do ecosystems exist? Am. Nat. **118**(2), 284–287 (1981)

C.F. Jordan, *Conservation: Replacing Quantity with Quality as a Goal for Global Management* (Wiley, New York, 1995)

S.E. Jørgensen, J.C. Marques, S.N. Nielsen, *Integrated Environmental Management: A Transdisciplinary Approach* (CRC Press/Taylor and Francis Group, Boca Raton/London/New York, 2016)

S.A. Kauffman, *Investigations* (Oxford University Press, New York, 2000)

E. Masood, Battle over biodiversity. Nature **560**, 423–425 (2018)

MIT, Physics of living systems at MIT. (2020), https://cpls.scripts.mit.edu/wordpress/. Accessed 5 May 2021

F. Montagnini, C.F. Jordan, The role of insects in productivity decline of cassava (*Manihot esculenta Crantz*) in Amazonian slash and burn agriculture. Agrl. Ecosyst. Environ. **9**, 293–301 (1983)

J.C. Moore, P.C. deRuiter, *Energetic Food Webs: An Analysis of Real and Model Ecosystems* (Oxford University Press, 2012)
H.J. Morowitz, *Energy Flow in Biology* (Academic Press, 1968)
National Park Service. (2019), https://www.nps.gov/thro/learn/historyculture/theodore-roosevelt-and-conservation.htm. Accessed 27 Dec 2020
Nature Editorial, The value of biodiversity is not the same as its price. An assessment of nature's contribution to economic growth must listen to diverse voices. Nature **573**, 463–464 (2019)
Nature News. Top UN panel paints bleak picture of world's ecosystems. Nature 28 March 2018. (2018), https://www.nature.com/articles/d41586-018-03891-1. Accessed 27 Dec 2020
N.N. Nielsen, B.D. Fath, J.C. Marques, F. Müller, B. Patten, R.E. Ulanowicz, S.E. Jørgensen, E. Tiezzi, *A New Ecology Systems Perspective* (Elsevier, Amsterdam, 2020)
E.P. Odum, *Fundamentals of Ecology* (Saunders, Philadelphia, 1959)
H.T. Odum, *Environment, Power and Society* (Wiley, New York, 1971)
J. Ono, J.W.H. Yong, K. Takayama, et al., *Bruguiera hainesii*, a critically endangered mangrove species, is a hybrid between B.cylindrica and B. gymnorhiza (Rhizpphoraceae). Conserv. Genet. **17**, 1137–1144 (2016)
B. Oskin, Criteria for 'Red List' of endangered ecosystems released. (2013), https://www.livescience.com/29445-iucn-red-list-ecosystems-detailed.html. Accessed 30 Dec 2020
G. Peeples, Little Tennessee River recognized for native fish conservation. (2015), https://www.fws.gov/southeast/articles/little-tennessee-river-recognized-for-native-fish-conservation/. Accessed 30 Dec 2020
Z.J.B. Plater, *The Snail Darter: How Pork-Barrel Politics Endangered a Little Fish and Killed a River* (Yale University Press, New Haven, 2013)
N. Qualtieri, 21 conservation quotes from president Theodore Roosevelt. (2019), https://gearjunkie.com/conservation-quotes-president-theodore-roosevelt. Accessed 1 Jan 2021
T. Roosevelt. *Theodore Roosevelt: An Autobiography. Chapter IV, In Cowboy Land; Chapter VIII, Appendix A, Conservation; Chapter XI. The Natural Resources of the Nation.* 2016 edition (Open Road Integrated Media Inc., New York, 1913)
R. Schiffman, Fighting wildfires with fire. Wall. Str. J. **2020**, C4 (2020)
E.D. Schneider, D. Sagan, *Into the Cool* (University of Chicago Press, Chicago, 2005)
R.E. Schultes, A. Hofmann, C. Ratsch, *Plants of the Gods: Their Sacred, Healing and Hallucinogenic Powers* (Healing Arts Press, Rochester Vermont, 1992)
Union of Concerned Scientists, The IPCC: Who are they and why do their climate reports matter? (2018a), https://www.ucsusa.org/global-warming/science-and-impacts/science/ipcc-backgrounder.html#.W43c1OhKiUk. Accessed 5 Jan 2021
Union of Concerned Scientists, Scientists agree: Global warming is happening and humans are the primary cause. (2018b), https://www.ucsusa.org/global-warming/science-and-impacts/science/scientists-agree-global-warming-happening-humans-primary-cause#.W43jRehKiUm. Accessed 5 Jan 2021
J.E. Watson, D.A. Keith, B.B.N. Strassburg, O. Venter, B. Williams, E. Nicholson, Set a global target for ecosystems. Nature **578**(7705), 360–362 (2020)
B. H. Weber, D. J. Depew, J. D. Smith (eds.), *Entropy, Information, and Evolution* (MIT Press, Cambridge, 1988)
R. Whitmore, R.C. Brusca, P. Gonzalez-Zamorano, R. Mendoza-Salgado, E.S. Amador-Silva, G. Holquin, C.C. McIvor, F. Galvan-Magana, J.A. Seminoff, J.L. de la Luz, The ecological importance of mangroves in Baja California Sur: conservation implications for an endangered ecosystem, in *Biodiversity, ecosystems, and conservation in Northern Mexico*, ed. by J. E. Cartron, G. Ceballos, R. S. Felger, (Oxford University Press, Oxford, 2005), pp. 298–333
World Ocean Review. The rapid disappearance of species (2010), https://worldoceanreview.com/en/wor-1/marine-ecosystem/biodiversity/. Accessed 6 Jan 2021
World Wildlife Foundation. Protecting species (2020), https://www.worldwildlife.org/initiatives/protecting-species. Accessed 6 Jan 2021

Chapter 2
What Can We Learn by Studying Ecosystems that We Can't Learn from Studying Populations?

2.1 The Predator-Prey Conundrum

A key question in ecology has been how species persist over time when there are pairs of species in which one is an exploiter and the other its victim (Hastings 2020). According to the Lotka-Volterra model (Lotka 1910), the prey consumption rate by a predator is directly proportional to the prey abundance. This means that predator feeding is limited only by the amount of prey in the environment. Prey numbers first increase, giving the predator more food. The subsequent increase in predators would lead to a decline in prey. Predator numbers would then decline. This gives the prey an opportunity to increase their population, thereby restarting the cycle. However, it has proved unexpectedly challenging to demonstrate this type of predator–prey cycle in simple controlled systems in the laboratory (Blasius et al. 2020).

2.2 The Serengeti Ecosystem

2.2.1 Evolution in the "Ecological Theater"

The Serengeti ecosystem of Tanzania and Kenya has been of special interest to ecologists because predator interaction with herbivores grazing in the savanna seems to be an adaptation that conveys stability to the populations. Efforts to understand stability in the Serengeti and other ecosystems have taken various approaches. One has been the "trophic cascade" theory, a view that looks at predator effects propagating downward thorough food webs across multiple trophic levels (Ripple et al. 2016). Within a trophic cascade, there are "keystone species". These are species that if they were removed, the food webs would change drastically (Paine 1969). In the Serengeti, large herbivores such as wildebeests and top predators such

C. F. Jordan, *Evolution from a Thermodynamic Perspective*,
https://doi.org/10.1007/978-3-030-85186-6_2

as lions have been designated as "keystone species" (Carroll 2016). Speaking teleologically, the trophic cascade approach explains how lions control the populations of herbivores to keep them from destroying the grasses on which the food chain depends. Of course the lions aren't cognizant of the fact that if they don't keep the wildebeest population under control, they will soon be out of food. They just do it because they are hungry.

2.2.2 *Predator-Prey Interactions Tell Only Part of the Story*

The cascade approach tells us only what we have suspected since Aldo Leopold published "Deer irruptions" in 1943 (Binkley et al. 2006): Top predators control animal populations. It doesn't tell us why persistent predator–prey cycles can't be duplicated in laboratory experiments. The reason is that persistent cycles depend on many other interactions besides predator-prey, particularly decomposer-autotroph and autotroph-herbivore that change according to season and that vary as animals migrate between ecosystems. Laboratory studies duplicate only part of this system. All the parts are necessary for deviation dampening feedback that is the ultimate cause of both population and ecosystem stability (McNaughton 1985).

2.2.3 *Evolution in the "Thermodynamic Theater"*

The Serengeti plains of East Africa make up a landscape, and the ecosystems they contain vary within the landscape. Grazing sequences by different species at the landscape scale enables stability at the ecosystem level (Bell 1971). Every year, millions of large mammals migrate across the Serengeti plains of East Africa to follow rainfall patterns and nutritious forage (World Wildlife Foundation 2019). By the onset of the dry season (late May), the grasses on these plains have either dried out or been eaten down to stubble, and water is scarce (Schaller 1972). This triggers the massive migration of wildebeest and zebra, later followed by Thomson's gazelle and eland from the plains to the Acacia-Commiphora woodlands where foliage is still green (Campbell and Borner 1995, World Conservation Monitoring Centre 2020). At the beginning of the rainy season, these animals return to the plains and complete the cycle.

2.2.3.1 Ruminants

Ruminants are one of the most diverse and successful groups of mammals. Their ability to acquire nutrition from plants that contain chemical or mechanical defenses is due to fermenting of the leaves through microbial actions in a rumen that facilitates digestion. This may underlie their success in competition with other mammalian species (Zahn 2019). Breakdown of recalcitrant plant material by microbes in the rumen results in excrement in which nutrients are soluble or almost so. Plant material returned to the soil via excrement of ruminants is broken down further by dung beetles (Yamada et al. 2007) and ultimately by bacteria. Once nutrients are released from organic compounds, they are transferred to plants via arbuscular mycorrhizal fungi. Although they account for less than 1% of the total biomass, the nutritional benefit provided by arbuscular mycorrhizal fungi can increase the biomass of macro-organisms in the Serengeti by up to 48% (Stevens et al. 2018). Veldhuis et al. (2018) have suggested that mechanisms that facilitate faster availability of nutrients are a type of autocatalytic stimulation that results in more rapid plant growth. Increased plant productivity resulting from autocatalysis reinforces the evolutionary advantage of species such as ruminate wildebeests and gemsbok antelope, as these nomadic grazers follows rainfall patterns throughout the Serengeti plains (McNaughton 1985).

2.2.3.2 Adaptation of Ruminants in the Serengeti

Leaves of shrubs and trees in the Acacia woodlands are protected against herbivory by sclerophyllous leaves. This means they have a high ratio of carbon to nitrogen, a characteristic that renders them unappealing to non-ruminate species because it makes leaves difficult to digest (Bryant et al. 1983, Tuomi et al. 1988). Sclerophylly not only inhibits digestion, it also inhibits microbial activity and thus decomposition. When leaves die and are shed, nutrient release is inhibited, recycling into new plant growth is slow, and productivity is lower than would occur without these defenses. Ruminants such as Grant's gazelle have adapted to this defense with the aid of microbes that decompose the coarse material inside the animal's rumen making it easier to digest. Once digested plant material is defecated, it decomposes more readily and the recycling of nutrients is accelerated.

Another defense against herbivory in the Serengeti is high silica content of grasses (McNaughton et al. 1985). Grasses tend to have high concentrations of silica in their leaves that retards mechanical breakdown of chlorenchyma cells and reduces their digestibility (Hunt et al. 2008). Silica also inhibits nutrient recycling. Wildebeest and topi that feed on a wide variety of hard, silica-rich grasses (Kingdon 1997) have high-crowned teeth, which are capable of grinding tough plant tissues (Romer 1959) preparing them for the rumen.

2.2.3.3 Productivity in the Serengeti

McNaughton (1983) found that the grasses found in grazed patches are more productive than ungrazed patches. The Kenya Long-term Exclosure Experiment showed that herbivore dung was an important reason (Young 2020). These patches are able to support large concentrations of herbivores (Hiernaux and Turner 1996). Animal wastes alone however, produce patchy effects, and so cannot explain benefits on a landscape scale (Augustine and Frank 2001). There is another interaction between herbivores and plants that results in optimum performance of the landscape. Many grasses show highest productivity when leaves are of intermediate length. Below this length, grazing deprives roots of carbohydrates necessary for growth. Above this, the tops of leaves begin to dry and lose nutrition. Studies on the migratory grazing of ruminants on the plains of Africa (Savory and Butterfield 2016) showed that continuous migration results in light grazing pressure that actually stimulates growth of grasses because it maintains leaves at the optimum length for rapid growth. Continuous migration also prevents elimination of the most nutritious grasses. When grazing animals are restricted to a particular area, their selective grazing on the most nutritious grasses gradually eliminates these species, and gives a competitive advantage to less desirable species. Migration allows recovery of the favored species (Ritchie et al. 1998). Fire is also important in maintaining Serengeti grasslands by burning off thorny bushes and trees (University of York 2019) and eliminating dead grass stalks thereby allowing more light to penetrate to the apical meristem of grasses.

2.2.3.4 Fitness Results from Synchronous Evolution

Through the simultaneous evolution of all these factors, the performance (power output or productivity) of the landscape system is theoretically maximized, a measure of "fitness" in general (Odum 1995) and Serengeti in particular (McNaughton 1976; McNaughton et al. 1997; Dyer et al. 1986). Evidence that mortality of some wildebeests is due to undernutrition instead of predation (Mduma et al. 2001) suggests that the system could become less "fit" due to lion hunting by the Maasai tribes-people to defend their cattle.

2.2.3.5 What Have We Learned?

So now we have the answer to the question as to why predator-prey cycles cannot be demonstrated in simple controlled systems in the laboratory. The cycles are part of a complex of systems that require feedback between several levels of organization; the landscape, the ecosystems, the animal and plant populations, and the soil microbial communities. All these elements must be in place for conservation of species to be effective.

References

D.J. Augustine, D.A. Frank, Effects of migratory grazers on spatial heterogeneity of soil nitrogen properties in a grassland ecosystem. Ecology **82**, 3149–3162 (2001)

R.H.V. Bell, A grazing ecosystem in the Serengeti. Sci. Am. **224**, 86–93 (1971)

D. Binkley, M.M. Moore, W.H. Romme, P.M. Brown, Was Aldo Leopold right about the Kaibab deer herd? Ecosystems **9**, 227–241 (2006)

B. Blasius, L. Rudolf, G. Weithoff, U. Gaedke, G.F. Fussman, Long-term cyclic persistence in an experimental predator–prey system. Nature **577**(7789), 226–230 (2020)

J.P. Bryant, S. Chapin, D.R. Klein, Carbon/nutrient balance of boreal plants in relation to vertebrate herbivory. Oikos **40**(3), 357–368 (1983)

K. Campbell, M. Borner, Population trends and distribution of Serengeti herbivores: Implications for management, in *Serengeti II: Dynamics, Management and Conservation of an Ecosystem*, ed. by A. R. E. Sinclair, P. Arcese, (University of Chicago Press, Chicago, 1995), pp. 117–145

S.G. Carroll, *Serengeti Rules: The Quest to Discover how Life Works and Why it Matters* (Princeton University Press, Princeton, 2016)

M.I. Dyer, D.I. DeAngelis, W.M. Post, A model of herbivore feedback on plant productivity. Math. Biosci. **79**, 171–184 (1986)

A. Hastings, Long-term predator–prey cycles finally achieved in the lab. Nature **577**(7789), 172–173 (2020)

P. Hiernaux, M.D. Turner, The effect of clipping on growth and nutrient uptake of Sahelian annual rangelands. J. Appl. Ecol. **33**, 387–399 (1996)

J.W. Hunt, A.P. Dean, R.E. Webster, G.N. Johnson, A.R. Ennos, A novel mechanisms by which silica defends grasses against herbivory. Ann. Bot. **102**(4), 653–656 (2008)

J. Kingdon, *The Kingdon Field Guide to African Mammals* (Academic Press Ltd., London, 1997)

A.J. Lotka, Contribution to the theory of periodic reaction. J. Phys. Chem. **14**(3), 271–274 (1910)

S.J. McNaughton, Serengeti migratory wildebeest: Facilitation of energy flow by grazing. Science **81**, 92–94 (1976)

S.J. McNaughton, Serengeti grassland ecology: The role of composite environmental factors and contingency in community organization. Ecol. Monogr. **53**, 291–320 (1983)

S.J. McNaughton, Ecology of a grazing ecosystem. The Serengeti. Ecol. Monogr. **55**(3), 259–294 (1985)

S.J. McNaughton, J.L. Tarrants, M.M. McNaughton, R.H. Davis, Silica as a defense against herbivory and a growth promoter in African grasses. Ecology **66**(2), 528–535 (1985)

S.J. McNaughton, F.F. Banyikwa, M.M. McNaughton, Promotion of the cycling of diet-enhancing nutrients by African grazers. Science **278**, 1798–1800 (1997)

S.A.R. Mduma, A.R.E. Sinclair, R. Hilborn, Food regulates the Serengeti wildebeest: A 40-year record. J. Anim. Ecol. **68**(6), 1101–1122 (2001)

H.T. Odum, Self-organization and maximum power, in *Maximum Power*, ed. by C. A. S. Hall, (University Press of Colorado, 1995), pp. 311–330

R.T. Paine, A note on trophic complexity and community stability. Am. Nat. **103**(929), 91–93 (1969)

W.J. Ripple, J.A. Estes, O. J. Schmitz, V. Constant, M.J. Kaylor, A. Lenz, J, L. Motley, K.E. Self, D. S. Taylor, C. Wolf. What is a Trophic Cascade? Trends in Ecology and Evolution **31**(11), 842–849 (2016). https://doi.org/10.1016/j.tree.2016.08.010

M.E. Ritchie, D. Tilman, J.M.H. Knops, Herbivore effects on plant and nitrogen dynamics in oak savanna. Ecology **79**, 165–177 (1998)

A.S. Romer, *The Vertebrate Story* (University of Chicago Press, Chicago, 1959)

A. Savory, J. Butterfield, *Holistic Management*, 3rd edn. (Island Press, Washington, DC, 2016)

G.B. Schaller, *The Serengeti Lion. A Study of Predator-Prey Relations* (University of Chicago Press, Chicago, 1972)

M. Stevens, J. Propster, G.W.T. Wilson, A. Abraham, C. Ridenour, C. Doughty, N.C. Johnson, N. Wurzburger, Mycorrhizal symbioses influence the trophic structure of the Serengeti. J. Ecol. **106**(2), 536–546 (2018)

J. Tuomi, P. Niemela, F.S. Chapin III, J.P. Bryant, S. Siren, Defensive responses of trees in relation to their carbon/nutrient balance, in *Mechanisms of Woody Plant Defenses Against Insects: Search for Pattern*, ed. by W. J. Mattson, J. Levieux, C. Bernard-Dagan, (Springer, 1988), pp. 57–72

University of York, Declining fire threatens Serengeti ecosystem. (2019), https://www.york.ac.uk/news-and-events/news/2019/research/declining-fire-threatens-serengeti-ecosystem/#:~:text=A%20study%20of%20the%20Serengeti,to%20the%20Serengeti%2DMara%20ecosystem. Accessed 5 Jan 2021

M.P. Veldhuis, M.P. Berg, M. Loreau, H. Olff, Ecological autocatalysis: A central principle in ecosystem organization? Ecol. Monogr. **88**(3), 304–319 (2018)

World Conservation Monitoring Centre, Serengeti National Park, Tanzania. (2020), http://www.thesalmons.org/lynn/wh-wcmc/Tanzania%20-%20Serengeti.pdf. Accessed 6 Jan 2021

World Wildlife Foundation, Eastern Africa: The greater Serengeti grassland ecosystem in northern Tanzania. (2019), https://www.worldwildlife.org/ecoregions/at0714. Accessed 6 Jan 2021

D. Yamada, O. Imura, K. Shi, T. Shibuya, Effect of tunneler dung beetles on cattle dung deposition, soil nutrients, and herbage growth. Grassl. Sci. **53**, 121–129 (2007)

T. P. Young, Kenya Long-term exclosure experiment. (2020), https://tpyoung.ucdavis.edu/klee. Accessed 6 Jan 2021

L.M. Zahn, Resolving ruminants. Science **364**(6446), 1150–1151 (2019)

Chapter 3
A Thermodynamic Definition of Ecosystems

3.1 Ecosystems in the Twentieth Century

In his book Fundamentals of Ecology, Odum (1959) described an ecosystem in light of ecological understanding in mid-twentieth century. *An ecosystem is a community of living organisms in conjunction with the nonliving components of their environment, interacting as a system. These biotic and abiotic components are linked together through nutrient cycles and energy flows.* The ecosystem concept was not accepted by many biologists at the time, in part because of rigid academic boundaries. Many universities had a course in animal ecology in the zoology department that focused on animal populations, and a course in plant ecology in the botany department concerned with plant communities. Microbiology was in a different department, and soil studies were in an entirely separate college of agriculture. There was little interaction between groups. In the 1960s, ecology began to change as a result of atmospheric testing of nuclear weapons. Because of concern over radioactive fallout and its effect on health and the environment, the U.S. Atomic Energy Commission sponsored research projects to determine the environmental fate of radioisotopes in several different ecosystem types (Jordan 1986; Golley 1993). An unexpected outcome of these projects was the recognition that cycles of elements defined an ecosystem. If you track radioactive fallout through the environment, you can't avoid becoming an ecosystem scientist.

3.1.1 Cycling of Strontium-90

The cycling of strontium-90 following its fallout from atmospheric tests of nuclear weapons tests in the 1960s was a unique circumstance that showed the interconnectedness of all parts of an ecosystem. The secondary pulse that moved through

C. F. Jordan, *Evolution from a Thermodynamic Perspective*,
https://doi.org/10.1007/978-3-030-85186-6_3

the system 12–16 years after the initial input in a tropical rain forest was evidence of feedback within the system (Appendix 1, Fig. 1.2).

3.1.2 Cesium-137 in Food Chains

Many patterns and mechanisms of radioisotope cycling in ecosystems were presented in the Proceedings of the First National Symposium on Radioecology (Schultz and Klement 1963) through the Fourth National Symposium on Radioecology (Cushing 1975). Cycles differed among isotopes because of differences in how each is held in the soil and in biomass. Atmospheric fallout of cesium-137 (^{137}Cs) entered food chains in Alaska following adsorption to the surface of mosses and lichens (Hanson 1967). When caribou ate these plants and indigenous people ate the caribou, the isotope was passed up through the food chain. Concentrations increased by a factor of two at each successive trophic level. In the year 2000, radiocesium was still detectable in caribou in Canada, Alaska, and Greenland (Macdonald et al. 2007). The long residence time in the ecosystem may be attributed to slow growth of the mosses and lichens and their affinity for ^{137}Cs as well as to recycling.

3.1.3 Recycling of Isotopes in Norwegian Sheep

The explosion of the Russian Chernobyl nuclear reactor in 1986 was another unique incident that injected radioactivity into the atmosphere. It allowed the tracing of fallout in ecosystems throughout Europe. Almost 30 years after the incident, radiation levels in some Norwegian sheep still exceeded thresholds allowed for consumption of their meat. At least a few sheep are expected to miss radiation thresholds for at least another 50 years (Orange 2013). Since several generations of sheep and grass have occurred since the initial fallout, contamination of the sheep could only have occurred through cycling of the radioactive isotopes through the ecosystem. Unlike the long-lived mosses and lichens that contaminated Alaskan caribou, grasses and forbs eaten by sheep are short lived, thus all radioactivity has resulted from a continuous cycle of isotope ingestion by sheep, transfer to soil following defecation, uptake of isotopes from soil by plants, and recycling through ingestion of new plants by a new generation of sheep.

3.2 Ecological Energetics

The field of ecological energetics emerged in the 1960s and laid the groundwork for a thermodynamic theory of ecology. Many of the studies were made by physiologists interested in differences in respiratory energy loss as a function of organisms' biomass or differences between taxonomic groups (Zar 1968; Reichle 1968; Gates 1965). Interests of ecologists included: energy budgets of natural ecosystems (Odum 1957; Goldman 1968) and successional ecosystems (Woodwell and Whittaker 1968; Odum 1960); efficiencies of energy transfer (Slobodkin 1960; Turner 1970; Kozlovsky 1968); energy storage (Lindeman 1942; Golley 1961; Odum et al. 1965); the energetic balance of particular species in an ecosystem context (Richman 1958; Smalley 1960; Teal 1962; Englemann 1968); and many methods papers included in Wiegert's (1976) book on benchmark papers in Ecological Energetics.

3.2.1 Is it Time to Bury the Ecosystem Concept?

Lindeman's (1942) trophic-dynamic aspect of ecology was an important advance but that was almost 80 years ago. Since the 1970s, studies of nutrient transfer between system compartments led to an understanding of chemical element recycling (e.g. Bormann and Likens 1979). Has anything more been added to the ecosystem concept? One could add many concepts such as above-ground-below -ground interactions, autocatalytic cycles and process-level interactions such as nitrogen dynamics in ecosystems, but these all are explanations of how energy flow and nutrient cycling works in both natural and managed ecosystems. They are "bricks" that add to the edifice of the ecosystem concept.

The ecosystem concept has had a great impact on society, particularly in regard to sustainability. Public awareness of the importance of energy conservation and the benefits of recycling are now part of a global culture. Nevertheless, the ecosystem is merely a paradigm. You can't lay your hands on it or take a picture of it. It is merely a product of the human mind's limited ability to understand the complexity of the real world. It has provided great insights about the natural world, but like all paradigms, it tends to become outmoded. The concept has flourished for half a century, but in 2001, R.V. O'Neill asked "Are ecosystem studies losing their momentum? Is it time to bury the ecosystem concept?" (O'Neill 2001).

"Probably not", he concluded. Applications of concepts derived from other branches of science have invigorated ecosystem science (Chapman et al. 2016). Autocatalysis as a central principle in ecosystem organization is a concept borrowed from biochemistry (Veldhuis et al. 2018). Ecosystem modelers have used cybernetic principles developed by engineers to analyze ecosystem control and stability (Nielsen et al. 2020). Hierarchy theory (Allen and Starr 1982; O'Neill et al. 1986) is now important in ecosystem control. Schneider (1988) and Wicken (1988) suggested that "entropy production" is the source of organization in living systems. These concepts come together in a thermodynamic theme of ecosystem ecology, but we need a new definition of the ecosystem that reflects the scientific insights gained in the last half century.

3.2.2 A Thermodynamic Definition of Life

Perhaps what is needed first is a new definition of life, a definition that recognizes living organisms as open systems which during the process of growth and metabolism use solar energy to counter the entropy that degrades all systems. Macklem and Seely (2010) recognized this in their Twenty-first Century definition of life. Life is a *"self-contained, self-regulating, self-organizing, self-reproducing, interconnected, open thermodynamic network of component parts which performs work, existing in a complex regime which combines stability and adaptability in the phase transition between order and chaos, as a plant, animal, fungus, or microbe."*

3.2.3 A Thermodynamic Definition of Ecosystems

Nielsen et al. (2020) in their book "A New Ecology Systems Perspective" develop the ecosystem principles that underlie a thermodynamic theory of evolution. The principles can be cogently summarized in a thermodynamic definition of ecosystems by a minor modification of Macklem and Seely's definition of life. An ecosystem is a *self-contained, self-regulating, self-organizing, interconnected, open thermodynamic network of component parts which evolves as a complex regime that combines stability and adaptability in the phase transition between order and chaos, through activities of plants, animals, fungi, and microbes.* The only part omitted is "self-reproducing", but ecological succession is a form of ecosystem reproduction, so that term could be included also. The twenty-first century ecosystem concept includes only energy-containing components i.e. living biomass or recently living biomass like soil organic matter. Mineral soil, part of the twentieth century concept is omitted because it does not contain embedded energy. It influences nutrient recycling but it is not part of the internal energy feedback system.

3.2.4 The Phase Transition Between Order and Chaos

Energy flow through species interactions keeps ecosystems organized. The incentive of species is to obtain energy. Herbivores eat autotrophs, carnivores eat herbivores, and decomposers use energy from all three. Interactions are not random. As long as there is an energy supply, interactions follow well-defined pathways, but well-defined pathways easily become unrecognizable following disturbances to the system. Because of entropy, there is the tendency of species within ecosystems to become disorganized like school children dismissed for recess or a platoon of soldiers given passes for liberty. That leads to chaos. To counter entropy and maintain stability, there has to be a continuous source of energy input that maintains relationships between individuals or species, relationships that maintain or restore organization and order. For school children the source is teachers, for soldiers it is sergeants. In ecosystems, the source is the energy captured by plants and used to support species interactions. It must be sufficient not only to maintain plant metabolism, but to overcome entropy in the ecosystem. The transition phase between order and chaos is the battleground between energy to keep the system organized and entropy that causes system disintegration.

References

T.F.H. Allen, T.B. Starr, *Hierarchy: Perspectives for Ecological Complexity* (University of Chicago Press, Chicago, 1982)

F.H. Bormann, G.E. Likens, *Pattern and Process in a Forested Ecosystem* (Springer, New York, 1979)

E.J. Chapman, D.L. Childers, J.J. Vallino, How the second law of thermodynamics has informed ecosystem ecology through its history. Bioscience **66**(1), 27–39 (2016)

C. E. Cushing (ed.), *Proceedings of the Fourth National Symposium on Radioecology Corvallis, Oregon, May 12–14, 1975* (Dowden, Hutchinson and Ross, Stroudsburg, 1975)

M.D. Englemann, The role of soil arthropods in community energetics. Am. Zool. **8**(1), 61.69 (1968)

D.M. Gates, Energy, plants, and ecology. Ecology **46**(1, 2), 1–13 (1965)

C.R. Goldman, Aquatic primary production. Am. Zool. **8**(1), 31–42 (1968)

F.B. Golley, Energy values of ecological materials. Ecology **42**(3), 581–584 (1961)

F.B. Golley, *A History of the Ecosystem Concept in Ecology* (Yale University Press, New Haven, 1993)

W.C. Hanson, Cesium-137 in Alaskan lichens, caribou and Eskimos. Health Phys. **13**(4), 383–389 (1967)

C.F. Jordan, Ecological effects of nuclear radiation, in *Ecological Knowledge and Environmental Problem Solving: Concepts and Case Studies*, ed. by Committee on the Applications of Ecological Theory to Environmental Problems, Commission on Life Sciences, National Research Council, (National Academy Press, Washington, DC, 1986), pp. 331–344

D.G. Kozlovsky, A critical evaluation of the trophic level concept: 1. Ecological efficiencies. Ecology **49**(1), 48–60 (1968)

R.L. Lindeman, The trophic-dynamic aspect of ecology. Ecology **23**, 399–418 (1942)

C.R. Macdonald, B.T. Elkin, B.L. Tracy, Radiocesium in caribou and reindeer in northern Canada, Alaska and Greenland from 1958 to 2000. J. Environ. Radioactiv. **93**, 1–25 (2007)

P.T. Macklem, A. Seely, Towards a definition of life. Perspect. Biol. Med. **53**(3), 330–340 (2010). https://doi.org/10.1353/pbm.0.0167

N.N. Nielsen, B.D. Fath, J.C. Marques, F. Müller, B. Patten, R.E. Ulanowicz, S.E. Jørgensen, E. Tiezzi, *A New Ecology Systems Perspective* (Elsevier, Amsterdam, 2020)

R.V. O'Neill, Is it time to bury the ecosystem concept? (With full military honors, of course!). Ecology **82**(12), 3275–3284 (2001)

R.V. O'Neill, D.L. Deangelis, J.B. Waide, T.F.H. Allen, *A Hierarchical Concept of Ecosystems* (Princeton University Press, Princeton, 1986)

H.T. Odum, Trophic structure and productivity of Silver Spring, Florida. Ecol. Monogr. **27**, 55–112 (1957)

E.P. Odum, *Fundamentals of Ecology* (Saunders, 1959)

E.P. Odum, Organic production and turnover in old field succession. Ecology **41**(1), 34–49 (1960)

E.P. Odum, S.C. Marshall, T. Marples, The caloric content of migrating birds. Ecology **46**(6), 901–904 (1965)

R. Orange, Record low number of radioactive sheep. (2013), https://www.thelocal.no/20130923/chernobyl-radiation-in-norway-sheep-hits-new-low. Accessed 15 June 2016

D.E. Reichle, Relation of body size to food intake, oxygen consumption, and trace element metabolism in forest floor arthropods. Ecology **49**(3), 538–542 (1968)
S. Richman, The transformation of energy by *Daphnia pulex*. Ecol. Monogr. **28**(3), 273–291 (1958)
E.D. Schneider, Thermodynamics, ecological succession, and natural selection, in *Entropy, Information, and Evolution*, ed. by B. H. Weber, D. J. Depew, J. D. Smith, (MIT Press, Cambridge, MA, 1988), pp. 107–138
V. Schultz, A. W. Klement (eds.), *Proceedings of the First National Symposium on Radioecology Held at Colorado State University, Fort Collins, Colorado, September 10–15, 1961* (Reinhold Publishing Corporation, 1963)
L.B. Slobodkin, Ecological energy relationships at the population level. Am. Nat. **94**(876), 213–236 (1960)
A.E. Smalley, Energy flow of a salt marsh grasshopper population. Ecology **41**(4), 672–677 (1960)
J.M. Teal, Energy flow in the salt marsh ecosystem of Georgia. Ecology **43**(4), 614–624 (1962)
F.B. Turner, The ecological efficiency of consumer populations. Ecology **51**(4), 741–742 (1970)
M.P. Veldhuis, M.P. Berg, M. Loreau, H. Olff, Ecological autocatalysis: A central principle in ecosystem organization? Ecol. Monogr. **88**(3), 304–319 (2018)
J.S. Wicken, Thermodynamics, evolution, and emergence ingredients for a new synthesis, in *Entropy, Information, and Evolution*, ed. by B. H. Weber, D. J. Depew, J. D. Smith, (MIT Press, Cambridge, MA, 1988), pp. 138–169
R.G. Wiegert, *Ecological Energetics. Benchmark Papers in Ecology 4* (Dowden Hutchison & Ross, Stroudsburg, 1976)
G.M. Woodwell, R.H. Whittaker, Primary production in terrestrial ecosystems. Am. Zool. **8**(1), 19–30 (1968)
J.H. Zar, Standard metabolism comparisons between orders of birds. Condor. **703**, 278 (1968)

Chapter 4
Thermodynamic Characteristics of Ecosystems

4.1 Equilibrium

The term "equilibrium" has been used in two ways to describe mature ecosystems. In the past, failure to distinguish between the two has caused ecologists to misinterpret the nature of ecosystems.

4.1.1 The Equilibrium Law

The equilibrium law or Le Chatelier's principle says that when a system is disturbed, it will adjust to diminish the change that has been made to it. For biological systems, that means that any change in the environment prompts an opposing reaction that returns the system to its original state. In that sense, mature ecosystems are in stable equilibrium (White et al. 1992), (Fig. 4.1B). "Unstable equilibrium" means that equilibrium at the end of the year is not the same as equilibrium at the beginning (Fig. 4.1d). Successional ecosystems are in dynamic equilibrium when the balance between input and output is increasing (Fig. 4.1c). Static equilibrium implies a steady state (Fig. 4.1a). Ecologists once considered mature ecosystems to be in equilibrium because they seemed to obey Le Chatelier's principle. Despite changes in the environment during the course of a year, biomass at the end of the year is the same as the beginning.

C. F. Jordan, *Evolution from a Thermodynamic Perspective*,
https://doi.org/10.1007/978-3-030-85186-6_4

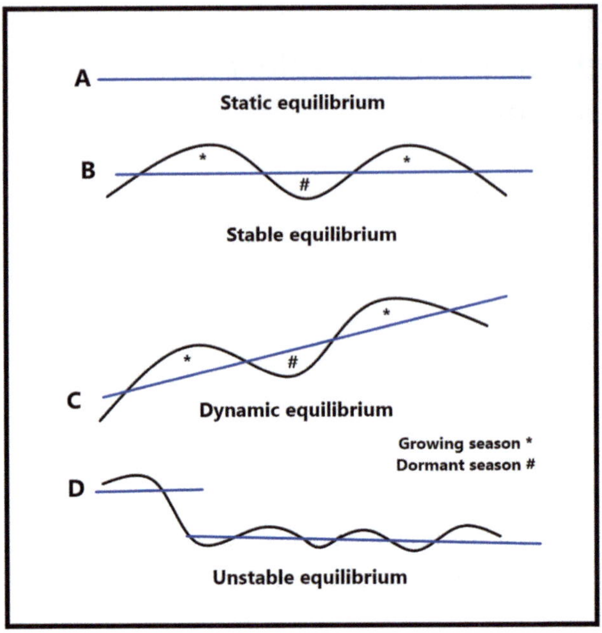

Fig. 4.1 Trends of energy content of ecosystems under stable, dynamic, and unstable conditions. Black lines represent seasonal patterns of ecosystem energy content. Blue lines represent long-term yearly averages. The deviation of the black line from the blue line represents bounded limits within which energy flow can vary without the ecosystem becoming unstable. Plants and animals exhibit a similar pattern of energy dynamics, except the wave frequency can be hours, days, weeks, or months. Dynamic equilibrium exists when they are growing, and stable equilibrium when they are mature

4.1.2 Thermodynamic Equilibrium

Thermodynamics deals with the transformation of energy from one form to another (Drake 2019). Physicists say that a black box isolated from the environment is in thermodynamic equilibrium. It is a closed system. There are no energy transformations within the box and there is no energy exchange between the system and its environment. Atoms are randomly bumping against each other so there is no direction of energy flow (Prigogine 1978). Ecologists now realize that organisms are not in thermodynamic equilibrium because molecules in their metabolic systems are continually transferring energy. Ecosystems are not in thermodynamic equilibrium because species are arranged to continually transform and transfer energy.

4.2 Open Thermodynamic Systems

An open thermodynamic system receives energy from a source outside itself, transforms it to heat while doing "work" and returns dispersed heat energy to a sink beyond its boundaries.

4.2.1 *Ecosystems Are Thermodynamically Open Non-Equilibrium Systems*

Ecologists now understand that living systems are thermodynamically open, non-equilibrium systems. They are thermodynamically open because they receive energy from the sun and disperse heat while metabolizing. There is exchange of energy between the system and its environment and there are energy transformations within the system. A horse is an open thermodynamic system. It receives energy from the sun in the form of oats, uses the energy to pull a wagon and as it sweats, discharges energy as heat to the environment. Although the biomass of the horse remained constant, energy flowed through the system. The Serengeti Is an open thermodynamic system on a landscape level. It receives energy through photosynthesis of grasses, and uses it to support the food chain of grazing herbivores, predators, and microbes that transform energy stored in biomass into heat.

4.2.2 *Work Is Performed by Non-equilibrium Systems*

In physics, work transfers energy from one place to another, or from one form to another. Work means that molecules within a system are not moving randomly, but are transferring energy in a very specific direction. In living systems, "work" is accomplished when solar energy is transformed to chemical energy that is used to fuel metabolic systems that keep systems alive. The energy flowing through the system keeps the system organized, and in the process of working to maintain organization, solar energy is transformed to heat. In a species, proteins are not randomly distributed in chromosomes, but are ordered in a particular sequence. As proteins degrade, they must be repaired or replaced, an energy demanding process.

In an ecosystem, species are not randomly distributed, but are arranged in a thermodynamic structure. Building and maintaining ordered structure requires work to counter entropy. For each trophic level beyond autotrophs, the energy comes from the energy embedded in the biomass of prey species. As herbivores eat autotrophs and predators eat herbivores, work is accomplished and ecosystem structure is maintained. In successional ecosystems, work is performed by building biomass. In mature ecosystems, biomass remains approximately constant because heat loss (entropy) balances work. (O'Neill 2001; O'Neill et al. 1982; Moore and deRuiter 2012). If the external source were to cease, the systems would disintegrate.

4.2.3 *Advantage of a Thermodynamically Open System*

England (2020) raised the question as to why animals are thermodynamically open systems, when a closed system like a crystal would be more stable and much less energetically expensive. He then answered, "the difficulty with a closed system is that it leaves no way of being structurally sturdy and (at the same time being) capable of withstanding pushes from the environment without also essentially being frozen into one rigid shape. What life typically needs is to combine structural integrity with the ability to rapidly reorganize itself in response to new information about the environment, and the way this is accomplished is through sustained energy flow." The same answer can be given as to why ecosystems are thermodynamically open systems. They must continually reorganize themselves in response to new information about the environment.

4.3 Ecosystems Are Entropic

Living organisms are entropic systems. Like machines, they tend to stop if they are not fed, and they wear out if they are not continually repaired. In the process, life's source of energy (concentrated solar energy) is degraded into diffuse heat energy. If life doesn't continually receive energy to replace what has been degraded, they will die. In this sense, life can be defined as the fight against entropy. Ecosystems too tend to degrade when their flow of energy is disrupted by volcanically produced aerosols, disease that kills autotrophs, loggers that destroy soil organic matter, and hunters that kill predators.

4.4 Ecosystems Are Cybernetic

4.4.1 Cybernetic Systems

Feedback is the essence of cybernetic systems. Feedback is defined as "the return to the input of a part of the output". Conventional science is based on the concept that cause produces an effect. Cybernetic science is based on the concept that after a cause produces an effect, the effect produces a new cause.

In cybernetic systems, stability is controlled by negative (deviation dampening) feedback. The effect of a deviation from the current state by one part of the system is compensated for by a reaction in another part of the system. Control is a circular chain of influence in which a system component is influenced indirectly by changes it has induced several steps in the past. To understand step "n" of the cycle, you have to understand what happened in steps "n-1, n-2 etc."

4.4.2 Economic Systems Are Cybernetic

Feedback is a key concept in economics (Golec 2004). Our understanding of feedback in economics can help us understand the role of feedback in ecology. In the economic system, stable growth is maintained by feedback between the Federal Reserve Bank, businesses, and consumers. To maintain the desired growth rate of the economy, the Federal Reserve Bank adjusts interest rates to local banks based on the rate that money is flowing through the economy. If growth is too slow, interest rates are lowered so that borrowing is cheaper. Local banks can then lower their interest rates, and companies that borrow can charge less for their goods and services. Consumer decisions to spend are based in part on prices of goods and services offered by companies. With lower prices, spending increases and the economy (theoretically) revives. Change in economic systems is perpetual and occurs as a result of continuous changes in the money flow through the system. While taking my elementary economics course circa 1950s, I had to read the text book twice to understand step *n*, because step *n* depended on what happened during previous steps. The problem for students trying to understand thermodynamic feedback in ecosystems is the same as the one that exists for students in elementary economics. They may have to read this book twice.

4.4.3 The Ecosystem Feedback Function

Change in ecosystems is perpetual, and occurs as a result of continuous changes in the energy flow entering the ecosystem and in biomass of species comprising the ecosystem. The feedback cycle is illustrated in Fig. 4.2.

Energy from the sun is transformed into chemical energy by autotrophs through photosynthesis. That energy is used to build biomass of plants. Energy embedded in biomass is passed through the food chain through herbivores, carnivores and detritovores. As they digest their food, energy stored in the consumed biomass is converted to glucose that is then transported through their bodies and converted to adenosine triphosphate in cells (Knowles 1980). Cells use that energy for several functions: to increase biomass; to propel the organism in which they are embedded; other metabolic functions. During these functions, chemical energy is converted to heat and lost to the atmosphere.

Biomass as it moves through the food chain contains not only energy but also embedded nutrients. By the time that detritovores in a natural system finish consuming the biomass from higher trophic levels, all the energy captured by photosynthesis has been depleted. In contrast to energy, only a very small proportion of nutrients have been lost during each transition of the trophic chain. Most of the nutrients remain available for recycling to autotrophs. It is the recycling of nutrients that closes the feedback loop. Nutrient movement from detritovores to autotrophs is the feedback link that controls energy flow through ecosystems. In ecosystems with rich soil, destruction of the feedback link does not immediately decrease productivity. Feedback is not critical. In such cases, exogenous factors such as temperature and precipitation control nutrient recycling and productivity.

Box 4.1
Some agricultural ecosystems are on soils rich in nutrients. Destruction of the feedback system by plowing does not initially decrease productivity because the soil is a nutrient reservoir. Eventually however, the reservoir becomes depleted and nutrients become limiting. Fertilizers become increasingly important for continued productivity

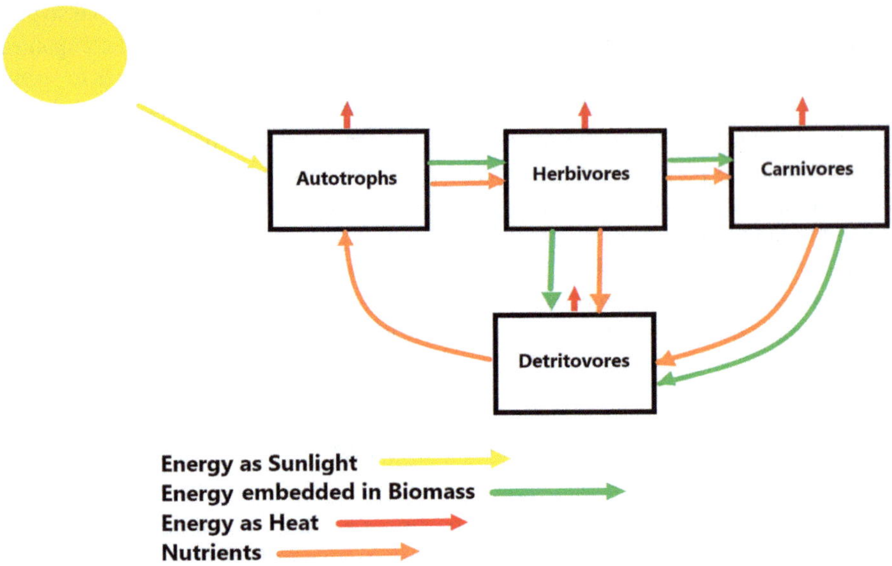

Fig. 4.2 An ecosystem-level model of energy flow through a terrestrial ecosystem, and feedback provided by nutrient recycling

In marine ecosystems decomposition of dead organic matter is not always complete, and some energy also is recycled along with nutrients by bacterio plankton to complete the feedback cycle. Biomass is recycled with nutrients in freshwater ecosystems also. *Utricularia* (a genus of carnivorous plants) serves as a substrate for periphyton growth, which in turn is grazed by zooplankton. Zooplankton provides nourishment to the *Utricularia* (Ulanowicz 1995).

4.4.4 Indirect vs. Direct Feedback

Feedback in Fig. 4.2 is indirect. An increase in herbivore population and biomass initially results in a decrease in autotroph biomass. Predators respond to the increase in herbivores by expanding their own population and eating more herbivores. More biomass is subsequently delivered to decomposers that increase the rate of nutrient delivery to autotrophs. Autotroph production expands, thereby re-establishing stability to the system.

Mutualisms exemplify direct feedback. An example is the interaction between flowering plants and pollinators. The plant benefits from pollinator activities, and the pollinator benefits from the energy supplied by the nectar. Mutualisms evolved as a result of commensalism in which species A benefitted from species B, but to encourage this benefit, species B began to provide a benefit to species A. Mutual benefits resulted in a positive feedback loop.

4.4.5 Deviation Dampening and Amplifying Feedback

Changes in energy transformations in one part of the system are compensated for by changes in another part. If the population of herbivores increases beyond the capacity of autotrophs to support them, populations of carnivores will increase and restore stability to the system. These adjustments constitute deviation *dampening* feedback. Systems remain stable over the long term because their biomass remains constant.

As a result of deviation *amplifying* feedback, biomass and power output of ecosystems increases during succession as feedback amplifies changes. During each cycle of nutrients through the system, energy output of plants increases. As energy output to the ecosystem increases, the herbivore population increases. An increase in predators soon follows. Increase in biomass of these consumers as well as the plants themselves leads to greater biomass and energy reaching the decomposers. Greater activity of decomposers leads to greater recycling of nutrients. This in turn affects the primary productivity of the plants that supply the organic matter (Carrillo et al. 2011).

When autotroph populations reach their maximum power output, the strategy of succession then changes to maintaining stability through deviation *dampening* feedback. When an ecosystem reaches maturity, interactions in the system (teleologically speaking) change to promote stability. If there is an increase in herbivore populations beyond the capability of autotrophs to support them, plants evolve chemical defenses against the herbivores. Predation pressure and disease also act to suppress populations that increase at a rate uncontrolled by other means.

The deviation dampening and amplifying feedback cycle also occurred during evolution of ecosystems but on a time scale of millions of years. The cycle has repeated at least five times during geological history as a result of extinction events caused by changes in energy flow through the environment.

4.4.6 Set Points

Engelberg and Boyarsky (1979) claimed that ecosystems are not cybernetic because they lack a fixed set point. Ecosystems do have a fixed set point, but the bounded limits around the set point are greater than those around the set points for a thermostat. The temperature change that triggers a thermostat to turn on the heat or air conditioning may be two degrees, one degree, or less, but it is not zero. The amount of change required to activate the system represents the bounded limits for temperature in the room. The amount of time it takes to regain the set point is the lag time.

The bounded limits for energy flow through an ecosystem can be visualized as the distance that the black line in Fig. 4.1b deviates from the blue line. Deviation can vary widely around the set point because of the length of time it takes for a system to respond. The cybernetic nature of ecosystems depends on the recycling of nutrients from organic matter to autotrophs. As energy flow changes following changes in environmental conditions, the feedback mechanism operates to stabilize the system by increasing or decreasing the energy flow through species populations. It is the slow response that causes relatively large bounded limits around the set point. While the temperature of a room responds to thermostatic control quickly and the set point is regained within a matter of minutes, adjustment of energy flow in ecosystems always lags behind the seasonal change in temperature and precipitation. Ecosystems respond slowly to seasonal changes and it may take a year to regain the original set point for ecosystem operation.

4.5 Ecosystems Are Autocatalytic

A single chemical reaction is said to be autocatalytic if one of the reaction products is also a catalyst for the same or a coupled reaction (Steinfeld et al. 1999). A set of chemical reactions can be said to be "collectively autocatalytic" if a number of those reactions produce, as reaction products, catalysts for enough of the other reactions that the entire set of chemical reactions is self-sustaining given an input of energy (Hordijk 2013). Veldhuis et al. (2018) envision subsystems within ecosystems to be collectively autocatalytic. "Interacting populations of different species in an ecosystem can form an autocatalytic loop if each species produces the resources needed by the next species in the loop, in such a configuration that the whole set of species is self-sustaining and even self-promoting given sufficient input of energy and essential carbon and nutrients." For example, soil fauna alter the effects of litter composition on nitrogen cycling. Increased nitrogen cycling stimulates growth of plants that then produce increasing amounts of litter. Increased litter means more energy for soil fauna (Carrillo et al. 2011).

Autocatalysis occurs in entire ecosystems as in Fig. 4.2 when nutrients released by detritovores interact with energy captured by photosynthesis to produce carbohydrates in amounts greater than needed for plant metabolism. This results in an increase in plant biomass, an increase in the availability of energy for herbivores and predators that in turn increase the supply of energy and nutrients for decomposers. The greater supply of nutrients results in an increase in photosynthetic output. The cycle eventually becomes limited by the energy-capturing ability of plants.

4.6 Ecosystems Have Boundaries

To understand ecosystem stability, an ecosystem model must have tangible boundaries that are either physical or biological (Strayer et al. 2003). When ecologists study a terrestrial ecosystem, they often define the boundaries as being coincidental with a plant community. The definition is practical because vegetation usually coincides with a soil type. An example would be Warren Woods, a maple-beech community in Michigan defined by an Alfisol (Poulson and Platt 1996). A major dispute occurred among ecologists in the mid Twentieth Century regarding whether community boundaries actually exist. On the one side was the Clementsian (1916) view that plant species existed in discrete communities that could be delineated. On the other side, Gleason's (1939) view was that vegetation exists along environmental gradients, and species are distributed along this "continuum" with species overlapping between communities of trees that supposedly are ecosystems. Appendix 2, Fig. 1 illustrates how both points of view are represented in a single landscape.

A definition of an ecosystem boundary could be that it is the line where there are more interactions between species within the boundary than across the boundary, as would occur in a pond ecosystem. That is a problem for modeling ecosystem dynamics, because top predators are an important component for stability, and their range can encompass several ecosystem types. In Appendix 2, Fig. 1, jaguars and ocelots range across all seven of the ecosystems portrayed. Ecosystem modelers get around that problem however by assuming that when one predator leaves a particular ecosystem, another takes its place. The Yellowstone region of North America consists of several different ecosystem types, and wolves, cougars and grizzly bears have roamed among the types. There is evidence of that these top predators have sequentially controlled trophic dynamics in each type (Ripple and Beschta 2004). Similar evidence has come from the Serengeti region in Africa where top predators can vary. An ecosystem can be modeled with either lions, leopards, or cheetahs as the top predator (Carroll 2016). Ecosystem models do not require that the same predator to be the single top predator all year long. While the species may change, the function remains the same.

4.7 Ecosystems Are Hierarchical

Because ecosystem ecologists themselves define the boundary of the system they are analyzing, critics of the ecosystem concept say it is arbitrary and therefore not real. That may be true, but it does not mean the concept is not useful. Political boundaries between counties, states, and nations also are arbitrary, but that does not mean the concept of counties, states and nations is not useful. Like ecosystems, they exist in a nested series. Their boundaries are helpful in resolving economic and political problems that vary depending on scale.

4.7.1 Hierarchy in Physical Systems

Feedback systems often exist in hierarchies where stability at all hierarchical levels depends on both bottom up and top-down control. Systems higher in a hierarchy exert external control over lower systems, while information uploaded from the lower system ensures stability of the higher-level system. An air traffic control system is a good example of hierarchical feedback control. The lower-level system is that of an airplane's autopilot that compares actual instrument readings on speed and altitude with the speed and altitude settings specified in the autopilot. If they do not match, feedback information in the form of electronic signals causes the autopilot to adjust fuel flow and elevator angle until actual and specified speed and altitude are congruent. This control is internal to the autopilot system which is embedded in a higher-level feedback system consisting the pilot, the air traffic control center and the controllers that use radar to determine the aircraft's location, altitude, speed, and direction. A computer analyzes the positions of all aircraft in the region, and a controller radios instructions back to the pilot as to whether the plane should maintain its current speed and altitude. If a change is needed due to storms, traffic congestion or other factors, the controller sends instructions to the pilot who then inputs this new information into the autopilot system. The autopilot is controlled externally as well as internally. In this system, feedback is deviation dampening, that is, feedback decreases the deviation between actual and desired performance. In contrast, if an inexperienced pilot tries to correct an impending stall by pulling up on the controls, a deviation amplifying system ensues in which the plane enters a what is called a "death spiral". The beginning of a stall causes the pilot to pull up the nose, which in turn increases the severity of the stall. Each deviation increases the next deviation, eventually causing the system to crash.

4.7.2 Hierarchy in Ecological Systems

Ecological systems exist in a nested series (Allen and Starr 1982). A forest ecosystem is embedded in a landscape system. The system that supports a wolf might consist of a landscape that contains several ecosystem types – old growth forest, successional forest, grassland, marsh. For an albatross, its system would be the whole earth. As the boundaries expand, the ecosystem concept becomes a whole-earth ecosystem. Working in the other direction, a decaying log is a micro-system nested in a forest macro-ecosystem depicted in Fig. 4.2. Interactions between species in non-living organic matter such as a decomposing log comprise a subsystem illustrated by Fig. 4.3.

4.7.3 Common Currencies

To understand how ecosystems function, ecologists make conceptual and mathematical models, which are a type of hypothesis. How does a researcher conceptualize an ecosystem so that it can be modeled? The model has to be a model of something. In Appendix 1, strontium is the common currency. In tropical rain forests, nitrogen is often fixed by blue-green algae and lichens that live on the surface of leaves. Here, nitrogen is the common currency for the leaf sub-system. The something that ties together all parts of all ecosystems is energy. Energy is the common currency that flows within ecosystems, and between systems and sub-systems just as money is the common currency that flows through and between economic systems. In Fig. 4.3, energy is the common currency that connects the species. Bacteria get part of their energy from plant material. The bacteria are fed upon by predaceous nematodes which are fed upon by predacious arthropods. When arthropods die, they become part of the organic material fed upon by bacteria. Within an ecosystem, there are dozens or more of such sub-system loops.

4.7.4 Macro- and Micro-system Models

Economists have macro-economic models that look at money flow through the entire economy from the federal reserve to regional banks, businesses, and consumers in the aggregate. Money flow is increased by lowering interest rates for money borrowed from the Federal Reserve and decreased by raising the rates. Control of the economy based on macro-models is top-down control. Economists also have micro-economic models that deal with how people make economic decisions. Micro-economics focuses on issues that affect individuals and companies. It includes consumer theory and theory of the firm. Hundreds of micro-economies influence the economy bottom-up through the "Invisible Hand of the Marketplace" (Smith 1776) that regulates supply and demand. Buyers affect money flow by deciding to save their money when prices are too high, or to increase spending when prices are low. Sellers affect money flow by lowering prices when sales are slow and raising prices when they cannot supply the demand.

Understanding of ecosystems also occurs at two levels, the micro- and the macro. The micro-level model in Fig. 4.3 deals with the decomposer function of an ecosystem with each species as an entity in the model. Veldhuis et al. (2018) stressed the importance of micro-systems as follows: "*This loop of species* cannot *simply be described by classic predator-prey interactions* (as in Fig. 4.2), *which would miss the elements of ecosystem engineering*".

There are many types of interactions in microbial loop cycles (Clarholm 1985). Mycorrhizal associations drive plant populations through increased nutrient uptake (Tedersoo et al. 2020). Root herbivores shape communities of above-ground herbivores via changes in plant nutrients (Johnson et al. 2013). Litter quality differences among plant species determines in turn the community structure of decomposer fauna that consumes the litter (Wardle et al. 2006). Plant communities are affected by some or all of these interactions (Megias and Möller 2010). In an analysis of ecosystem control, these interactions are pooled. While pooling obscures many fascinating ecological details, an attempt to include them all in a single model obscures efforts to understand control at the level of the entire ecosystem.

In contrast, the macro-level (ecosystem level) predicts the stability of the entire system resulting from all species in each trophic level treated in the aggregate (Venzon et al. 2001). The micro-level model of a decomposer trophic web cannot answer a question about how stability is controlled in a forest ecosystem, and an

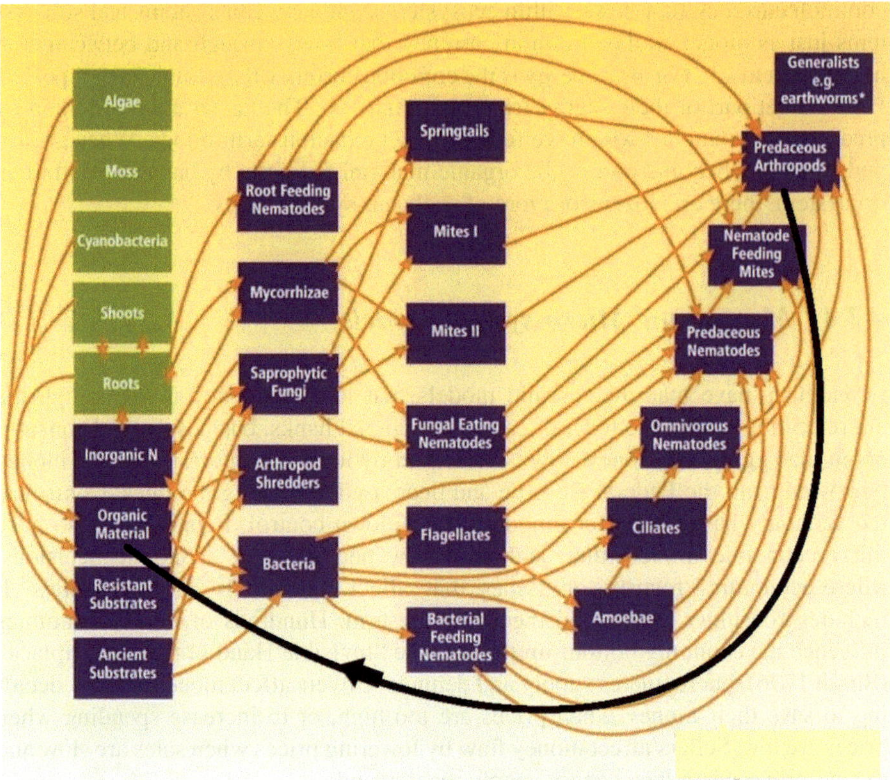

Fig. 4.3 A food web in decomposing organic matter. The black arrow added to the original figure delineates one of the many feedback loops in a trophic web on a forest floor. That loop is where bacteria are fed upon by predaceous nematodes which are fed upon by predacious arthropods. When the arthropods die, they become part of the organic material fed upon by bacteria. (Adapted from Ingham 2019)

ecosystem model that examines stability cannot answer questions about the evolution of energy transfer from nectar producing plants to hummingbirds (Suarez and Welch 2017). Macro-ecological models and micro-ecological models represent studies at different hierarchical levels, each with their own independent questions and objectives.

The Long-Term Ecological Research Program (LTER Network 2019) has provided funding to scientists who are willing to carry out their research at the same place and the same time as many other researchers. Most of these studies are not of ecosystem properties but of processes at lower hierarchical levels. Although it was hoped that these process level models could be integrated bottom-up into a model of ecosystem properties that could be used for ecosystem management, it was a false hope. As modelers incorporate more phenomena, the model's predictions typically become less accurate. Every parameter measured has a statistical distribution, and it is unlikely that an average value for a parameter will reflect the true value. The errors for each sub-model are compounded into the comprehensive model rendering the total error to be unknown (Saltelli et al. 2020). Attempts to understand ecosystem properties and control must be based on top-down models that constrain the dynamics of micro-system models. The failure to incorporate top-down control in an early IBP (International Biological Program) ecosystem model of the short-grass prairie resulted in uncontrolled build-up of bison manure (Van Dyne 1971 personal communication). There were no macro-system constraints on micro-system models of bison defecation.

Measurements of the energy dynamics of all sub-systems are not needed to answer questions at the macro-ecosystem level. All that is needed for a model of stability for the short-grass prairie is the rate at which energy moves through each trophic level: the grass, the bison and other herbivores that eat the grass, hunters that thin the bison herd, and the decomposers that break down grass litter and bison manure.

Resource managers have to know how to carry out studies that will help them get answers they need within a short time frame and within an allocated budget. This leads to the question, what parameters should an ecosystem ecologist measure to understand ecosystem control and stability in management situations, and how precisely should he or she measure them? The answer is, only those relevant to the question being asked, and only precisely enough to answer the question being asked. Appendix 3 gives several examples.

4.7.5 Why an Ecosystem Model that Includes Everything Is Not Possible

In the 1980s, I attended several meetings where the objective was to determine how population ecology, community ecology and ecosystem ecology could all be incorporated into a single comprehensive model. Since nested systems are connected by energy flow, it appeared to be a straightforward task. To increase the understanding of stability of a forest ecosystem model, it might seem that a sub-model of decomposer organisms in the decaying logs should be part of the main model because more detail would result in a better simulation. This turned out not to be true. Real world ecosystems (as opposed to their mathematical counterparts) are enormously complex associations of species which interact in diverse ways. As a matter of practical necessity, ecosystem ecologists can rarely specify, much less quantify all of the interactions in an ecosystem. Consequently to answer questions about ecosystem stability and sustainability, ecosystem models generally consider fewer than the total number of interacting species" (Schaffer 1981). This is the "it ignores my favorite species" complaint that population ecologists make about ecosystem models. The complaint reflects a lack of understanding that distinct levels of interest call for distinct levels of analysis. Studies at one hierarchical level have different paradigms, different questions and different types of data than studies of other levels.

Theoretically it is possible for a population ecologist to help with an ecosystem model by quantifying the energy flowthrough a population, but that is not a study that will obtain her grants or ensure her tenure.

4.7.6 A Nested Marine Community

There are no distinct ecosystems with discernable borders in open oceans except for corals. The intertidal zone with its community of organisms (Paine 1966) are not ecosystems, but simply communities that are links in the oceanic trophic chain. In marine systems, Sonnewald et al. (2020) identified what they called global marine ecological provinces based on plankton community structure and nutrient flux. A province consists of a "coherent combination of biogeochemical and ecological factors" aggregated into a cluster. These clusters are associated with upwelling, where nutrients required by the plankton are dissolved or suspended in the water. Control of such provinces is through exogenous factors such as oligotrophic gyres and regions of upwelling influenced by trade winds.

Within oceans there are dozens of trophic webs such as that in Fig. 4.4 that get their energy from the collective autotrophic output of the larger ecosystem in which this sub-system is embedded. The autotrophs in the larger system are photosynthetic

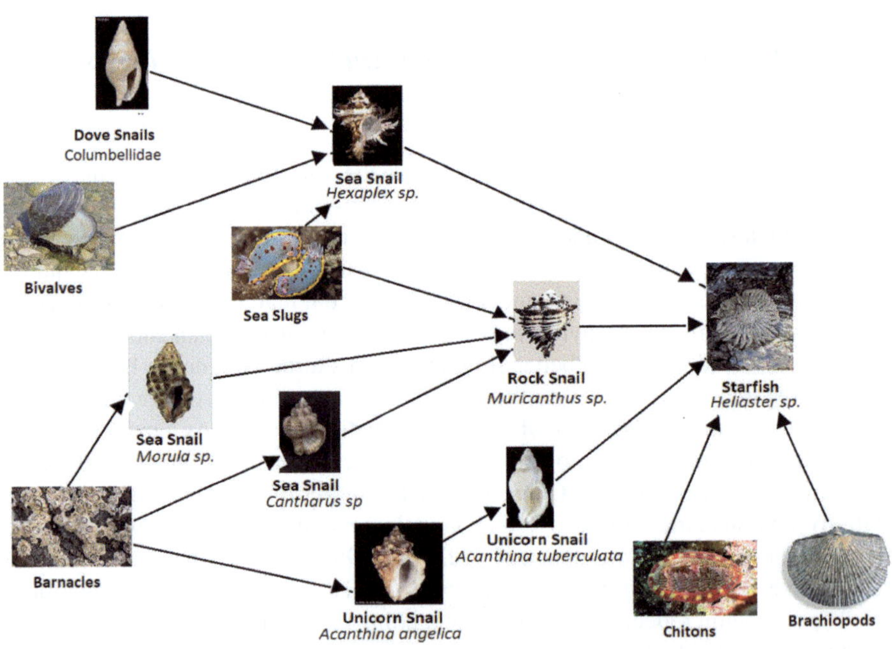

Fig. 4.4 An intertidal food web that is subsumed by the overall marine ecosystem. Relationships are from by Paine (1966)

plankton consumed by the filter feeders that are fed upon in turn by starfish. When the predator starfish are consumed by other organisms such as otters, the nutrients and energy rejoin the larger trophic chain topped off by predators such as sharks. Nutrients egested by these higher trophic levels are taken up by plankton that then become the nutrient and energy source for the lowest trophic level in Fig. 4.4.

4.8 Ecosystems Are Deterministic

Chaos theory recognizes that within complex systems such as ecosystems, there are underlying patterns, constant feedback loops, repetition, self-similarity, fractals, and self-organization (Fractal Foundation 2020). The theory says that the resulting chaos makes it impossible to predict the ultimate fate of a complex system: the outcome of any process is sensitive to its starting point (Rickles et al. 2007). When viewed from the population perspective, ecosystems are chaotic and development is stochastic. There are too many variables to predict which way populations will proceed during succession and evolution. The time when succession starts is only one. If an agricultural field is abandoned in June, the plant community that appears will be different from one that starts in August. Other factors include the types of insects, birds and mammals that bring in seeds, and the direction that the wind is blowing. In contrast when viewed from the thermodynamic point of view, ecosystems are deterministic.

A deterministic system is a system in which no randomness is involved in the development of future states. A deterministic model will always produce the same output. Ecosystems are deterministic because development of ecosystem structure and trophic levels is guided by the thermodynamic principles that control ecosystem dynamics. Evolution of species populations and their interactions is stochastic, but evolution of the ecosystem in which the species exist is deterministic. Regardless of species and starting point, the ecosystem will increase in energy flow, biomass, and species diversity.

Chaos theory says that the flap of a butterfly's wings in the Amazon could cause a tornado in Texas. That is impossible. Populations and ecosystems operate on two distinct levels of organization. The larger, more powerful level controls the dynamics of the lower, subordinate level. The energetics of insect populations are ultimately controlled by the dynamics of the Amazon rain forest. Energy flow through an encompassing hierarchical level constrains energy flow in a nested level. Hierarchical control prevents the flap of an anxious butterfly's wing to alter significantly the energy flow through the rain forest (Upadhyay 2009). Events in the Amazon forest can affect tornadoes in Texas only if the entire structure of the forest is modified. Most of the rain in the Amazon region comes from water transpired by the forests (Wright et al. 2017). Transpiration also cools the earth. Clear cutting of the Amazon forest might cause a desert in Brazil resulting in atmospheric changes that cause a tornado in Texas.

4.9 Ecosystems Are Information Rich

4.9.1 An Engineering Definition of Information

For engineers, a cybernetic system is a group of interacting entities through which information is transmitted in a closed feedback loop (Åström and Murray 2008). Although "information" is the general term for what is conveyed through feedback control (Wiener 2013), the nature of "information" varies depending on the system. For systems in general, information is that which is produced by feedback about the gap between the actual level and the reference level of a system parameter. This information is used to alter the gap in some way (Ramaprasad 1983). In engineering systems, information is derived from measurements of parameters such as temperature, pressure, speed, light intensity, sound, and chemical concentration.

4.9.2 Information to Facilitate Exchange

An ecosystem is an energy marketplace, where exchanges of energy and nutrients are controlled by information about whether species in an adjacent trophic level are increasing or decreasing in biomass. Maintenance of ecosystem stability occurs through an interplay of internal material cycles and energy flows, under informational controls that generate self-organizing feedback (Patten and Odum 1981). Just as the balance between supply and demand determines rate of money flow in the economy, the balance between energy supply and demand determines rate of energy flow (ecosystem productivity). In economic and ecological systems, rate is information.

4.9.3 High Energy Information

"Rate" is a type of information used in many different systems. High energy information comes from direct rate measurements of energy flow. Electric meters measure the rate at which electricity is used and this information is used by the power company to bill users. The rate at which vehicles travel along highways is important for traffic control. In mechanical systems, there is always some type of gauge or speedometer that registers rate. In economic systems, Gross Domestic Product is a measure of the rate that money flows through the economy. In ecological systems, productivity is a measure of rate that energy flows through the ecosystem.

In trophic webs, information is the rate of energy flow through species interactions. The receiving species in each interaction reacts to information defined as the rate that it is receiving energy-containing reduced carbon compounds. If the rate is declining, population biomass declines, and the rate that energy is transmitted to the next population also declines. If the rate is increasing, the reverse occurs. There is a continual adjustment in each species population determined by the rate that energy becomes available to that population. High energy information passed between species and through an ecosystem adjusts energy flow to keep the system stable. Patten and Odum (1981) considered this informationally nonspecific interaction to be a "brute force" exchange of information.

4.9.4 Low Energy Information

In cybernetic systems such as thermostat control and flight control systems, high energy flow (burning of fuel in a furnace or jet engine) is controlled by low energy information (electric signals controlled by switches). The energy involved in transmission of feedback information is small compared to the amount of energy that is regulated. In ecology, information that is transferred between organisms through sight, smell and sound, is low energy information. Chemical cues, another type of information, regulate the behavior of many types of animals from invertebrates through mammals, and also stimulate reactions in plants (Hay 2009). These chemically mediated interactions strongly affect high energy flows including population structure and community organization through foraging strategies, feeding choices, mutualistic associations, selection of mates and habitats, competitive interactions, and transfer of energy and nutrients between species (Aartsma et al. 2019).

4.9.5 Information Theory

Species richness is simply the number of species in an ecosystem. A diversity index (Shannon 1948) considers not only the number of species, but also the number of individuals in each species. In an old growth forest, there may be thousands of plant and animal species, one of which may be the northern spotted owl. Some species are represented by many individuals, but the spotted owl may be represented by only one individual. Through the use of the diversity index, information theory reduces the uncertainty that the owl will occur in future states of the ecosystem (Ulanowicz 2009). We can be fairly certain that following disturbance to the system, the spotted owl will be one of the first to become extinct.

Information can be further refined by quantifying the ways that energy can flow through a system. By weighting the interactions by the amount of energy flow through each interaction, we have measure of the probability that interaction will occur in the future (Ulanowicz 2001). This reduction of uncertainty constrains the almost infinite number of mathematically possible interactions to the number that would actually occur, based on the probability of each interaction (Fath 2015). One probable route of energy flow would be from a mouse to an owl. An improbable route would be from an owl to a mouse. Margalef (1963, 1968) argued that the present information content of a successional system sets limits or patterns for future states of the system.

4.9.6 Genetic Information

Brooks and Wiley (1986) looked at information in the context of evolution where information is the complexity of the genetic structure of species. Their question was, how does genetic information increase? That is a basic question for evolutionary biologists.

4.10 Ecosystems Are Non-teleological

To facilitate understanding of complex ideas, ecologists sometimes use teleologic expressions to avoid the awkward but scientifically correct way of writing. For example, the often-cited paper "The strategy of ecosystem development" (Odum 1969) does not mean that ecosystems have a strategic plan for development. It is just a clear way of expressing an idea. A scientifically correct title would be "The ontogeny of ecosystem development", but that is a title appropriate for a strictly disciplinary audience. Odum's goal was to reach a more general audience. In this book when I say "the goal of evolution is maximizing power output", I don't mean that species decided among themselves that this is an important and worthwhile thing to do. Species do not have evolutionary goals. A scientifically correct way of putting this would be to say, "species are selected on the basis of whether the characteristics that best enable their winning the competition for survival involve maximizing power output." Sentences like that can be bewildering. When you encounter statements in the following chapters that are teleologic, please keep in mind that teleology was used only to enhance communication and streamline development of a pertinent idea.

4.11 Criticisms of Ecosystem Models

A frequent criticism of ecosystem models is that they are oversimplifications of reality. This criticism was valid when it was thought that to have a realistic model of an ecosystem, all the species in the system had to be included. Now that we understand that ecosystems are hierarchical, we realize that simplification through separation and aggregation of population and community level data is necessary to understand the properties of each hierarchical level (Jørgensen and Fath 2006). In nature, top-down and bottom-up control are operating simultaneously. Hundreds of sub-systems are operating simultaneously and all contributing to bottom-up control.

Theoretically it would be possible for there to be one system model that included all sub-system models and their metabolic sub-sub-systems each contributing to the continual feedback from higher levels to lower levels and vice-versa, but that would require a model as complex and dynamic as nature itself.

Scientists working at each hierarchical level have distinct questions and particular data requirements. While thermodynamic models predict ecosystem level responses, they cannot make any predictions about the emergence and extinction of a particular species or sub-system. That is not their purpose. This difference in purpose between hierarchical levels is not universally understood by ecologists. Betts et al. (2019), in a review criticized ecosystem models of forest fragmentation because they fail to predict the fate of particular species. Predictions of species eruptions or extinctions require lower level, micro-ecological models that receive input from ecosystem level models that inform as to whether energy flow is increasing or decreasing.

References

Y. Aartsma, B. Leroy, W. van der Werf, M. Dicke, E.H. Poelman, F.J.J.A. Bianchi, Intraspecific variation in herbivore-induced plant volatiles influences the spatial range of plant–parasitoid interactions. Oikos **128**(1), 77–86 (2019)

T.F.H. Allen, T.B. Starr, *Hierarchy: Perspectives for Ecological Complexity* (University of Chicago Press, Chicago, 1982)

K.J. Åström, R.M. Murray, *Feedback Systems: An Introduction for Scientists and Engineers* (Princeton University Press, Princeton, 2008)

M.G. Betts, C. Wolf, M. Pfeifer, et al., Extinction filters mediate the global effects of habitat fragmentation on animals. Science **366**, 1236–1239 (2019)

D.R. Brooks, W.O. Wiley, *Evolution as Entropy: Toward a Unified Theory of Biology* (University of Chicago Press, Chicago, 1986)

Y. Carrillo, B.A. Ball, M.A. Bradford, C.F. Jordan, M. Molina, Soil fauna alter the effects of litter composition on nitrogen cycling in a mineral soil. Soil. Biol. Biochem. **43**, 1440–1449 (2011)

S.G. Carroll, *Serengeti Rules: The Quest to Discover How Life Works and Why it Matters* (Princeton University Press, Princeton, 2016)

M. Clarholm, Interactions of bacteria, protozoa and plants leading to mineralization of soil nitrogen. Soil Biol. Biochem. **17**, 181–187 (1985)

F.E. Clements, *Plant Succession: An Analysis of the Development of Vegetation* (Carnegie Institution of Washington, Washington, DC, 1916)

G.W.F. Drake, Thermodynamics. Encyclopedia Britannica. (2019), https://www.britannica.com/science/thermodynamics. Accessed 17 Dec 2020

J. Engelberg, L. Boyarsky, The non-cybernetic nature of ecosystems. Am. Nat. **114**, 317–324 (1979)

J.L. England, *Every Life Is on Fire: How Thermodynamics Explains the Origins of Living Things* (Basic Books, New York, 2020)

B.D. Fath, Quantifying economic and ecological sustainability. Ocean. Coast. Manage. **108**, 13–19 (2015)

Fractal Foundation, What is chaos theory. (2020), https://fractalfoundation.org/resources/what-is-chaos-theory/. Accessed 19 Dec 2020

H.A. Gleason, The individualistic concept of the plant association. Am. Midl. Nat. **21**(1), 92–110 (1939)

P. Golec, Feedback. (2004), http://www.economicswebinstitute.org/glossary/feedback.htm. Accessed 19 Dec 2020
M.E. Hay, Marine chemical ecology: Chemical signals and cues structure marine populations, communities, and ecosystems. Annu. Rev. Mar. Sci. **1**, 193–212 (2009). https://doi.org/10.1146/annurev.marine.010908.163708
W. Hordijk, Autocatalytic sets: From the origin of life to the economy. Bioscience **63**(11), 877–881 (2013)
E.R. Ingham, Soil Biology Primer. (2019), https://www.nrcs.usda.gov/wps/portal/nrcs/detailfull/soils/health/biology/?cid=nrcs142p2_053865. Accessed 21 Dec 2020
S.N. Johnson, C. Mitchell, J.W. McNicol, J. Thompson, A.J. Karley, Downstairs drivers—root herbivores shape communities of above – ground herbivores and natural enemies via changes in plant nutrients. J. Anim. Ecol. **82**(5), 1021–1030 (2013)
S.E. Jørgensen, B.D. Fath, Examination of ecological networks. Ecol. Model. **196**(3, 4), 283–288 (2006). https://doi.org/10.1016/jecolmodel.2006.02.029
J.R. Knowles, Enzyme-catalyzed phosphoryl transfer reactions. Annu. Rev. Biochem. **49**, 877–919 (1980). https://doi.org/10.1146/annurev.bi.49.070180.004305. PMID 6250450
LTER Network, (2019) https://lternet.edu/. Accessed 8 Mar 2021
R. Margalef, On certain unifying principles in ecology. Am. Natt. **97**(897), 357–374 (1963)
R. Margalef, *Perspectives in Ecological Theory* (University of Chicago Press, Chicago, 1968)
A.G. Megias, C. Müller, Root herbivores and detritovores shape above-ground multitrophic assemblage through plant-mediated effects. J. Anim. Ecol. **79**(4), 923–931 (2010)
J.C. Moore, P.C. deRuiter, *Energetic Food Webs: An Analysis of Real and Model Ecosystems* (Oxford University Press, 2012)
R.V. O'Neill, Is it time to bury the ecosystem concept? (With full military honors, of course!). Ecology **82**(12), 3275–3284 (2001)
R.V. O'Neill, R.H. Gardner, D. Weller, Chaotic models as representations of ecological systems. Am. Nat. **120**, 259–263 (1982)
E.P. Odum, The strategy of ecosystem development. Science **164**, 262–270 (1969)
R.T. Paine, Food web complexity and species diversity. Am. Nat. **100**(910), 65–75 (1966)
B.C. Patten, E.P. Odum, The cybernetic nature of ecosystems. Am. Nat. **118**, 886–895 (1981)
T.L. Poulson, W.J. Platt, Replacement patterns of beech and sugar maple in Warren Woods, Michigan. Ecology **77**(4), 1234–1253 (1996)
L. Prigogine, Time structure and fluctuations. Science **201**(4358), 777–785 (1978)
A. Ramaprasad, On the definition of feedback. Behav. Sci. **28**(1), 4–13 (1983). https://doi.org/10.1002/bs.3830280103
D. Rickles, P. Hawe, A. Shiell, A simple guide to chaos and complexity. J. Epidemiol. Comm. H. **61**(11), 933–937 (2007). https://doi.org/10.1136/jech.2006.054254
W.J. Ripple, R.L. Beschta, Wolves, elk, willows and trophic cascades in the upper Gallatin range of southwestern Montana, USA. Forest. Ecol. Manag. **184**, 299–313 (2004)
A. Saltelli et al., Five ways to ensure that models serve society: A manifesto. Nature **582**(7813) (2020)
W.M. Schaffer, Ecological abstraction: The consequences of reduced dimensionality in ecological models. Ecol. Monogr. **51**(4), 383–401 (1981)
C. Shannon, A mathematical theory of communications. Bell. Syst. Technol. J. **27**, 379–423 (1948)
A. Smith, *An Inquiry into the Nature and Causes of the Wealth of Nations* (London, 1776)
M. Sonnewald, S. Dukiewicz, C. Hill, G. Forget, Elucidating ecological complexity: Unsupervised learning determines global marine eco-provinces. Sci. Adv. **6**(22), eaay4740 (2020). https://doi.org/10.1126/sciadv.aay4740
J.I. Steinfeld, J.S. Francisco, W.L. Hase, *Chemical Kinetics and Dynamics*, 2nd edn. (Prentice-Hall, New York, 1999), pp. 151–152
D.L. Strayer, M.E. Power, W.F. Fagan, S.T.A. Pickett, J. Belnap, A classification of ecological boundaries. Bioscience **53**(8), 723–729 (2003)
R.K. Suarez, K.C. Welch, Sugar metabolism in hummingbirds and nectar bats. Nutrients **9**(7), 743 (2017). https://doi.org/10.3390/nu9070743

L. Tedersoo, M. Bahram, M. Zobel, How mycorrhizal associations drive plant populations and community biology. Science **367**(6480), 867–876 (2020)

R.E. Ulanowicz, *Utricularia's* secret—The advantage of positive feedback in oligotrophic environments. Ecol. Model. **79**, 49–57 (1995)

R.E. Ulanowicz, Information theory in ecology. Comput. Chem. **25**(4), 393–399 (2001)

R.E. Ulanowicz, The dual nature of ecosystem dynamics. Ecol. Model. **220**, 1886–1892 (2009)

R.K. Upadhyay, Dynamics of an ecological model living on the edge of chaos. Appl. Math. Comput. **210**(2), 455–464 (2009). https://doi.org/10.1016/j.amc.2009.01.006

M.P. Veldhuis, M.P. Berg, M. Loreau, H. Olff, Ecological autocatalysis: A central principle in ecosystem organization? Ecol. Monogr. **88**(3), 304–319 (2018)

M. Venzon, A. Pallini, A. Janssen, Interactions mediated by predators in arthropod food webs. Neotrop. Entomol. **30**(1), 1–9 (2001)

D.A. Wardle, G.W. Yeates, G.M. Barker, K.I. Bonner, The influence of plant litter diversity on decomposer abundance and diversity. Soil Biol. Biochem. **38**, 1052–1062 (2006)

I.D. White, D.N. Mottershead, S. Harrison, *Environmental Systems: An Introductory Text* (Chapman and Hall, London, 1992)

N. Wiener, *Cybernetics, or Control and Communication in the Animal and Machine* (Martino Publishing, Mansfield Centre, 2013)

J.S. Wright, R. Fu, J.R. Worden, S. Chakraborty, N.E. Clinton, C. Risi, Y. Sun, L. Yin, Rainforest-initiated wet season onset over the southern Amazon. PNAS (2017). https://doi.org/10.1073/pnas.1621516114. Accessed 6 Jan 2021

References

Chapter 5
Ecosystem Control: A Top-Down View

5.1 Two Ways to Look at Systems

Systems can be evaluated in two ways: by analyzing the parts of the system (the bottom-up view) and by measuring performance of the system (the top-down view). The benefits of each approach become apparent when, for example, there is a need to evaluate a young man who wants to be a football player. A physician would take the reductionistic approach and analyze parts of the man's system: his heart condition, blood pressure, body mass, and blood chemistry. In contrast, a coach would take the holistic approach and measure the prospect's speed, his ability to catch a pass, and his ability to tackle a ball carrier. An analytic attempt to predict the performance of an ecosystem is as futile as a medical doctor trying to predict a football players success on the field. Both holistic and analytic viewpoints are needed to fully understand and manage all types of systems.

5.2 Composing and Decomposing Trophic Webs

Ecosystems consist of two types of trophic webs or communities, a composing web and a decomposing web. A *composing* ecological community is one where energy is used to build ecosystem structure. A *decomposing* community is one where the structure is broken down. Composing and decomposing communities are tied together in ecosystems by a flow of energy and a feedback loop of nutrients (Fig. 5.1). Composing webs described by Lindeman (1942) consist of autotrophs such as plants, and consumers, meaning herbivores and carnivores. They are composing webs because the ecosystem structure is increasing. Their biomass is built from the energy contained in the reduced carbon compounds of the biomass in the preceding level. Lindeman estimated that roughly 10% of the energy at one level in

C. F. Jordan, *Evolution from a Thermodynamic Perspective*,
https://doi.org/10.1007/978-3-030-85186-6_5

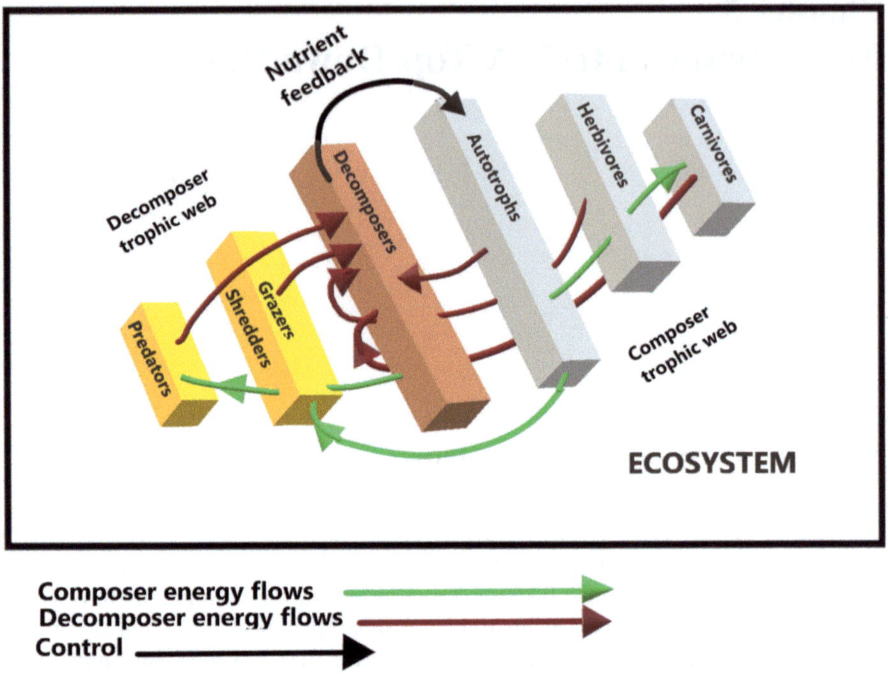

Fig. 5.1 Endogenous control over trophic communities via nutrient feedback in a terrestrial ecosystem. Each trophic level consists of species populations that are controlled by interactions with species in adjacent trophic levels. Functional groups of species such as herbivores are the bars that comprise trophic levels. A composer energy flow arrow from autotrophs to carnivores through herbivores indicates that herbivores are a step in the composer food chain

the composer food chain is incorporated in the biomass of the next. The other 90% is lost (Turner 1970). That "loss" means that biomass energy returns directly to decomposers instead of being passed along through the rest of the food chain. Until recently, top predators were usually considered the final step in energy transfer through trophic communities.

5.2.1 Decomposers in Soil Organic Matter

Although trophic webs in the soil have long been studied by soil ecologists (Hendrix et al. 1986), the importance of decomposer trophic webs has only recently been recognized as important for ecosystem control. Decomposers are at the base of what is called the below-ground trophic web in terrestrial ecosystems, although some of the species may occur in the litter layer above the soil, or even completely above ground. Decomposer webs receive energy from the composer webs (Moore and de Ruiter 2012). The transfer often begins when an organism dies, but could begin while it is still alive. In the case of animal tissue, the web usually begins with microbes. For wood, fungi are the initial decomposers. For leaf litter, which constitute the greatest transfer of energy between composing and decomposing webs, shredding and grazing arthropods begin the decomposer chain followed by bacteria. The decomposer trophic web also contains composer flows as when bacteria are fed upon by protozoa, nematodes and predatory arthropods. When predaceous arthropods and other consumers die, the energy from their bodies then re-enters the decomposing chain (Fig. 4.3, Chap. 4).

5.2.2 Decomposers in Marshes and Mangroves

The high productivity of fresh and saltwater marshes and of mangroves was not recognized until the studies of Teal (1962), Odum (1970), and Odum and Heald (1972). In these systems as in terrestrial systems, release of nutrients from decaying organic matter (sometimes called detritus) is a key to the high productivity of plants and the animals that consume the plants. In both cases, herbivores and omnivores consume plant material and pass along nutrients in a form easily assimilated by plants. In terrestrial systems, we usually think of conspicuous vertebrates that eat living vegetation as the consumers and micro-arthropods and bacteria as the decomposers (Fig. 5.1). But large herbivores such as deer can be thought of as primary decomposers, because they are the first to begin breaking down the structure of plant biomass. In marshes and mangroves, inconspicuous arthropods are thought of as part of a decomposer food web because they eat dead leaves and roots. They are breaking down, not building up ecosystem structure. However, they are part of a secondary composing food chain because they compose the decomposing food web. Even bacteria are composers in they make up part of the decomposing food chain. They are the last stage though, because that is where the energy that supports life runs out. Bacteria are the ultimate consumers as well as ultimate decomposers.

Box 5.1
When we eat meat, vegetables, and potatoes, we think of ourselves as ultimate consumers. We are not. It is the bacteria in our gut that ultimately consumes the food, depletes the energy, and releases vitamins, nutrients and energy in a form that we can use. We just help break down the food structure

5.3 Control of Systems

5.3.1 Top-Down vs. Bottom-Up

Political systems are controlled top-down by government. Bottom-up control in democratic systems is by voters. Economic systems are controlled top-down by the Federal Reserve Bank and bottom-up by consumers and small business. Top-down control in ecosystems means energy flow through species populations is regulated by feedback at the decomposer-autotroph interface. Bottom-up control means species populations control energy flow through ecosystems as species respond to feedback by increasing or decreasing energy flow through their populations. In this chapter, we look at ecosystem control from a holistic, top-down point of view. In Chap. 6, we look at ecosystem control from an analytic, bottom-up perspective.

5.3.2 Top-Down Exogenous Control

Energy flow through natural ecosystems is controlled top-down both exogenously and endogenously. Exogenous control is by environmental conditions including precipitation, soil fertility, temperature and available sunlight. They affect the rate at which solar energy is captured by plants and the rate at which that energy is introduced into the ecosystem's feedback system.

5.3.3 Exogenous Impacts and Stability

Rate of energy flow through ecosystems is influenced by periodic exogenous factors such as seasonal changes in temperature and rainfall. Periodic exogenous factors do not destabilize ecosystems, because species have adapted to the daily and seasonal changes in energy input. Amplitude of variations in energy flow through ecosystems due to periodic factors constitute bounded limits, that is the amount of variation in energy flow that can occur without ecosystems becoming unstable. In terrestrial systems, precipitation sets bounded limits for energy flow. Unlike temperature variations which on a seasonal basis are quite regular, precipitation can have large seasonal and yearly variations which result in large variations in annual energy flow. The widest variation that an ecosystem can tolerate without degrading (where energy-out due to metabolism is greater than energy-in from photosynthesis) sets the bounded limits for ecosystem stability (See Fig. 4.1b).

Because of variations in exogenous inputs, ecosystems are never in a steady state where energy content of the system is constant. There are always diurnal and seasonal variations in energy flow through, an ecosystem, resulting in pulses of energy flow and productivity. In deserts and high latitude/altitude regions, seasonal pulses can last only a few months or less. Energy flow through these ecosystems decreases when water is lacking, or temperature inhibits metabolic processes of plants. Nevertheless, the systems remain stable, because species have adapted to these pulses by shutting down metabolism when conditions are too dry or too cold.

Non-periodic environmental changes such as caused by climate change, asteroid impact, volcanic eruptions, and elimination of predators destabilizes ecosystems. When disturbances interrupt energy flow to an extent greater than that to which they are adapted, it is not possible to predict the fate of individual species.

Román-Palacios and Wiens (2020) in their examination of species extinction and survival in response to climate change showed "the importance of maximum temperatures for causing species extinction, and niche shifts for allowing their survival." Since we cannot predict a species' reaction to these variables, their response is stochastic.

In contrast, thermodynamic models are deterministic and can predict the trajectory of ecosystem resilience following catastrophic disturbance (Fath et al. 2004). Following a perturbation that exceeds an ecosystem's bounded limits, power output and species diversity will decrease until new species adapt to the altered environment. Diversity and power output will then gradually increase until photosynthetic capture of energy becomes limiting.

5.3.4 Top-Down Endogenous Control

Exogeneous impacts are dampened by endogenous feedback. Feedback controls ecosystem stability, resilience and resistance. We can enter the feedback cycle at the point where solar energy is fixed in plants by photosynthesis. From plants, energy continues as biomass (reduced carbon compounds) through species, trophic webs, and ecosystems. Some of the energy is dissipated as metabolic heat as it flows through the trophic chains, but a large portion moves directly to decomposers through leaf-fall or when animals die due to disease or old age. By the end of the decomposer trophic chain, energy is completely depleted. Biomass has been oxidized to carbon dioxide, and the released energy has been returned to the atmosphere as heat. Although nutrients move together with biomass through trophic webs, nutrients are still present in the system and are recycled (Loreau et al. 2004). For nutrients to be recycled to autotrophs in terrestrial systems, they first have to be released by microbes from the organic matter in which the nutrients are embedded. It is at this junction between decomposers and autotrophs that top-down endogenous control of feedback occurs (Fig. 5.1) (Moore et al. 2003). Feedback is controlled top-down by the autocatalytic reaction between energy captured by photosynthesis and nutrients, a reaction that synthesizes new biomass. The energy in biomass then moves through food chains from autotrophs to downstream populations where bottom-up control takes place.

Box 5.2

Top-down and bottom-up control in community ecology have different meanings than they do in ecosystem control. In community ecology, top-down control means that populations within trophic webs are controlled by top predators. Bottom-up control means that trophic webs are controlled by plants that produce secondary metabolites that defend against herbivores (Hunter and Price 1992; Dyer 2014)

Decomposition of an autotroph by a decomposer and recycling of nutrients from a decomposer to the autotroph was the very first feedback cycle that occurred in the very first ecosystem 2.7 billion years ago and set the pattern for ecosystem control as food chains and food webs evolved (Chap. 12). Animals came into existence merely to facilitate decomposition and accelerate nutrient feedback. Energy flow from autotrophs to decomposers is still the route through which most of energy and nutrients flow.

5.4 Endogenous Control Through Nutrient Recycling

5.4.1 Autocatalysis

In Chap. 4, autocatalysis in ecosystems was said to occur when interacting populations of different species form a loop whereby each species produces the resources needed by the next species in the loop, in such a configuration that the whole set of species is self-promoting and self-sustaining given sufficient input of energy and essential carbon and nutrients. Veldhuis et al. (2018) proposed that interactions in microbial loop cycles promote such autocatalysis.

Loop type autocatalysis can explain how an ecosystem maintains energy and nutrient flow, but does not explain the increase in flows that occurs during succession and evolution. To understand that, we can look at enzyme kinetics. In reactions catalyzed by enzymes, substrates undergo chemical change in which chemical bonds are broken and new ones formed to synthesize products. In ecosystems, glucose produced by photosynthesis and nutrients released from decomposing organic matter are reactants. Inside the cells of plants their reaction is catalyzed by enzymes to form new plant material in the form of leaves and wood that will become energy sources for other trophic levels in the ecosystem. When an increase in nutrient availability allows an increase in biomass production, all trophic levels of the system subsequently increase in biomass. This results in nutrient-containing biomass being returned to decomposers at an increased rate. Nutrients are released from dead biomass by extracellular enzymes secreted by bacteria and fungi. The enzymes break down the organic matter on which the microbes feed, thereby releasing the nutrients bound up in that organic matter and making them available to plants for re-use in synthesizing new biomass (Bardgett and Wardle 2010). The increase in nutrient availability results in a further increase in the rate of biomass production. The ecosystem grows autocatalytically. Biomass will continue to increase until such time as the photosynthetic capacity of the autotrophs becomes limiting.

5.4.2 Control of Microbial Activity

Microbial activity determines the rate at which nutrients are made available for uptake by plants and ultimately for autocatalytic reaction with glucose. The rate that microbes break down organic matter depends on several factors. On a global scale, climate is the factor that influences decomposition (Steidinger et al. 2019). On a local scale, decomposition is influenced by soil fertility and biota in addition to climate (Jenny 1941). Microbes excrete nutrients that are in excess of their own requirements. Excreted nutrients are in forms that are biologically available (Ayers et al. 2007).

5.4.3 *Inhibition of Microbial Activity by Leaf Sclerophylly*

Decomposition is influenced by adaptations of plants growing on nutrient-poor soils or where nutrient cycling is inhibited by local climate. Conservation of nutrients is important because of their scarcity in these environments. Trees often have sclerophyllous leaves (Medina et al. 1990), an adaptation that gives leaves a long life, up to 50 months (Reich et al. 1991). Sclerophyllous leaves contain high concentrations of cellulose and lignin (Ågren and Andersson 2012), resulting in C/N (carbon/nitrogen) ratios of 58:1 to 88:1 (McGroddy et al. 2004). That ratio is too high for optimum decomposition after abscission. A C/N ratio of between 25 and 30: 1 is optimal for decomposition (USDA 2019). In the cloud forests that occupy high altitudes on tropical mountains, sclerophylly is common. Because of the high C/N ratio in leaf litter, microbial activity is inhibited, litter is slow to decompose and nutrients are sequestered in a layer of undecomposed humus. Because nutrient recycling is inhibited, photosynthesis is further reduced, trees are stunted, and forests grow only a few meters high (Bruijnzeel and Proctor 1995). Similar "caatinga" forests appear on lowland white sands of the Amazon Basin, as well as on the top of "tepuis", blocks of Precambrian quartz sandstone (Cuevas 2001). Decomposition of conifer needles is also inhibited by a high C/N ratio at high latitudes where low temperatures inhibit microbial activity (Monk 1966).

5.4.4 *Inhibition of Microbial Activity by Chemical Defenses*

Chemical defenses in plants protect the plants against herbivores. When attacked by the tobacco budworm *Heliothis virescens*, tobacco, cotton, and maize plants can emit a particular blend of chemicals that attracts the parasitic wasp *Cardiochiles nigriceps*, which lays its eggs in the budworm larvae (Demoraes et al. 1998). Chemical defenses also inhibit number, type, and activity of decomposer organisms (FAO 2019). Sour manioc (*Manihot esculenta*), the roots of which are an important source of carbohydrate for humans in tropical regions, can be left in the ground without rotting for long periods of time. It produces hydrocyanic acid precursors that protect the leaves from attack by insects. Presumably that is the reason the roots decompose so slowly. Slow decomposition means the roots can be left in the ground until it is convenient to harvest them.

In spruce-fir forests of high latitudes, there is a steady accumulation of organic matter on the forest floor (Paré et al. 1993) due to slow decomposition caused by tannins in the spruce needles. As a result, nutrients become sequestered in the litter layer, and regenerating forests experience a decline in growth rate after about 70 years (Erdle and Baskerville 1986). Nutrient limitations also cause an increase in susceptibility of the forest to an outbreak of spruce budworm (*Choristoneura spp.*) that severely injures the trees (Candau et al. 2018). As they die back, the increased litter on the forest floor increases the potential for fire. The resulting ash contains concentrated nutrients (except nitrogen and sulfur) that are readily available for recycling as the forest repeats the cycle (Houle et al. 2009).

5.4.5 Inhibition of Microbial Activity by Ecological Stoichiometry

Decomposition also depends on the balance of multiple chemical elements particularly carbon, nitrogen and phosphorus in trophic levels (Zhang et al. 2003; Elser et al. 2008). "Redfield stoichiometry" refers to the consistent atomic ratio of carbon, nitrogen and phosphorus found in marine phytoplankton and throughout the deep oceans (Redfield 1958). Optimal stoichiometry occurs when the nutrients in decomposing organic matter are in the relative concentrations that are required for optimal decomposer stoichiometry (Sinsabaugh and Follstad Shah 2012). That enhances nutrient recycling. Ratios have been examined in various terrestrial plant tissues (McGroddy et al. 2004) and in soil microbes (Cleveland and Liptzin 2007). Often the nitrogen-phosphorus ratio of dead plant material does not reflect the relative availability of N and P that is optimum for microbes (Garrish et al. 2010). Decomposer organisms often require them in different ratios than provided by their resources (Swift et al. 1979). On a global scale, C:N:P ratios in the nonliving soil organic matter (186:13:1) do not match well the ratios soil microbial biomass (60:7:1) (Cleveland and Liptzin 2007). The high ratios of C and N to P in soil organic matter may retard the release of nutrients (Moe et al. 2005), since decomposer organisms require a higher level of P.

5.4.6 The Synchrony Principle

Release of nutrients through decomposition and uptake of nutrients by plants in natural ecosystems are said to be well synchronized according to the" Synchrony Principle" (Palm et al. 2001) because changes in temperature and moisture affect rates of both photosynthesis and decomposition in the same way. In warmer, wetter environment, both photosynthesis and decomposition are higher than in colder, drier environments. Because of chemical defenses in leaves to protect against herbivory, nutrient ratios may not be optimum for decomposition. There are other mechanisms through which nutrient recycling is facilitated.

5.4.7 The Decay Law

Where nutrients are sequestered in the soil organic matter and nutrient availability retards decomposition below the rate required to supply growing trees, decomposition is controlled by the Decay Law (Olson 1963). Decomposition depends on a decay constant k in Eq. 5.1.

$$\frac{X}{Xo} = e - kt \tag{5.1}$$

Xo: the original amount of biomass
X: the biomass remaining after time *t*
e: the base of natural logarithms
k: the decomposition constant

Equation 5.1. The fraction of biomass remaining after time *t* as a function of the decay constant *k*.

The decay law means that loss rate of soil organic matter is a function of the organic matter remaining in or on the soil at any given moment (Jordan 1985). As mass increases linearly, rate of decomposition increases exponentially, so there can never be an infinite buildup of biomass on the forest floor. Although decomposition per gram of litter and hums of the forest floor is low, absolute decomposition per square meter can be high because there is a thick accumulation of organic matter. Since microbial biomass is an important component of soil organic matter (Crowther et al. 2019), when activity level of soil micro-organisms is low, total soil organic matter and the accompanying microbial biomass increase (teleologically speaking) to keep nutrients flowing and to keep the system stable. Greater biomass of decomposers compensates for their lower activity. This prevents the mat of organic materials on the floor of a nutrient-poor rain forest from exceeding 30–40 cm (Jordan and Escalante 1980). Microbial activity in peat bogs of Northern Europe also is low, but because of the decay law, build-up does not increase indefinitely (Lew et al. 2019).

In aquatic systems, decomposition of some organic matter by microbes takes place inside the gut of animals. In these systems, herbivorous fish consume autotrophic plankton and most nutrients are quickly rendered soluble and are excreted into the water. Recycling also occurs via drifting decomposers. Dissolved organic nitrogen may bypass drifting decomposers, and move directly from fish to autotrophs (Berman and Bronk 2003).

5.4.8 Direct Nutrient Cycling

"Direct nutrient cycling" is a mechanism to translocate nutrients directly from decomposing leaves to roots. It occurs in nutrient-poor ecosystems when decomposition of organic matter is too slow to supply the needs of trees or shrubs. Direct nutrient recycling can occur when nutrients move from decomposing leaves to roots via ectomycorrhiza (Went and Stark 1968). To test this hypothesis, Herrera et al. (1978) performed an experiment in a tropical rain forest on nutrient-poor Oxisols in the Amazon region of Venezuela. They labeled leaves with ^{32}P and placed them in growth chambers on the forest floor. Intact roots from the living root mat were inserted in a lateral hole in the chamber. Eight weeks later, the leaves and the root length that had grown attached to them were harvested and preserved. An electron micrograph of a mycorrhizal hyphae bridge showed ^{32}P in the decomposing leaf, the mycorrhiza, and the living root (Herrera et al.1978).The sequence of leaf decomposition as mycorrhizae attach to leaves and digest the biomass is shown in Fig. 5.2.

Many ectomycorrhizal fungi are active producers of phytase and phosphatase enzymes (Leake and Read 1997), and this aids them in obtaining both P and N from a range of organic sources, including partially decayed tree litter (Read and Perez-Moreno 2003). Direct nutrient transfer speeds up the recycling process. In the absence of ectomycorrhizal mycelium, partially decayed tree litter only very slowly releases inorganic phosphorus (Bending and Read 1995).

5.4.9 The Role of Animals

Sclerophyllous leaves often occur on plants living on nutrient poor soils as a way of conserving nutrients by protecting leaves against decomposition (Medina and Cuevas 1994) while still on trees where decomposition is not wanted (teleology again). It also protects leaf litter on the soil surface where decomposition *is* wanted. Consumption of sclerophyllous plant leaves by herbivores overcomes the defense and facilitates nutrient recycling. Herbivores thus are another type of decomposer.

Fig. 5.2 The sequence of leaf litter decomposition in the Amazon rain forest of Venezuela. From left to right: 1. Root attachment to freshly fallen leaf. 2. Initial stage of decomposition. 3. Leaf in advanced stage of decomposition showing root shell where tissue between veins is missing. 4. Center of figure shows a white fungal “ghost” where leaf tissue once existed. (Photos by C.F. Jordan)

In fact, all animals are decomposers because their gut microbiome speeds up recycling of nutrients back to autotrophs. The reason animals came into existence was to facilitate nutrient recycling and increase ecosystem productivity. Ecosystems that include animals have greater power output than those without, and therefore have been more successful.

The first ecosystem consisted only of autotrophs and decomposers. It functioned satisfactorily, at least until the autotrophs began producing chemical compounds that deterred decomposers. Why did some types of eukaryotic cells become consumers? It was a convenient way for them to get energy. It was beneficial for the ecosystem because it accelerated nutrient recycling. Incorporating animals into ecosystems was beneficial for both the animals and the ecosystem. That is why animals persisted, and why evolution produced so many kinds of species. The more species in an ecosystem, the more efficiently nutrients are recycled.

5.5 Marine Systems

5.5.1 Nutrient and Energy Recycling

Figure 5.3 illustrates the feedback system of energy and nutrients in marine environments.

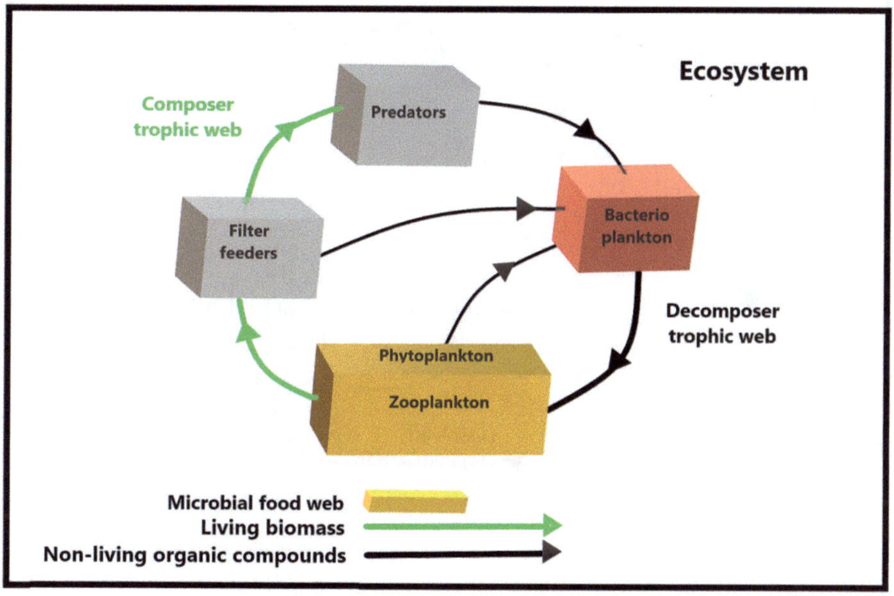

Fig. 5.3 Simplified model of energy and nutrient flow through a marine system. Details of the microbial food web are in Fig. 5.4

In marine systems, organic compounds resulting from various sources enter bacterioplankton (Van den Meersche et al. 2004). Sources include the leakage or exudation of fixed carbon from phytoplankton, sudden cell senescence, sloppy feeding by zooplankton, the excretion of waste products by aquatic animals, death of aquatic animals, and the breakdown or dissolution of organic particles from terrestrial plants and soils. Bacterioplankton degrade these compounds into dissolved organic carbon (Fig. 5.4).

Bacterioplankton are the major consumers of biomass energy in the sea (Pomeroy 1974). They are the top-down endogenous control of nutrient and organic matter cycling in marine systems. At the bacterioplankton stage, the cycles and flows differ from terrestrial ecosystems. In terrestrial ecosystems, decomposers deplete any energy that is remaining in the biomass that they consume. Only nutrients recycle into the composing trophic web. In ocean systems, some energy containing organic

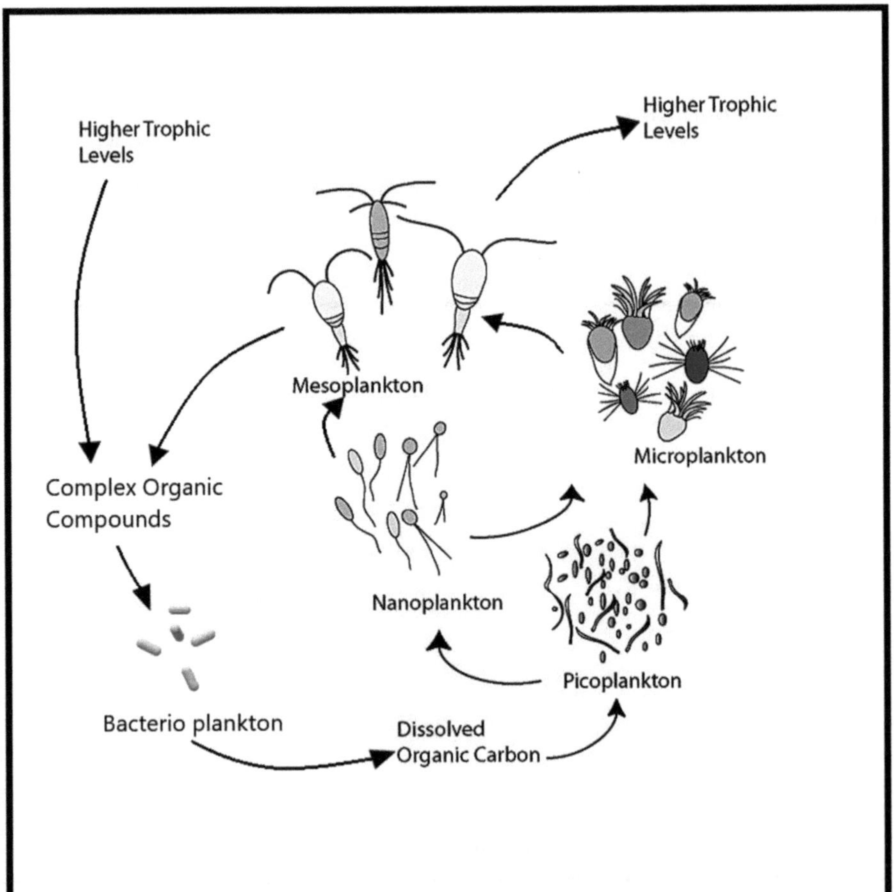

Fig. 5.4 Microbial feedback loop in marine systems

matter remains in the form of dissolved organic carbon after passing through bacterioplankton (Giovannoni 2017). This is a recycled energy source for organisms in the microbial food web. New energy is added to the system through photosynthesis by phytoplankton. From plankton, energy and nutrients move up a composing food web in the form of living biomass, with filter feeders such as sardines and mackerel analogous to herbivores in terrestrial systems, and predators such as sharks analogous to wolves in terrestrial systems. Like herbivores in terrestrial systems, fish and marine mammals facilitate decomposition of autotrophs and thereby increase nutrient cycling in oceans.

5.5.2 Exogenous Control

Exogenous Control of marine communities emerges from oligotrophic gyres influenced by trade winds and associated with nutrient upwelling (Sonnewald et al. 2020).

5.6 Control in Lakes

In oligotrophic (nutrient poor) lakes, nutrients move together with dissolved organic carbon into the microbial food web. Control is endosomatic similar to that in food chains of terrestrial systems. In eutrophic (nutrient-rich) bodies of water, control is exosomatic. Growth of plankton is stimulated by the concentration of dissolved nutrients in the water. If rate of biomass production of plankton is greater than the rate that filter feeders consume the plankton, decomposition of dead plankton depletes the water of oxygen, causing hypoxic zones that hamper the function of organisms higher in the trophic chain. Excess biomass production occurs when nitrogen and phosphorus runoff from agricultural fertilizers enrich nutrient levels in lakes, bays, and around the mouths of rivers (Breitburg et al. 2009).

5.7 Control in Managed Ecosystems

In industrial farm systems, productivity is maintained with fossil fuel subsidies, primarily in the form of fertilizers and pesticides. Fertilizer runoff, erosion, and pesticide drift cause the ecological problems of agriculture listed in Appendix 4. To avoid these problems, management for sustainability mimics control in natural systems. Management maintains energy flow and nutrient recycling through feedback at the decomposer-autotroph interface.

Soil erosion was severe following clear-cut logging of Southern Appalachian forests in the nineteenth century (Yarnell 1998). In the Piedmont, rivers still run red during flood stage from soil eroded from cotton fields laid bare in the same epoch.

Restoration of feedback cycles is through systems in which soil organic matter is rebuilt and nutrients again are recycled.

References

G.I. Ågren, F.O. Andersson, *Terrestrial Ecosystem Ecology* (Cambridge University Press, Cambridge, 2012)

E.N.J. Ayers, K.M. Dromph, R. Cook, N. Ostle, R.D. Bardgett, The influence of below ground herbivory and defoliation of a legume on nitrogen transfer to neighboring plants. Funct. Ecol. **21**, 256–263 (2007)

R.D. Bardgett, D.A. Wardle, *Aboveground – Belowground Linkages: Biotic Interactions, Ecosystem Processes, and Global Change* (Oxford University Press, Oxford, 2010)

G.D. Bending, D.J. Read, The structure and function of the vegetative mycelium of ectomycorrhizal plants. V. The foraging behavior of ectomycorrhizal mycelium and the translocation of nutrients from exploited litter. New Phytol. **130**, 401–409 (1995)

T. Berman, D.A. Bronk, Dissolved organic nitrogen: A dynamic participant in aquatic ecosystems. Aquat. Microb. Ecol. **31**, 279–305 (2003)

D.L. Breitburg, D.W. Hondorp, L.A. Davias, R.J. Diaz, Hypoxia, nitrogen, and fisheries: Integrating effects across local and global landscapes. Annu. Rev. Mar. Sci. **1**, 329–349 (2009)

L.A. Bruijnzeel, J. Proctor, Hydrology and Biogeochemistry of Tropical Montane Cloud Forests: What Do We Really Know? in *Tropical Montane Cloud Forests. Ecological Studies (Analysis and Synthesis)*, ed. by L. S. Hamilton, J. O. Juvik, F. N. Scatena, vol. 110, (Springer, New York, 1995)

J. Candau, R.A. Fleming, X. Wang, Ecological patterns of spruce budworm – Wildfire interactions in Central Canada's forests. Forests **9**(3) (2018). https://doi.org/10.3390/f9030137

C.C. Cleveland, D. Liptzin, C:N:P stoichiometry in soil: Is there a "Redfield ratio" for the microbial biomass? Biogeochemistry **85**, 235–252 (2007)

T.W. Crowther, J. van den Hoogen, J. Wan, M.A. Mayes, A.D. Keiser, L. Mo, C. Averill, D.S. Maynard, The global soil community and its influence on biogeochemistry. Science **365**(6455), eaav0550 (2019). https://doi.org/10.1126/science.aav0550

E. Cuevas, Soil versus biological controls on nutrient cycling in Terra Firme forests, in *The Biogeochemistry of the Amazon Basin*, ed. by M. E. McClain, R. L. Victoria, J. E. Richey, (Oxford University Press, Oxford, 2001), pp. 52–67

C.M. Demoraes, W.J. Lewis, P.W. Pare, H.T. Alborn, J.H. Tumlinson, Herbivore-infested plants selectively attract parasitoids. Nature **393**, 570–573 (1998)

A. Dyer, *Chasing the Red Queen: The Evolutionary Race Between Agricultural Pests and Poisons* (Island Press, Washington, DC, 2014)

J.J. Elser, R.W. Sterner, E. Gorokhova, W.F. Fagan, T.A. Markow, J.B. Cotner, J.F. Harrison, S.E. Hobbie, G.M. Odell, L.W. Weider, Biological stoichiometry from genes to ecosystems. Ecol. Lett. **3**(6) (2008). https://doi.org/10.1111/j.1461-0248.2000.00185.x

T.A. Erdle, G.L. Baskerville, Optimizing timber yields in New Brunswick forests, in *Ecological Knowledge and Environmental Problem Solving*, ed. by Committee on the Applications of Ecological Theory to Environmental Problems, (National Academy Press, Washington, DC, 1986), pp. 275–300

FAO, Organic matter decomposition and the soil food web. (2019), http://www.fao.org/3/a0100e/a0100e05.htm. Accessed 17 Dec 2020

B.D. Fath, S. Eørgensen, B. Patten, M. Straškraba, Ecosystem growth and development. Biosystems **77**(1–3), 213–228 (2004)

V. Garrish, L.A. Cernusak, K. Winter, B.L. Turner, Nitrogen to phosphorus ratio of plant biomass versis soil solution in a tropical pioneer tree, *Ficus insipida*. J. Exp. Bot. **61**(13), 3735–3748 (2010)
S.J. Giovannoni, SAR11 Bacteria: The most abundant plankton in the oceans. Annu. Rev. Mar. Sci. **9**, 231–255 (2017)
P.F. Hendrix, R.W. Parmelee, D.A. Crossley, D.C. Coleman, E.P. Odum, P.M. Groffman, Detritus food webs in conventional and no-tillage agroecosystems. Bioscience **36**, 374–380 (1986)
R. Herrera, T. Merida, N. Stark, C.F. Jordan, Direct phosphorus transfer from leaf litter to roots. Naturwissenschaften **65**, S.208 (1978)
D. Houle, L. Duchesne, R. Boutin, Effect of a spruce budworm outbreak on element export below the rooting zone: A case study for a balsam fir forest. Ann. Forest. Sci. **66**, 707 (2009). https://doi.org/10.1051/forest/2009057
M.D. Hunter, P.W. Price, Playing chutes and ladders: Heterogeneity and the relative roles of bottom-up and top-down forces in natural communities. Ecology **73**(3), 724–732 (1992)
H. Jenny, *Factors of Soil Formation* (McGraw Hill, New York, 1941)
C.F. Jordan, *Nutrient Cycling in Tropical Forest Ecosystems* (Wiley, Chichester, 1985)
C.F. Jordan, G. Escalante, Root productivity in an Amazonian rain forest. Ecology **61**, 14–18 (1980)
J.R. Leake, D.J. Read, Mycorrhizal fungi in terrestrial habitats, in *Wicklow DT*, ed. by B. Söderström, (The Mycota IV. Environmental and microbial relationships, Springer, Heidelberg, 1997), pp. 281–301
S. Lew, K. Glińska-Lewczuk, M. Lew, The effects of environmental parameters on the microbial activity in peat-bog lakes. PLoS One **2019** (2019) Accessed 26 Dec 2020
R.L. Lindeman, The trophic-dynamic aspect of ecology. Ecology **23**, 399–418 (1942). https://doi.org/10.2307/1930126
M. Loreau, C. de Mazancourt, R.D. Holt, Ecosystem Evolution and Conservation, in *Evolutionary Conservation Biology*, ed. by R. Ferrière, U. Dieckmann, D. Couvet, (Cambridge University Press, 2004), pp. 327–343
M.E. McGroddy, T. Daufresne, L.O. Hedin, Scaling of C:N:P stoichiometry in forests worldwide: Implications of terrestrial Redfield-type ratios. Ecology (9), 85, 2390–2401 (2004)
E. Medina, E. Cuevas, Mineral nutrition: Humid tropical forests, in *Progress in Botany*, ed. by H. D. Behnke, U. Lüttge, K. Esser, J. W. Kadereit, M. Runge, vol. 55, (Springer, Berlin/Heidelberg, 1994), pp. 115–129. https://doi.org/10.1007/978-3-642-78568-9_6
E. Medina, V. Garcia, E. Cuevas, Sclerophylly and oligotrophic environments: Relationships between leaf structure, mineral nutrient content, and drought resistance in tropical rain forests of the Upper Rio Negro Region. Biotropica **22**(1), 51–64 (1990)
S.J. Moe, R.S. Stelzer, M.R. Forman, W.S. Harpole, T. Daufresne, T. Yoshida, Recent advances in ecological stoichiometry: Insights for population and community ecology. Oikos **109**, 29–39 (2005)
C.D. Monk, An ecological significance of evergreeness. Ecology **47**, 504–505 (1966)
J.C. Moore, P.C. de Ruiter, *Energetic Food Webs: An Analysis of Real and Model Ecosystems* (Oxford University Press, Oxford, 2012)
J.C. Moore, K. McCann, H. Setälä, P.C. de Ruiter, Top-down is bottom-up. Does predation in the rhizosphere regulate aboveground dynamics? Ecology **84**, 846–857 (2003)
W.E. Odum, Pathways of Energy Flow in a South Florida Estuary. Dissertation. University of Miami (1970)
W.E. Odum, E.J. Heald, Trophic analysis of an estuarine mangrove community. Bull. Mar. Sci. **22**, 671–738 (1972)
J.S. Olson, Energy storage and the balance of producers and decomposers in ecological systems. Ecology **44**, 322–332 (1963)
C.A. Palm, K.E. Giller, P.L. Mafongoya, M.J. Swift, Management of organic matter in the tropics: translating theory into practice, in *Managing Organic Matter in Tropical Soils: Scope and Limitations. Developments in Plant and Soil Sciences. vol. 93*, ed. by

C. Martius, H. Tiessen, P. L. G. Vlek, (Springer, Dordrecht, 2001), pp. 63–75. https://doi.org/10.1007/978-94-017-2172-1_7
D. Paré, Y. Bergeron, C. Camiré, Changes in the forest floor of Canadian southern boreal forest after disturbance. J. Veg. Sci. **4**, 6 (1993). https://doi.org/10.2307/3235619
L.R. Pomeroy, The ocean's food web, a changing paradigm. Bioscience **24**(9), 499–504 (1974)
D.J. Read, J. Perez-Moreno, Mycorrhizas and nutrient cycling in ecosystems – A journey towards relevance? New Phytol. **157**, 475–492 (2003)
A. Redfield, The biological control of chemical factors in the environment. Am. Sci. **46**(3), 205–221 (1958)
P.B. Reich, C. Uhl, M.B. Walters, D.S. Ellsworth, Leaf lifespan as a determinant of leaf structure and function among 23 Amazonian tree species. Oecologia **86**, 16–24 (1991)
C. Román-Palacios, J.J. Wiens, Recent responses to climate change reveal the drivers of species extinction and survival. PNAS **117**(8), 4211–4217 (2020)
R.L. Sinsabaugh, J.J. Follstad Shah, Ecoenzymatic stoichiometry and ecological theory. Ann. Rev. Ecol. Evol. S. **43**, 313–343 (2012). https://doi.org/10.1146/annurev-ecolsys-071112-124414
M. Sonnewald, S. Dukiewicz, C. Hill, G. Forget, Elucidating ecological complexity: Unsupervised learning determines global marine eco-provinces. Sci Adv **6**(22), eaay4740 (2020). https://doi.org/10.1126/sciadv.aay4740
B.S. Steidinger, T.W. Crowther, J. Liang, et al., Climatic controls of decomposition drives the global biogeography of forest-tree symbioses. Nature **569**(7756), 404–408 (2019)
M.J. Swift, O.W. Heal, J.M. Anderson, *Decomposition in Terrestrial Ecosystems. Studies in Ecology Volume 5* (University of California Press, Berkeley, 1979)
J.M. Teal, Energy flow in the salt marsh ecosystem of Georgia. Ecology **43**(4), 614–624 (1962)
F.B. Turner, The ecological efficiency of consumer populations. Ecology **51**(4), 741–742 (1970)
USDA, Natural Resources Conservation Service. Carbon to Nitrogen Ratios in Cropping Systems. (2019), https://www.nrcs.usda.gov/Internet/FSE_DOCUMENTS/nrcseprd331820.pdf. Accessed 5 Jan 2021
K. Van den Meersche, J. Middelburn, K. Soetaert, P. van Rijswijk, H.T.S. Boschker, C.H.R. Heip, Carbon-nitrogen coupling and algal-bacterial interactions during an experimental bloom: Modeling a 13C tracer experiment. Limnol. Oceanogr. **49**(3), 862–878 (2004). https://doi.org/10.4319/lo.2004.49.3.0862
M.P. Veldhuis, M.P. Berg, M. Loreau, H. Olff, Ecological autocatalysis: A central principle in ecosystem organization? Ecol. Monogr. **88**(3), 304–319 (2018)
F.W. Went, N. Stark. Mycorrhiza. BioScience **18**, 1035–1039 (1968)
S.L. Yarnell, *The Southern Appalachians: A History of the Landscape*, General Technical Report SRS-18 (U.S. Dept. of Agriculture Forest Service. Southern Research Station, 1998)
L. Zhang, B. Youngei, H. Xingguo, Application of N:P stoichiometry to ecology studies. Acta Bot. Sin. **45**(9), 1009–1018 (2003)

Chapter 6
Ecosystem Control: A Bottom-Up View

6.1 Species as Arbitrageurs of Energy

In economics, arbitrage is the process of exploiting differences in the price of an asset by simultaneously buying and selling it. Any change in supply or demand for particular goods or services is signaled by a change in price that increases or decreases the amount of money held by the arbitrageur. Bottom-up control in ecosystems occurs through species acting as arbitrageurs of energy. Species respond to changes in supply and demand for energy by a change in size of individuals within the population or a change in number of individuals in the population. Biomass of populations at any moment reflects the difference between: (1) the rate that individuals eat (or photosynthesize) and the rate they are eaten; (2) the difference between the population birth rate and death rate.

6.1.1 Relation Between Rate of Flow and Mass in Hydraulic Systems

The way that population biomass responds to changes in energy input is analogous to the way that water pressure in a cylinder responds to changes in depth of the water column. We can demonstrate that using a graduated cylinder with holes of equal size drilled every 4 cm along the length of the tube (Fig. 6.1). Water from a control valve is directed into the top of the cylinder at a rate that maintains the head 1 cm above the bottom hole. The flow rate out of the bottom hole equals rate of water input (Fig. 6.1a). Equilibrium is maintained because the pressure causing water discharge depends on the head of water above the hole. Then rate of water input is increased, and as the water level in the cylinder rises, the increased water pressure causes water to flow more quickly out of the bottom hole until equilibrium

C. F. Jordan, *Evolution from a Thermodynamic Perspective*,
https://doi.org/10.1007/978-3-030-85186-6_6

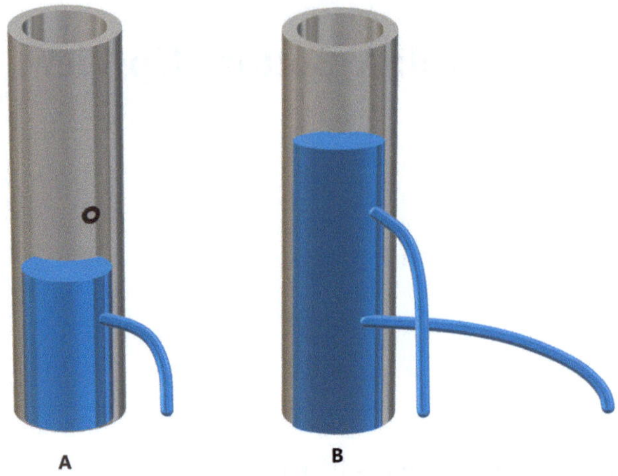

Fig. 6.1 Flow of water out of holes in a cylinder as a function of rate of water input into the cylinder. The head above a discharge point regulates the pressure that forces water out of that discharge point, thereby balancing output with input

is again achieved. If rate of input is increased until the water level rises above the second hole, the difference in pressure between the two holes can be seen by the distance that the water shoots out from each hole (Fig. 6.1b). Discharge rate at each hole is a function of the water pressure at each hole. As input of water to the cylinder increases, the head of water increases to keep the output balanced with input.

6.1.2 Relation Between Population Biomass and Rate of Energy Flow

Biomass of a species population responds to changes in energy flow just as mass of water in the cylinder responds to changes in water flow. The response is illustrated by predator-prey dynamics (Fig. 6.2) that have a theoretical foundation in the Lotka-Volterra predator- prey equations (Berryman 1992). The increase in biomass of a lower trophic level (hare) results in increased pressure for energy to exit. As prey population of an ecosystem rises, the predator population on that prey responds by increasing its population, A decrease in prey population has the opposite effect. As population of a species increases or decreases, the biomass of that species changes accordingly, so predator-prey interactions can also be expressed as the energy relations between predator and prey (Fig. 6.3).

When the rate of energy entering a population increases, biomass and its embedded energy increase. Pressure to equilibrate increases until the rate that energy moves out to the next trophic level matches the rate of input (Jørgensen and Fath 2006). Because of continuous feedback in the system, there is always either

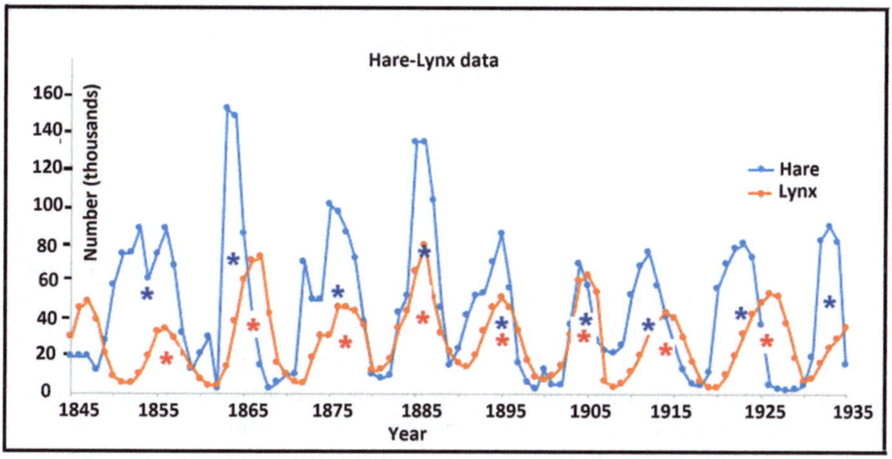

Fig. 6.2 Estimation of snowshoe hare and Canadian lynx populations in the Mackenzie River district of the North-West Canada for the period 1821–1934. Asterisks indicate average value for each 10-year interval. The data on which this figure is based was compiled by MacLulich (1937) from historic pelt-trading records of the Hudson Bay Company. (Data re-published by Zhang and Li 2007)

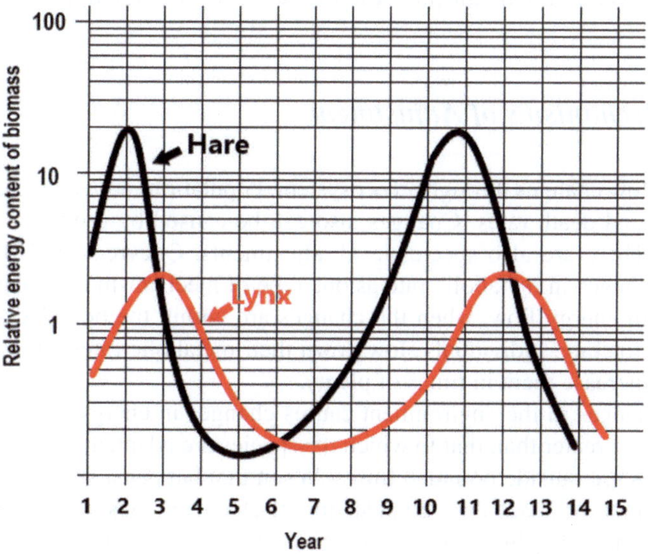

Fig. 6.3 The energy content of hare and lynx biomass populations follows the same dynamics as number of individuals. The pattern assumes that only 10% of the hare biomass becomes lynx biomass in accordance with Lindeman's (1942) trophic dynamic model that 90% of the energy in a trophic level is lost upon transfer to the next level

pressure to increase biomass or suction to decrease biomass. Energetic equilibrium in populations is never reached, but interacting species continually move asymptotically toward equilibrium (Yackinous 2015).

6.2 Equilibrium

According to the equilibrium law (Le Chatelier's principle), an ecological population is in stable equilibrium when its biomass responds to a stress by diminishing the change that has been made to it. Stable equilibrium depends on the time scale chosen for analysis. The blue and red lines of Fig. 6.2 indicate a lack of equilibrium over approximately 10-year intervals. The blue and red asterisks however suggest that over the course of a century, populations are close to stable equilibrium.

Responses of predators to changes in prey populations may not fully explain the cycles of the hare population in Fig. 6.2. Krebs et al. (2001) found that the hare-lynx cycle is produced by an interaction between predation and food supplies, and its biological impacts ripple across many species of predators and prey in the boreal forest. The question then becomes, what controls food supplies? The periodicity of the hare-lynx cycle coincides roughly with the sunspot cycle of 11 years. That theory could be valid only if sunspot variation causes a variation in plant productivity.

6.2.1 Mechanisms of Adjustment

As energy flow changes through an ecosystem, populations adjust by changing their birth rates, and death rates. Changes also can be caused by predation, disease, or starvation. There are also species-level adjustments. Quévreux and Brose (2018) proposed changes in metabolic rate as one type of mechanism to adjust to changes in ecosystem energy flow, when the changes are within the bounded limits that a system can tolerate. Individuals slow down their metabolic rates when facing starvation and increase them in times of plenty.

When changes in the environment causes changes in energy flow through ecosystems to be greater than that to which the species are adapted, we can say that the energy flows are outside bounded limits. When that happens, species will migrate, become extinct, or adapt through selection of characteristics better adapted to the new environment. Evolutionary adaptation is evident in the small white-tailed deer that exist in warm Key-West Florida that are a sub-population of the larger deer found in much colder New England. Smaller animals are better adapted to a warming climate because they have more surface area per unit body mass from which heat can radiate. Key-West deer eventually may become a new species. With a cooling climate, the opposite trend will occur. Another example is found among different species of bears. The smallest of bears, the sun bear, lives in the tropical forests of Southeast Asia and the largest bears, Kodiak and Polar, live in Alaska.

6.2.2 *Adjustments and Climate Change*

Terrer et al. (2021) argued that climate change models that predict carbon storage in both biomass and soil organic matter will increase as atmospheric CO_2 increases are incorrect. In a meta-analysis of forest biomass and carbon stock changes resulting from increased CO_2, they found that high accumulation of carbon in biomass was associated with soil carbon loss, whereas low biomass accumulation was associated with soil carbon gain. They proposed that plant nutrient-acquisition strategies are responsible. When a forest ecosystem in steady state equilibrium is subjected to increased CO_2, biomass of the forest increases but as a result, there is an increase in "mining" for soil nutrients, as ectomycorrhizal fungi facilitate an increase in breakdown of soil organic matter and release of stored nutrients. Cybernetic theory predicts what will happen next. As soil nutrients are depleted, biomass increase will be slowed and the system tends again toward steady state. When biomass accumulation begins to drop, "mining" decreases and soil organic carbon builds up.

6.2.3 *Bird Populations*

Cycles of population expansions and contractions have been a common feature of many bird species during the Quaternary period, likely coinciding with climate cycles (Nadachowska-Brzyska et al. 2015). When climate got colder during the ice ages and energy flow in the biosphere decreased, some species survived by migrating South, but others that remained in original habitat adapted evolutionarily to the colder regime, perhaps by increasing body insulation.

6.2.4 *Dis-equilibrium*

In contrast to the stable oscillations of the hare-lynx populations in Canada, wolf and moose populations, once in equilibrium on Isle Royale in Lake Superior (McLaren and Petersen 1994), are no longer stable because of the decline in wolf populations due to inbreeding (Mills 2016). From a thermodynamic perspective, population instability occurs when energy flow is disrupted or distorted beyond bounded limits. Loss of wolves on Isle Royale could cause the ecosystem to "collapse" because of the lack of control over herbivory. Without wolves to control the moose population, the moose will over-graze the trees and shrubs on which they depend for food, thereby killing the plants. Because they cannot emigrate, some or all of the moose will starve.

6.3 Population Instability vs. Ecosystem Instability

Individual species populations need not be stable and in equilibrium with an ecosystem's energy flux for the ecosystem to be stable. Energy flow through populations may vary while flow through ecosystems remains stable. Declines through one species population are often compensated by increases in another (Hooper et al. 2005). For example, the chestnut blight in eastern North America virtually wiped out the native American chestnut. Because there are five species of oaks in the same ecosystems, 72% of the pre-blight biomass had been replaced by the year 2000 (Diamond et al. 2000). Chestnut oak and red oak appear to have at least partially taken over the functional niche left vacant by the chestnut (Keever 1973).

When only one species makes up the dominant species in a mature ecosystem, its elimination can cause ecosystem instability. In the Southern Appalachian Mountains, hemlock populations make up the mature, "climax" forest. Survival of the hemlocks is threatened by sap-sucking woolly adelgids. If introduced predator beetles can stop the adelgids, they can provide resilience for the hemlocks (Patterson 2016). If not, the ecosystem will regress to an earlier stage of succession.

Fig. 6.4 Bottomland forest on Spring Valley Ecofarms in the Georgia Piedmont. Privet has been cleared out in front of the photographer. Dense privet monopolizing the understory is in the background. Unless controlled with herbicides, privet will resprout in the cleared area. (Photo by C.F. Jordan)

Ecosystem instability can be caused when an aggressive invasive species prevents reproduction of native populations in a trophic level. Stability of moist forest ecosystems of the Southeastern U.S. is threatened by the invasion of Chinese privet (*Ligustrum sinense*) that outcompetes seedlings for light energy by forming a dense understory, according to The Georgia Forestry Commission (Associated Press 2018) (Fig. 6.4). No species has appeared that can outcompete Chinese privet in the understory of moist forests, and there is yet no biological control. Only mechanical or herbicide control can eliminate it, but this is practical only in small areas.

6.4 Control by Interactions: Direct vs. Indirect

6.4.1 Indirect Interactions

Indirect interactions result from energy flow that connects species that are not adjacent, but are connected only through feedback. Indirect interaction means that any change in energy flow through a species caused by seasonal change or non-seasonal disturbance promulgates a change in energy and nutrient flow through direct species interactions downstream from the modified interaction. This chain of influence causes a change in ecosystem-level nutrient recycling from decomposers to autotrophs. Consequently, there is a change in energy flow to food webs that recalibrates the original species interactions toward a new equilibrium (Gellner and McCann 2016). This fine-tuning feedback results in "self-organization" that allows more efficient energy flow and nutrient recycling during changes in energy input to the system caused by climate change or other disturbances to the system.

Indirect interactions can be either beneficial or disadvantageous for the ecosystem. Beneficial indirect interactions can be deviation amplifying during succession and evolution when an ecosystem is increasing in biomass, or deviation dampening when it reaches maturity and is transitioning to stability. Disadvantageous indirect interactions often occur in human-managed ecosystems. Spraying a crop with pesticides has an indirect negative effect because those pests that evolve a strong resistance to pesticides survive and become harder to control.

6.5 Direct Interactions

Indirect interactions through feedback explains only part of the story of stability and resilience of populations and ecosystems. Direct interactions tell the rest. Direct interactions occur between two species within the same trophic level (competition), or between two adjacent trophic levels (predator-prey, mutualisms). They are "strong" interactions (Curtis et al. 2015) or brute force interactions (Patten and Odum 1981). Direct interactions are both the cause of and the result of feedback.

Adjustment of species populations to changes in energy flow through ecosystems is through direct interactions.

6.5.1 Predator – Prey

Predator – prey interactions have been classified as positive/negative because the interaction while benefitting the predator, harms at least one individual of the population being eaten. Harper (1977) asked how species could survive when they are involved in interactions that are detrimental for the species themselves. Interactions that are detrimental to individual organisms at the species level can be beneficial at the ecosystem level because they promote ecosystem stability (Mougi and Kondoh 2012; Gellner and McCann 2016). Stability of ecosystems is more important for survival of populations than is survival of a particular individual of that species. Wolves will kill an individual moose but that kill helps, rather than hinders the moose population as a whole by preventing overgrazing and starvation, and also benefits the ecosystem by ensuring that the moose population does not expand beyond the carrying capacity of the system (Slobodkin 1968). Although the sacrifice is unintentional, moose populations sacrifice individuals to wolves for the greater good of the population and the ecosystem.

Herbivore-plant interactions are also a predator-prey interaction that in the short-term may harm the plant, but in the long-term benefit ecosystems. Plants have evolved secondary metabolites that defend plants against disease (Futuyma and Slatkin 1983), but these compounds also inhibit bacterial decomposition of the leaves, thereby inhibiting decomposition and nutrient recycling. Some mammalian herbivores have evolved mechanisms that detoxify and eliminate defensive compounds (Kimball and Provenza 2003). The mammalian digestive system breaks down the hard-to-decompose leaves so they can be used as an energy source by the animal. Upon egestion, the manure now free of toxic metabolites, is quickly broken down through the action of bacteria in the soil. Rapid decomposition accelerates nutrient release and recycling, thereby benefitting the ecosystem. Herbivory is more important for ecosystem stability than is short-term energy balance in autotrophs.

6.5.2 Mutualisms

Mutualisms have been classified as a positive/positive interaction because they benefit both species involved. Egrets ride on the backs of large mammals picking parasites from their fur or skin to nourish themselves, which, in turn, frees these animals from the parasite burden. Egrets also are more sensitive to changes in their surroundings and more vigilant than their steeds, making them good at early warnings when it comes to predators or other dangers. Egrets and their hosts are more productive because of their mutualism. The interaction between trees and birds that eat the tree's fruit is a mutualism that is beneficial for both the plants and the birds (Howe and Smallwood 1982). The animals benefit through nutrition, and the plants benefit by having their seeds dispersed away from the parent plant. Seedlings clumped around a parent are more likely to be affected by disease and predation (Connell 1970).

One of the most important beneficial interaction in natural and managed ecosystems is the mutualistic relation between mycorrhizal fungi and higher plants (Martin et al. 2018). The ability of plants to scavenge nutrients from poor soils often is a limiting factor for their growth. In contrast, mycorrhizal fungi are very efficient at obtaining nutrients dispersed in the soil. Their very fine hyphae can penetrate very dense soil. What fungi don't do very well is obtain energy. A scenario for the development of the plant – mycorrhizal association could be that at one point in evolutionary time, a fungal body attached itself to a root hair that was leaking energy-containing organic compounds. The fungus adapted to the favorable environment, but the interaction (commensalism) was tenuous at first because the tree received no positive feedback. But when a nutrient rich fungus leaked phosphorus to a phosphorus deficient tree, the makings of a positive feedback loop were formed. Chemical signaling was established. The mutualism increased plant growth (Bolan 1991). The interaction between Rhizobia bacteria that live in the root nodules of legumes is another extremely important mutualism that increases the soil fertility of natural ecosystems as well as croplands. The benefit of these mutualisms at the ecosystem level is that as net primary productivity increases, the increase in energy flow (power output) increases the complexity and stability of ecosystems (Patten 2019).

6.5.3 *Competition*

Competition has been classified as negative/negative because it stresses interacting individuals and interacting species, but it has a long-term benefit for ecosystems because competition between sub-populations within a large population leads to formation of new species. An increase in species richness within a trophic level increases stability by providing alternative, redundant pathways of energy flow through that level (MacArthur 1955). Multiple species ensure that energy flow will continue, should one of the species become extinct.

Competition can start as a negative interaction as subspecies develop within a species, but the negative interaction disappears when the subspecies becomes a new species by either moving to a new habitat or developing a specialized energy source (thermodynamic niche, Appendix 5). Competition is the ultimate cause of diversity within trophic levels. Once negative interactions have disappeared, the positive results of increased species diversity due to competition become apparent.

6.5.3.1 Competition Leads to Complementarity and Formation of Thermodynamic Niches

In a stand of trees all of which are the same species, individuals compete for the same resources at the same time. All have leaves have the same vertical distribution, and therefore are competing for light. Roots compete at the same depth for nutrients and water.

Competition leads to formation of sub-populations that survive by adapting to part of the energy flow that is unused by the parent population. In a diverse ecosystem, there are different spatial and temporal demands for resources (Tilman and Downing 1994; Fichtner et al. 2020). Differential growth strategies among species in the same trophic level cause each species to develop its own thermodynamic niche (Appendix 5). An increase in species diversity within a trophic level increases the efficiency with which resources are captured or used (Huang et al. 2018). Differential patterns of leaf distribution among tree species ensure a more efficient use of solar radiation (Fig. 6.5). Differential patterns of root distribution ensure a

Fig. 6.5 Layering of leaves in a tropical rain forest. Composite assembled from Learning about Rainforests (2019)

more efficient use of soil nutrients and water (Cuevas et al. 1991). Ultimately though, competition is for energy. If a tree species can garner enough energy, it can outcompete other species for water and nutrients.

6.5.3.2 Competition in Terrestrial and Marine Systems

Autotrophic competition is important in terrestrial systems. In forest ecosystems, maximum production results from competition between species of trees with large biomass and slow turnover. Trees are part of a trophic pyramid with most of the biomass at the bottom. In marine systems, high productivity is achieved through a large number of autotrophic individuals (phytoplankton) with small biomass and rapid turnover. There is no apparent competition between individuals or species of phytoplankton. Physical factors such as wind and upwellings drive nutrients toward the water surface where they can be utilized by the rapidly reproducing autotrophs. They are part of a trophic pyramid with maximum biomass at the herbivore trophic level (Trebilco et al. 2013). In marine systems, competition is important between consumers. Maximum power in marine systems results from competition among herbivorous fish and some species of whales to consume the phytoplankton, and from competition between predators to catch filter feeding fish (Vanni et al. 2009). Grassland ecosystems are similar to marine systems in that maximum biomass is not at the autotroph level but at the herbivore level. Competition between individual herbivores for grass leads to selection of individuals with the greatest power output, but competition between species leads to development of distinct thermodynamic micro-niches (Appendix 5).

6.5.3.3 Ecosystem Competition

Early in evolutionary history when ecosystems resembled the microbial loops in the oceans of today, those loops with the highest power output were the ones that survived. High power output gave an ecosystem a competitive advantage. Although today's ecosystems such as the seven ecosystems in Appendix 2, Fig. 1 do not appear to be competing head-to-head, the species within each system are competing. Species competition increases the power output of ecosystems. New more powerful species may replace older less powerful ones, with the result of a higher power output of rehabilitated ecosystems. The new combination of species would be a new ecosystem that out-competed the one from which it was derived. Think about a football coach who replaces smaller players on the team with more powerful players. The team is the same, but the new combination could theoretically out-compete the combination that it replaced.

6.5.3.4 Nature, Red in Tooth and Claw

Evolutionary theory by Charles Darwin and Alfred Russel Wallace conflicted with the nineteenth century's Christian belief of harmony in the natural world. The phrase "Nature, red in tooth and claw" is a line from a poem by Alfred Tennyson in 1850 in which he deplored scientific ideas about nature functioning impersonally and with no direct divine intervention. Such intervention, it was thought, would facilitate cooperation, not the predatory killing that occurs in nature. Recent discoveries of mutualisms in nature seemed to support the idea that cooperation has been more important in evolution than competition (Gatti 2016). This view contrasts with thermodynamic theory. Thermodynamic theory argues that cooperation facilitated by mutualisms increases the ability of participants in the mutualism to become better competitors for energy in the race within food webs to maximize power output. The combined power output of two species involved in a mutualistic interaction is greater than the sum of the power outputs of the species acting independently. The two species have "joined forces" so to speak to become better competitors in the race for energy. It is like when two small armies join forces to defeat a more powerful army.

6.5.4 Decomposition

The interaction of bacteria and fungi with recently living biomass is beneficial from both the population and ecosystem perspectives. The decomposer populations use energy from the detritus to survive and to release nutrients bound in the detritus into a soluble form that can be taken up by roots of autotrophs, thereby producing the feedback that stabilizes the system.

6.5.5 Parasitism and Disease

Parasites are debilitating for the organisms affected, but in certain cases such as tsetse flies in the genus *Glossina*, they can defend the stability of the natural ecosystems in which they exist (Pearce 2000). Tsetse flies are obligate parasites that live by feeding on the blood of vertebrate animals. They have a prominent economic impact in sub-Saharan Africa as the biological vectors of trypanosomes, which cause human sleeping sickness and animal trypanosomiasis in domesticated cattle. Much forested land was cleared in the nineteenth century to develop pasture for cattle. Following accidental introduction of rinderpest in 1887, it was rapidly spread by tsetse flies. Most of the cattle in East Africa died off, and grassland reverted to woodlands, ideal habitat for tsetse flies (Jordan 1986). While millions of people died of sleeping sickness, wild mammal populations resistant to trypanosomiasis increased rapidly, and stability of energy and nutrient flow of the original ecosystem was presumably re-established. Disease benefitted the flow of energy and recycling if nutrients by eliminating cattle that disrupted these flows and cycles.

The interaction of bacteria and fungi with living organisms is negative for the living species, but it may be beneficial for stability of their population. When overpopulation threatens the resource base of a deer herd and there are no predators, bacterial and fungal diseases thin the populations (Pierce and Flinn 2018).

6.5.6 Commensalism and Amensalism

From an evolutionary perspective, commensalism and amensalism, are only short term interactions. Amensalism is a short-term relationship between organisms of different species in which one organism is inhibited or destroyed while the other organism remains unaffected. Black walnut exudes a chemical known as juglone from its roots and foliage, which builds-up in its local environment. The juglone is quite toxic to some other species of plants, but it is not toxic to others. Production of juglone may have evolved to inhibit disease or herbivory and its effect on other plants is just coincidental. The interaction disappears as plant species adapt.

Commensalism is an interaction between two organisms in which one benefits and the other derives neither benefit nor harm. It becomes permanent as the relationship evolves into a mutually beneficial interaction. The interaction between mycorrhizae and plants began as commensalism, but as feedback developed between species, it evolved into a mutualism.

6.5.7 Persistence of Negative Interactions

Why do interactions with negative components persist? Why are they not selected against? Interactions that are negative in the short term at the species level are positive at the ecosystem level in the long term. Selection for interactions that are beneficial for ecosystems is more important for survival of all life than selection that benefits a particular species in the short term. In the long-term, there are no negative interactions. All interactions are positive because they enhance the power output of ecosystems.

References

Associated Press, Asian privet, other plants are invading more Georgia ground. (2018), https://www.wabe.org/asian-privet-other-plants-are-invading-more-georgia-ground/. Accessed 7 Dec 2020

A. Berryman, The origins and evolution of predator-prey theory. Ecology **73**(5), 1530–1535 (1992)

N.S. Bolan, A critical review on the role of mycorrhizal fungi in the uptake of phosphorus by plants. Plant Soil **134**(2), 189–207 (1991)

J.H. Connell, On the role of natural enemies in preventing competitive exclusion in some marine animals and in rain forest trees, in *Dynamics of Populations*, ed. by P. J. Den Boer, G. R. Gradwell, (Wageningen Netherlands Center for Agricultural Publications, Wageningen, 1970), pp. 298–312

E. Cuevas, S. Brown, A.E. Lugo, Above-and belowground organic matter storage and production in a tropical pine plantation and a paired broadleaf secondary forest. Plant Soil **135**, 257–268 (1991)

R.J. Curtis, T.M. Brereton, R.L.H. Dennis, C. Carbone, N.J.B. Isaac, Butterfly abundance is determined by food availability and is mediated by species traits. J. Appl. Ecol. **52**(6), 1676–1684 (2015)

S.J. Diamond, R.H. Giles, R.L. Kirkpatrick, Hard mast production before and after the chestnut blight. South. J. Appl. For. **24**(4), 196–201 (2000)

A. Fichtner, F. Schnabel, H. Bruelheide, M. Kunz, K. Mausolf, A. Schuldt, W. Härdtle, G. von Oheimb, Neighborhood diversity mitigates drought impacts on tree growth. J. Ecol. (2020). https://doi.org/10.1111/1365-2745.13353

D.G. Futuyma, M. Slatkin, *Coevolution* (Sinauer Associates, Sunderland, 1983)

R.C. Gatti, A conceptual model of new hypothesis on the evolution of biodiversity. Biologia **2016**, 71. (3) (2016). https://doi.org/10.1515/biolog-2016-0032

G. Gellner, K.S. McCann, Consistent role of weak and strong interactions in high- and low-diversity trophic food webs. Nat. Commun. **7**, 11180 (2016)

J.L. Harper, The contributions of terrestrial plant studies to the development of the theory of ecology, in *The Changing Scenes in the Natural Sciences 1776–1976*, Special Publication 12, ed. by C. E. Goulden, (Academy of Natural Sciences, Philadelphia, 1977), pp. 139–157

D.U. Hooper, F.S. Chapin III, J.J. Ewel, et al., Effects of biodiversity on ecosystem functioning: A consensus of current knowledge. Ecol. Monogr. **75**(1), 3–35 (2005)

H.F. Howe, J. Smallwood, Ecology of seed dispersal. Annu. Rev. Ecol. Syst. **13**, 201–228 (1982)

Y. Huang, Y. Chen, N. Castro-Izaguirre, et al., Impacts of species richness on productivity in a large-scale subtropical forest experiment. Science **362**(6410), 80–83 (2018). https://doi.org/10.1126/science.aat6405

A.M. Jordan, *Trypanosomaisis Control and African Rural Development* (Longman, London/New York, 1986)

S.E. Jørgensen, B.D. Fath, Examination of ecological networks. Ecol. Model. **196**(3, 4), 283–288 (2006). https://doi.org/10.1016/jecolmodel.2006.02.029

C. Keever, Distribution of major forest species in Southeastern Pennsylvania. Ecol. Monogr. **43**, 303–327 (1973)

B.A. Kimball, F.D. Provenza Chemical defense and mammalian herbivores. USDA, National Wildlife Research Center – Staff Publications. 236. (2003), http://digitalcommons.unl.edu/icwdm_usdanwrc/236. Accessed 23 Dec 2020

C.J. Krebs, T. Boonstra, S. Boutin, A.R.E. Sinclair, What drives the 10-year cycle of snowshoe hares? Bioscience **51**, 25–35 (2001)

Learning About Rainforests, (2019). http://www.srl.caltech.edu/personnel/krubal/rainforest/Edit560s6/www/whlayers.html. Accessed 26 Dec 2020

R.L. Lindeman, The trophic-dynamic aspect of ecology. Ecology **23**, 399–418 (1942). https://doi.org/10.2307/1930126

R.H. MacArthur, Fluctuations of animal populations, and a measure of community stability. Ecology **36**, 53–536 (1955)

D.A. MacLulich, *Fluctuations in the Numbers of the Varying Hare (Lepus Americanus)*, University of Toronto Studies Biological Series 43 (University of Toronto Press, Toronto, 1937)

F.M. Martin, M.J. Harrison, S. Lennon, B. Lindahl, M. Opik, A. Polle, N. Requenq, M. Selosse, Cross-scale integration of mycorrhizal function. New Phytol. **220**(4), 941–946 (2018). https://doi.org/10.1111/nph.15493

B.E. McLaren, R.O. Petersen, Wolves, moose and tree rings on Isle Royale. Science **266**(5190), 1555–1558 (1994)

A. Mills, Two wolves remain on Ilse Royale. Michigan Tech News, April 19, 2016. (2016), https://www.mtu.edu/news/stories/2016/april/two-wolves-remain-isle-royale.html. Accessed 26 Dec 2020

A. Mougi, M. Kondoh, Diversity of interaction types and ecological community stability. Science **337**(6092), 349–351 (2012)

K. Nadachowska-Brzyska, C. Li, L. Smeds, G. Zhang, H. Ellegren, Temporal dynamics of avian populations during Pleistocene revealed by whole-genome sequences. Curr. Biol. **25**, 1375–1380 (2015). https://doi.org/10.1016/j.cub.2015.03.047

B.D. Patten, The wealth of ecosystems. How invisible hands (organisms, autonomy, biodiversity, connectivity) mold biological and environmental fitness in the economy of nature. Ecol. Indic. **100**, 4–10 (2019)

B.C. Patten, E.P. Odum, The cybernetic nature of ecosystems. Am. Nat. **118**, 886–895 (1981)

B. Patterson, An army of beetles could save dying hemlock forests. (2016), https://www.scientificamerican.com/article/an-army-of-beetles-could-save-dying-hemlock-forests/. Accessed 30 Dec 2020

F. Pearce, Inventing Africa. New. Sci. **167**(2251), 30 (2000)

R.A. Pierce, E. Flinn, Potential diseases and parasites of white-tailed deer in Missouri. Extension, University of Missouri. (2018), https://extension2.missouri.edu/g9489. Accessed 30 Dec 2020

P. Quévreux, U. Brose, Metabolic adjustment enhances food web stability. Oikos **128**(1), 54–63 (2018). https://doi.org/10.1111/oik.05422

L.B. Slobodkin, How to be a predator. Am. Zool. **8**(1), 43–51 (1968)

C. Terrer, R.P. Phillips, A. Hungate, et al., A trade-off between plant and soil carbon storage under elevated CO_2. Nature **591**, 599–603 (2021)

D. Tilman, J.A. Downing, Biodiversity and stability in grasslands. Nature **367**, 363–365 (1994)

R. Trebilco, J.K. Baum, A.K. Salomon, N.K. Dulvy, Ecosystem ecology: Size-based constraints on the pyramids of life. Trends Ecol. Evol. **28**(7), 423–431 (2013)

M. Vanni, J. Duncan, M. González, M.J. Horgan, Competition among Aquatic Organisms, in *Encyclopedia of Inland Waters*, ed. by G. E. Likens, vol. 1, (2009), pp. 395–404

W.S. Yackinous, *Understanding Complex Ecosystem Dynamics: A Systems and Engineering Perspective* (Elsevier, Amsterdam, 2015)

Z. Zhang, Z. Li, Factors affecting hare-lynx dynamics in the classic time series of the Hudson Bay, Canada. Clim. Res. **34**(2), 83–89 (2007)

Chapter 7
Ecosystem Stability

7.1 Background

Frederic Clements may have been the first ecologist to use the term ecological stability (Clements 1916). He observed that during the early phase of succession, communities had few species and composition changed rapidly, while in mature communities, species diversity was high and species composition varied little with time. This led to the widely accepted view that species diversity promotes ecosystem stability. That theory was questioned by May (1972, 1973, 1981). He tested stability of what he called an ecosystem model through the use of first-order partial derivatives of a vector-valued function. The model was based on positive, negative, or neutral interactions between populations. He found that contrary to prevailing ecological theory, there was no relationship between diversity and stability. This triggered decades of furious debate among ecologists, but the final verdict was that May was wrong (Hooper et al. 2005). May's model was closed to the environment, static and had no interactions with decomposers. It did not incorporate indirect feedback interactions, and did not differentiate between population stability and ecosystem stability. Ecosystems are now understood to be open, dynamic, and to contain constantly re-adjusting feedback loops that include decomposers. Jørgensen et al.'s (2016) criticism of May was more specific: *"His conclusion was correct according to his model, but the model was wrong."*

A conference in 1969 on diversity and stability in ecological systems (Woodwell and Smith 1969) included leading theorists of the time who presented various explanations for ecological stability. Theories centered on climate stability, fluctuating environments, competitive exclusion, and many variations on these themes. Lack of progress resulted in part from lack of agreement on the definition of stability. Since then, researchers have debated the meaning of ecological stability in species populations (Stoll and Prati 2001; Gilad 2008), in food webs (Slobodkin 1968; Murdoch 1966) and in ecosystems (Margalef 1968; Tilman 1996; Scheffer et al. 2002).

C. F. Jordan, *Evolution from a Thermodynamic Perspective*,
https://doi.org/10.1007/978-3-030-85186-6_7

Studies have focused resilience (Oliver et al. 2015), interactions within communities (Matson and Hunter 1992; Mougi and Kondoh 2012), and feedback within ecosystems (Patten and Odum 1981). Ives and Carpenter (2007) pointed to a number of empirical studies that demonstrated positive diversity-stability relationships. Jacquet et al.'s (2016) analysis of 116 quantitative food webs showed that intrinsic energetic organization of food webs is highly stabilizing and allows complex ecosystems to recover from perturbations. In contrast, a 20-year study of lake ecosystems, Kardol et al. (2018) found that the effects of species loss on community biomass decreased over time, suggesting that losing species did not affect biomass stability in the lakes. An interaction matrix described by Johnson et al. (2014) indicated that "trophic coherence" is a better predictor of stability than complexity. Trophic coherence means that species are not randomly distributed in an ecosystem, but are arranged according to their interactions. Grimm and Wissel (1997) made an inventory of 167 definitions of ecological stability used in the literature and found 70 different concepts. Pimm et al. (2019) have called for practical measurements to untangle the "bewildering array of terms under the heading of stability".

Part of the problem is that there is a difference between population stability, community stability, and ecosystem stability. Population stability can be considered in terms of the whether a population recovers partially or fully from a perturbation (Hamilton et al. 2009). Numbers of individual organisms in a particular population can decline but not enough so that the species is threatened with extinction. Community stability means the presence of the same set of species' interactions over long periods of time. Ecosystem stability means that individual populations can vary or become locally extinct but combined flows of energy and nutrients through all populations in the ecosystem remain constant or nearly so (Berlow et al. 2004). McNaughton (1977) proposed the ecosystem view when he suggested that *"The diversity-stability hypothesis --- simply states that species diversity mediates community functional stability through compensating interactions to environmental fluctuations among co-occurring species. Fluctuations in the abundances of species with different adaptive modes may be a mechanism stabilizing community function in a varying environment."*

7.2 A Thermodynamic Definition

In recent decades there have been advances in evolutionary theory, network science, cybernetics, food web theory, and soil ecology that have led to a better understanding of how energy flow and nutrient cycling form feedback loops that control ecosystem stability. These advances lead to a systems definition of ecosystem stability:

> Stability for a mature ecosystem means maintenance of nutrient and energy flows through species populations by deviation dampening feedback loops that keep flows within bounded limits.

Bounded limits for ecosystems are the extremes that energy flow through combined populations can vary without causing a "regime shift". These bounded limits are set by the usual daily and seasonal variations in energy flow to which individual species have become adapted. When deviation amplifying feedback causes energy flow to drop below the lower bounded limit, the system becomes unstable and changes to an alternative stable state (Lehtonen and Kokko 2012).

7.2.1 *Regime Shift*

Following severe disturbance to energy and nutrient recycling, an ecosystem may stabilize at a level of biomass and energy flow that is lower than that which existed before the disturbance. The transition is an ecological "regime shift" (Holling 1973). A number of examples occur in the literature. Fitzhugh (2018) reported that disturbance of ponderosa pine stands in California by fire or logging resulted in a degradation of the plant community to chaparral because of disturbance to the nutrient recycling mechanism. Chaparral is a community of dense shrubs on soil with low soil organic matter (Norton et al. 2019). During conversion, the flow of energy and nutrients through the ecosystem decreased but stabilized at a lower biomass and lower rate of flow. The point where the transition can no longer be reversed is called the "tipping point" (Dakos et al. 2019), because beyond this point, the system may not recover. Regime change following destruction of soil organic matter also has been observed forest ecosystems of Southeastern U.S. Brewer et al. (2012) showed that salvage logging following a tornado in Mississippi severely impacted recovery of successional species through removal of the organic matter in which nutrients were stored.

Mesquita et al. (2015) found that when Amazon rain forests are converted to cattle pastures by cutting and burning, soil organic matter is depleted. Overgrazing results in dieback of root systems that reduces the productivity of the grasses. As overgrazing continues, unpalatable shrubs invade the system. Cattle graze harder on the remaining grass that results in an even greater increase in shrubs. The feedback mechanism that recycles nutrients is damaged. Eventually grasses disappear. The abandoned pastures revert to stands dominated by a single genus *Vismia*, a slow-growing unpalatable competitor that can survive under poor soil quality. Even if cattle are removed, the shrubs remain dominant because of the decreased soil fertility.

In oligotrophic ponds and lakes, a regime shift through eutrophication can occur if nutrient concentrations in the water become too high as can occur from agricultural fertilizer runoff. Algae quickly respond to elevated nutrient levels. Communities change as rooted emergents such as cattails, floating plants like water lilies, and submerged plants like hydrilla become established.

7.2.2 Metastability

In physics, metastability is a stable state of a dynamical system other than the system's state of least potential energy. A ball resting in a hollow on a slope is an example of metastability. There is a force acting on the ball (the edge of the hollow) preventing it from attaining the state of least potential energy, the ground state at the bottom of the slope. In ecology, a stable state also is the state to which a system regresses when there are no forces acting on the system. A river valley could be in a ground state. If a beaver builds a dam and floods the valley, it enters a metastable condition. There is considerable potential energy in the water behind the dam. If a storm breaks the dam, the valley regresses to the ground state.

During boom-and-bust cycles of pine bark beetles, forest ecosystems alternate between metastable and ground states (Berryman et al. 1984). As beetles increase their population density to better access wood tissue partially digested by fungi, the system becomes metastable. Trees increase resin production for defense. Negative density dependence then reduces beetle populations and the forest returns to ground state (Goodsman et al. 2017).

7.2.3 Pulsed Stability

In ecosystems that experience seasonal change, energy input occurs in pulses (Odum 1971). Species have adapted to these pulses and become endangered when the pulses disappear.

Freshwater marshes such as the Florida Everglades are an example. (Odum 1959). The dry season draw-down speeds up aerobic decomposition of accumulated organic matter that results in nutrient solubilization followed by increased productivity during the wet. Wood storks have been adapted to this cycle. When water levels are falling, the small fish on which they feed are concentrated in small pools and are easy to catch. Regulation of water flow through the Everglades through canals and levees resulted in almost complete disappearance of these storks (Audubon Florida 2019). The comprehensive Everglades Restoration Plan authorized by Congress is intended to reestablish pulses of water through the marsh. Fires, another type of energy pulse that occurs during the Everglades' dry season prevents woody "hammock" species from spreading and converting marshlands to forest (National Park Service 2019).

An ecosystem in stable equilibrium as illustrated by Fig. 4.1b has pulses that occur in seasonal intervals. The pulses do not exceed the bounded limits of stability. Since seasonality exists in all ecosystems from the tropics to the poles, all natural ecosystems exhibit pulsed stability. Even in the continually wet tropics there are months when rainfall is less. Towards the end of the drier season, many trees exhibit flowering, an adaptation that results in seed release when there is optimum moisture for germination on the forest floor.

7.2.4 *Resistance and Resilience*

Resistance and resilience are characteristics of ecosystems closely related to stability and like stability, are the subject of much debate among ecologists (MacGillivray and Grime 1995; Oliver et al. 2015). Like stability, they are more precisely defined by a systems model. Ecological resistance means there are buffering mechanisms so that energy flow and nutrient recycling are scarcely affected by perturbations. High species diversity is a buffering mechanism. Resistant ecosystems are scarcely affected by the loss of a species when there are other species with similar functions available to compensate for the loss (Hooper et al. 2005). Ecological resilience means that once the energy and nutrient flow of an ecosystem are altered, the original flow is quickly restored. As in the case of resistance, high species diversity confers resilience.

In political systems, resistance to change results from a conservative bureaucracy that benefits from the status quo. It is an efficient but fragile system. Resilience results from a progressive system that is open to alternatives when the old ways no longer work very well. A political system that alternates between resistance and resilience works best. The same can be said for ecological systems.

7.3 Species Richness and Functional Stability

A high number of species in an ecosystem does not mean there is an optimal distribution of functions among the species. If there are 20 different herbivore species in an ecosystem but only one predator species, the potential for destabilization is much greater than if there are 16 herbivores and 5 predators, although species richness in both systems in the same. In an insect re-colonization experiment on mangrove islands off the Florida coast, Heatwole and Levins (1972) showed that stability depended more on how the diversity of functions in an ecosystem was distributed among species rather than simply on species richness.

In the Serengeti, there are three feline predators, lions, leopards and cheetahs (Planet Tiger 2016). However stability of energy flow through the system is increased when hyenas that have different functional characteristics feed upon the same herbivores. Hyenas and big cats in Africa cover the same ground, hunt the same prey, and scavenge the same remains of animals, but their hunting strategies differ. Hunting strategies by hyenas that act in packs may be more effective than solo strategies of big cats against some types of herbivores (Lehmann et al. 2017).

7.4 Species Richness and Cultural Values

Despite the recent increase in the rate of species extinction (IPBES 2019) and the documented losses of particular species in local ecosystems (Díaz et al. 2018), especially in marine systems (Blowes et al. 2019), the overall richness of many global ecosystems has changed little over the past 40 years (Hillebrand et al. 2018; Eriksson and Hillebrand 2019). There has been no systematic decrease in the total number of species (Dornelas et al. 2014). Most habitats undergo replacement of some or many species by other species with no general shift in the total number of species (Vellend et al. 2013). Some remote ecosystems show a systematic increase in average species richness over time, and declines are observed only in locations with strong human pressures (Elahi et al. 2015). A complete exchange of species composition over time without any change in the number of species commonly occurs in nature (Hillebrand et al. 2018), as species from one region move to fill a vacuum in another. Local species assemblages rapidly reorganize, with nearly a third of all species being replaced by new species every decade. These studies bring little comfort to population biologists who argue that the importance of species lies in the unique characteristics of particular species. Simply valuing species richness does not reflect the concerns of population biologists that a species that has great cultural value may be replaced by one with little cultural value (Diaz et al. 2018).

7.5 Keystone Species, and Population and Ecosystem Stability

Keystone species such as the wolves on Isle Royale are species that have an outsize impact on energy flow because of their control over other species. The term keystone species was coined by Paine (1969) following his studies of coral reefs. He wrote: "*Within complex systems, the species composition and physical appearance were greatly modified by the activities of a single native species high in the food web. These individual populations are the keystone of the community's structure, and the integrity of the community and its unaltered persistence through time is determined by their activities and abundances.*" Because top predators have such a dramatic effect on populations stability, they have often been classified as "apex" species in trophic cascades (Terborgh et al. 2001). Elimination of keystone species destabilizes other populations in the ecosystem, a phenomenon called a "trophic cascade" (Terborgh and Feeley 2010). Conservation of keystone species depends on the preservation of particular interactions in the ecosystem, while preservation of all interactions is vital to conservation of particular interactions (Carroll 2016).

"Keystone species" is a term that depends on the scale of the system being studied (appendix 6). An earthworm is a keystone species in the soil ecosystem, an elk in a forest ecosystem, and a wolf in a landscape system. Paine (1969) proposed that carnivorous starfish are a keystone species in rocky intertidal areas from Alaska to Baja California, and that their removal results in significant changes in the local trophic webs. However they are not keystone species for the Pacific Ocean, even though they are part of that system. Students sometimes think that the particular species they happen to study is a keystone species because through their focus on that species they uncover the many ways it interacts and influences other species in its local environment.

7.5.1 *Keystone Species in the Yellowstone Region of Wyoming*

Wolves were hunted in the Yellowstone region since the late 1800s because they preyed upon cattle that were easier to catch than elk. As a result, elk population swelled to the point where it destroyed the willows, aspen, and cottonwoods that grew in river bottoms and upon which the elk fed (Ripple and Beschta 2004). Consequently, soil erosion increased. Wolves were reintroduced to control the elk in 1995, but a major result was for the wolves to drive the elk to uplands where they were preyed upon by cougars and grizzly bears. This allowed recovery of the willows and cottonwoods, especially where beaver dams raised the water level (Boyce 2018). However, some researchers have been critical because "Yellowstone is still not 100% back to normal – and may never be" (Lackey 2018, referring to populations). That is because populations are stochastic. Ecosystems however, are deterministic. Thermodynamic theory however predicts that energy flow will be stabilized at close-to-previous levels.

References

Audubon Florida, Wood stork. (2019), https://fl.audubon.org/birds/wood-stork. Accessed 7 Dec 2020

E.L. Berlow, A.M. Neutel, J.E. Cohen, P.C. De Ruiter, B. Ebenman, M. Emmerson, J.W. Fox, V.A.A. Jansen, J.I. Jones, G.D. Kokkoris, D.O. Logofet, A.J. McKane, J.M. Montoya, O. Petchey, Interaction strengths in food webs: Issues and opportunities. J. Anim. Ecol. **73**, 585–598 (2004)

A.A. Berryman, N.C. Stenseth, D.J. Wollkind, Metastability of forest ecosystems infested by bark beetles. Res. Popul. Ecol. **26**, 13–29 (1984). https://doi.org/10.1007/BF02515505

S.A. Blowes, S.R. Supp, L.H. Antão, et al., The geography of biodiversity change in marine and terrestrial assemblages. Science **366**(6463), 339–345 (2019)

M. Boyce, Wolves for Yellowstone: Dynamics in time and space. J. Mammal. **99**(5), 1021–1031 (2018)

J.S. Brewer, C.A. Bertz, J.B. Cannon, J.D. Chesser, E.E. Maynard, Do natural disturbances or the forestry practices that follow them convert forests to early-successional communities? Ecol. Appl. **22**(2), 442–458 (2012)

S.G. Carroll, *Serengeti Rules: The Quest to Discover how Life Works and Why it Matters* (Princeton University Press, Princeton, 2016)

F.E. Clements, *Plant Succession; an Analysis of the Development of Vegetation* (Carnegie Institution of Washington, Washington, DC, 1916)

V. Dakos, B. Matthews, A.P. Hendry, J. Levine, N. Loeuille, J. Norberg, P. Nosil, M. Scheffer, L. De Meester, Ecosystem tipping points in an evolving world. Nat. Ecol. Evol. **3**, 355–362 (2019)

S. Díaz, U. Pascual, M. Stenseke, et al., Assessing nature's contributions to people. Science **359**(6373), 270–272 (2018)

S. Díaz, U. Pascual, M. Stenseke, et al., Assessing nature's contributions to people. Science **359**(6373), 270–272 (2018). https://doi.org/10.1126/science.aap8826

M. Dornelas, N.J. Gotelli, B. McGill, H. Shimadzul, F. Moyes, C. Sievers, A.E. Magurran, Assemblage time series reveal biodiversity change but not systematic loss. Science **344**(6181), 296–299 (2014)

R. Elahi, M.I. O'Connor, J.E.K. Byrnes, J. Dunic, B. Eriksson, M.J.S. Hensel, P.J. Kearns, Recent trends in local-scale marine biodiversity reflect community structure and human impacts. Curr. Biol. **25**(14), 1938–1943 (2015)

B.K. Eriksson, Hillebrand, Marine systems outpace terrestrial habitats in biodiversity erosion. Science **366**(6463), 308–309 (2019)

E.L. Fitzhugh, California wildlife habitat relationships system. California department of fish and game, California interagency wildlife task group: Ponderosa Pine, Vegetation. (2018), https://nrm.dfg.ca.gov/FileHandler.ashx?DocumentID=67322. Accessed 17 Dec 2020

O. Gilad, Competition and Competition Models, in *Encyclopedia of Ecology*, ed. by S. E. Jorgensen, B. D. Fath, (Elsevier, 2008), pp. 707–712

D.W. Goodsman, B.J. Cooke, M.A. Lewis, Positive and negative density-dependence and boom-bust dynamics in enemy-victim populations: A mountain pine beetle case study. Theor. Ecol. **10**, 255–267 (2017)

V. Grimm, C. Wissel, Babel, or the ecological stability discussions: An inventory and analysis of terminology and a guide for avoiding confusion. Oecologia **109**(3), 323–334 (1997)

M.J. Hamilton, O. Burger, J.P. DeLong, R.S. Walker, M.E. Moses, J.H. Brown, Population stability, cooperation, and the invasibility of the human species. PNAS **106**(30), 12255–12260 (2009)

H. Heatwole, R. Levins, Trophic structure stability and faunal change during recolonization. Ecology **53**, 531–534 (1972)

H. Hillebrand, B. Blasius, E.T. Borer, et al., Biodiversity change is uncoupled from species richness trends: Consequences for conservation and monitoring. J. Appl. Ecol. **55**(1), 169–184 (2018)

C.S. Holling, Resilience and stability of ecological systems. Annu. Rev. Ecol. Syst. **4**, 1–23 (1973)

D.U. Hooper, F.S. Chapin III, J.J. Ewel, et al., Effects of biodiversity on ecosystem functioning: A consensus of current knowledge. Ecol. Monogr. **75**(1), 3–35 (2005)

IPBES, Summary for policymakers of the global assessment report on biodiversity and ecosystem services, in *The Intergovernmental. Science-Policy Platform on Biodiversity and Ecosystem Services*, ed. by S. Díaz, J. Settele, Brondizio, et al., (IPBES Secretariat, Bonn, 2019). https://ec.europa.eu/environment/nature/pdf/Summary%20for%20Policymakers%20IPBES%20Global%20Assessment.pdf. Accessesd 21 Dec 2020

A.R. Ives, S.R. Carpenter, Stability and diversity of ecosystems. Science **317**(5834), 58–62 (2007)

C. Jacquet, C. Moritz, L. Morissette, P. Legagneux, F. Massol, P. Archambault, D. Gravel, No complexity–stability relationship in empirical ecosystems. Nat. Commun. **7**, 12573 (2016)

S. Johnson, V. Domınguez-Garcıa, L. Donetti, M.A. Munoz, Trophic coherence determines food-web stability. PNAS **111**, 17923–17928 (2014)

S.E. Jørgensen, S.N. Nielsen, B.D. Fath, Recent progress in systems ecology. Ecol. Model. **319**, 112–118 (2016)

P. Kardol, N. Fanin, D.A. Wardle, Long-term effects of species loss on community properties across contrasting ecosystems. Nature **557**, 710–713 (2018). https://doi.org/10.1038/s41586-018-0138-7

K. Lackey, Yellowstone's wolves are back, but they haven't restored the park's ecosystem. Here's why. USA Today; Sept. 7, 2018. (2018). https://www.usatoday.com/story/tech/science/2018/09/07/wolves-reintroduction-yellowstone-ecosystem/973658002/. Accessed 26 Dec 2020

D.S. Lehmann, T.M. Montgomery, S.M. Maclachlan, J.M. Parker, O.S. Spagnuolo, K.J. Vandewetering, P.S. Bills, K.E. Holekamp, Lions, hyenas and mobs (Oh my!). Curr. Zool. **63**(3), 313–322 (2017). https://doi.org/10.1093/cz/zow073

J. Lehtonen, H. Kokko, Positive feedback and alternative stable states in inbreeding, cooperation, sex roles and other evolutionary processes. Philos. Trans. Rl. Soc. B **367**(1586) (2012). https://doi.org/10.1098/rstb.2011.0177

C.W. MacGillivray, J.P. Grime, Testing predictions of the resistance and resilience of vegetation subjected to extreme events. Funct. Ecol. **9**, 640–649 (1995)

R. Margalef, *Perspectives in Ecological Theory* (University of Chicago Press, Chicago, 1968)

P.A. Matson, M.D. Hunter, Special feature: The relative contributions to top-down and bottom-up forces in population and community ecology. Ecology **73**(3), 723–723 (1992)

R.M. May, Will a large complex system be stable? Nature **238**, 413–414 (1972)

R.M. May, Qualitative stability in model ecosystems. Ecology **34**(3), 638–641 (1973)

R.M. May, *Theoretical Ecology*, 2nd edn. (Blackwell Scientific publications, Oxford/London, 1981)

S.J. McNaughton, Diversity and stability of ecological communities: A comment on the role of empiricism in ecology. Am. Nat. **111**(979), 515–525 (1977)

R. Mesquita, P.E. Massoca, C.C. Jakovac, T.V. Bentos, G.B. Williamson, Amazon rain forest succession: Stochasticity or land-use legacy? Bioscience **65**(9), 849–861 (2015)

A. Mougi, M. Kondoh, Diversity of interaction types and ecological community stability. Science **337**(6092), 349–351 (2012). https://doi.org/10.1126/science.1220529

W.W. Murdoch, Community structure, population control, and competition. Am. Nat. **100**, 219–226 (1966)

National Park Service, (2019), https://www.nps.gov/thro/learn/historyculture/theodore-roosevelt-and-conservation.htm. Accessed 30 Jan 2021

J.B. Norton, W.R. Horwath, U. Norton, Fire frequency effects on soil organic matter, vegetation, and runoff in chaparral shrublands of the Sierra Nevada foothills. (2019), www.uwyo.edu/esm/faculty-and-staff/jay-norton/research/asa2005-chap-pster.pdf. Accessed 27 Dec 2020

E.P. Odum, *Fundamentals of Ecology* (Saunders, Philadelphia, 1959)

H.T. Odum, *Environment, Power and Society* (Wiley, New York, 1971)

T.H. Oliver, M.S. Heard, N.J.B. Isaac, et al., Biodiversity and resilience of ecosystem functions. Trends Ecol. Evol. **30**(11), 673–684 (2015). https://doi.org/10.1016/j.tree.2015.08.009

R.T. Paine, A note on trophic complexity and community stability. Am. Nat. **103**(929), 91–93 (1969)

B.C. Patten, E.P. Odum, The cybernetic nature of ecosystems. Am. Nat. **118**, 886–895 (1981)

S.L. Pimm, I. Donohue, J.M. Montoya, M. Loreau, Measuring resilience is essential to understand it. Nat. Sustain. **2**, 895–897 (2019)

Planet Tiger, World …reasons for the disappearance of the big cats. (2016). http://www.planete-tigre.org/news-29.html. Accessed 30 Dec 2020

W.J. Ripple, R.L. Beschta, Wolves, elk, willows and trophic cascades in the upper Gallatin range of southwestern Montana, USA. For. Ecol. Manag. **184**, 299–313 (2004)

M. Scheffer, F. Westley, W.A. Brock, M. Holgran, Dynamic interaction of societies and ecosystems: Linking theories from ecology, economy, and sociology, in *Panarchy: Understanding Transformations in Human and Natural Systems*, ed. by L. Gunderson, C. S. Holling, (Island Press, Washington, DC, 2002), pp. 95–239

L.B. Slobodkin, How to be a predator. Am. Zool. **8**(1), 43–51 (1968)

P. Stoll, D. Prati, Intraspecific aggregation alters competitive interactions in experimental plant communities. Ecology **82**(2), 319–327 (2001)

J. Terborgh, K. Feeley, Propagation of trophic cascades via multiple pathways in tropical forests, in *Trophic Cascades: Predators, Prey, and the Changing Dynamics of Nature*, ed. by J. Terborgh, J. A. Estes, (Island Press, Washington, DC, 2010), pp. 125–140

J. Terborgh, L. Lopez, V.P. Nuñez, M. Rao, G. Shahabuddin, G. Orihuela, M. Riveros, R. Acanio, G.H. Adler, T.D. Lambert, L. Balbas, Ecological meltdown in predator-free forest fragments. Science **294**(5548), 1923–1926 (2001)

D. Tilman, Biodiversity: Population versus ecosystem stability. Ecology **77**, 350–363 (1996)

M. Vellend, L. Baeten, I.H. Myers-Smith, S.C. Elmendorf, R. Beauséjour, C.D. Brown, P. De Frenne, W.S. Verheyen, Global meta-analysis reveals no net change in local-scale plant biodiversity over time. 2013. PNAS **110**(48), 19456–19459 (2013)

G.M. Woodwell, H.H. Smith, *Diversity and Stability in Ecological Systems*, Brookhaven Symposia in Biology No. 22 (Brookhaven National Laboratory, Brookhaven, 1969)

Chapter 8
Case Studies of Ecosystem Control and Stability

8.1 Walden

In 1845, Henry David Thoreau built a small cabin on the shore of Walden Pond near Concord Massachusetts and spent 2 years recording the names of indigenous plants and animals and observing the seasonal changes in nature. It was a spiritual quest. Meaning in life, he felt, is not revealed through accumulation of material wealth, but through seeking ultimate truth through the study of nature. Thoreau enjoyed watching the spring thaw of Walden pond, and grew ecstatic as he witnessed the green rebirth of nature, the "indescribable innocence and beneficence of Nature" as he put it. His message was that there is a harmony in nature and that society could benefit by emulating that harmony. Nevertheless, there seems to be a contradiction in his thinking. At Walden, Thoreau observed a war between two races of ants, herons gobbling up tadpoles and other interactions hinting that nature is in reality "red in tooth and claw". Thoreau remarked "I love to see that Nature is so rife with life that myriads can be afforded to be sacrificed and suffered to prey on one another that tender organizations can be so serenely squashed out of existence like pulp."

8.1.1 "Harmony in Nature"

Why is it that nature seems harmonious at first glance? It is because everything is so well coordinated. Birds build their nests in the spring when food becomes abundant. As flowers bloom, insects come to transfer pollen. As trees leaf out, caterpillars emerge to fatten themselves. Dead animals disappear quickly because scavengers are always ready. But who or what coordinates all this harmony? A symphony conductor ensures harmony between musicians. Who does it for nature? The Christian bible in the book of Genesis says it is God who created harmony in nature. That is

C. F. Jordan, *Evolution from a Thermodynamic Perspective*,
https://doi.org/10.1007/978-3-030-85186-6_8

not an explanation. It is merely putting a label on a phenomenon for which there is no apparent explanation.

8.1.2 Feedback Produces Nature's "Harmony"

Although Lotka (1922) recognized energy flow as the integrating force in ecosystems, and Elton (1927) organized ecosystems into trophic levels, Lindeman (1942) was perhaps the first to show that energy flow did the organizing. The question then for ecologists was how this organization came about. It would have been impossible for a nineteenth century naturalist to walk into a forest as did Thoreau and decipher what coordinated all this "harmony". Even after the recognition that energy flow through plants, animals and decomposers constituted an ecosystem, the underlying mechanisms that produced "harmony" among the parts were not obvious.

Harmony is the composite product when individual musical voices group together to form a cohesive whole. That is how Thoreau interpreted nature at Walden Pond – a cohesive whole. Cohesive means well integrated. What integrates a system? What ties the pieces together and synchronizes their operation? A clue came from the emerging field of systems analysis. A well-integrated system is a stable system. Since we now have an objective definition of stability, we can propose the mechanism that promotes ecosystem stability. It is keeping flows of energy within bounded limits. It is through feedback that stability of populations and ecosystems is maintained. Feedback provides the cohesiveness that integrates the species so that the flows of energy and nutrients that supports the species are optimized and do not exceed bounded limits. Specifically, a species population adjusts its biomass in response to the energy flow through its ecosystem. An ecosystem adjusts its energy flow in response to the energy and nutrients supplied by species.

8.1.3 Feedback Mechanisms

To understand the mechanisms that underlie feedback, flows of energy and recycling of nutrients must be experimentally disturbed or interrupted. A perturbation is a disturbance that affects the functioning of one or more of an ecosystem's trophic levels, thereby disrupting the feedback cycle. The disturbance can be natural or deliberately imposed. Observations on how the ecosystem recovers or destabilizes and the accompanying species dynamics give clues as to the mechanisms that work to recalibrate the flows and cycles that control feedback.

There have been numerous perturbation studies of ecosystems. Some have been of experimentally produced disturbances, and others followed natural ones. Cases from tropical forest ecosystems are presented here because they are where I happened to spend most of my early professional career. It was fortuitous for gaining an understanding of ecosystem control and stability because they often have tighter feedback systems and depend less on exogenous input of nutrients, thereby allowing a focus on feedback as the endogenous mechanism that stabilizes ecosystems.

8.2 Perturbations in Amazonian Rain Forests

Soil organic matter is particularly important for stability in tropical rain forests because it contains the organisms that recycle nutrients. Because decomposition in the hot, wet climate is rapid, it is important to preserve soil organic matter so that nutrients are recycled before they are leached.

Soil organic matter remaining after a perturbation depends on the length, size, and intensity of the disturbance. To illustrate, Jordan (1987) compared reports from seven projects in the Amazon Basin that included studies of nutrient cycling in forests. When natural disturbances were short in duration, small in scale, and mild in intensity, ecosystems recovered. Stresses such as hurricanes and fires are of short duration, and recycling mechanisms often remain intact. Stresses such as shifting cultivation by indigenous peoples are limited in scale, so disturbed areas recover by organic matter input from the surrounding forest. Stresses such as disease may affect one population, but there are always other species to maintain the energy and nutrient flow. In contrast, human induced disturbances are often long in duration, large in scale, and severe in intensity. They often result in degradation to a lower level of productivity and biomass. Stresses that are of long duration include conversion of rain forest to pasture. Stresses that are of high intensity are characterized by conversion of forest to plantations for pulp, palm oil, and charcoal. Large scale disturbance to the Amazon rain forest is exemplified by the Trans-Amazon Colonization Project (Moran 1981). The critical factor in all these cases is degradation of the soil organic matter. When the soil organic matter is destroyed, the system may never recover to its previous state unless the recycling mechanisms are restored.

Charcoal found in soils of mature Amazon rain forest indicated fires occasionally occurred there for at least 6000 years, but did not permanently degrade the forests (Sanford et al. 1985). Fires set by indigenous cultivators during shifting cultivation do not destroy the forest because the area disturbed is small and heals rapidly. In contrast, when new roads are constructed into the rain forest either for colonization projects or for logging, slash and burn agriculture by colonists is the first step in degradation of rain forests. Colonists clear small-scale patches to grow staples but after a few years, productivity declines due to loss of soil fertility. Farms are not large enough to allow natural succession in part of them to restore soil organic matter. Farmers sell or abandon the cleared land to ranchers and move further into the forest. To maintain grass for their cattle, the ranchers burn the grass, and fires then spread into adjacent forests. In human-dominated landscapes of Eastern Brazil that once were covered by diverse rain forest species, the babassu palm, *Attalea speciosa*, became particularly dominant in areas abandoned from pasture and shifting cultivation (Campos et al. 2017).

8.3 Top-Down Control

Top-down control concerns the role of feedback in dampening the oscillations of energy flow through the ecosystem.

8.3.1 The San Carlos Project: A Small-scale, Low Intensity, Short Duration Disturbance

The San Carlos Project (Herrera et al. 1981) was part of the International Man and the Biosphere Program launched in 1971 by UNESCO to establish a scientific basis for the improvement of relationships between people and their environments (Golley 1993). The project concerned ecophysiology, energy flow, and nutrient dynamics of forests on very old soils classified as Oxisols and Spodosols in the Amazon region of Venezuela. Over millions of years, leaching removed most of the nutrient elements (Putzer 1984; Kronberg et al. 1979; Jordan 1985a). Forests of low stature called caatinga forests occur on the sandy Spodosols, and tierra firme forests of greater structure grow on Oxisols with higher clay content and lateritic concretions (Appendix 2, Fig. 2.1). The first case discussed here is of a forest on an Oxisol. Oxisols comprised about half of the soil types in the San Carlos region. These forests have been called "stressed ecosystems" but it is unclear if they really are "stressed" because the species have adapted to low nutrient availability (Jordan 1989).

8.3.1.1 Nutrient Recycling

Because of the low levels of nutrients in the soil, survival depends on efficient recycling of nutrients. A mechanism to facilitate efficiency is direct transfer of nutrients from leaf litter to roots. Most of the roots in the forest ecosystem together with leaf litter form a mat on top of the mineral soil (Jordan and Escalante 1980; Jordan 1982c) (Fig. 8.1).

Within that mat, ectomycorrhiza enable direct transfer of nutrients from decomposing leaf litter to roots (Went and Stark 1968; Herrera et al. 1978). Nutrients bound in the litter are released from the biomass as a result of mycorrhizal exudation of organic acids that catalyze nutrient solubilization (Leake and Read 1997). Extracellular enzymes secreted by bacteria also contribute to solubilization (Bardgett and Wardle 2010; Reed et al. 2011).

Fig. 8.1 Nutrient distribution (red dots) in ancient soils of the Amazon Basin (left), and in younger soils of higher latitudes (right). In the Amazon soils, nutrients are concentrated in the soil organic matter. In the younger soils, a large proportion of the nutrients are exchanged on the surface of clay minerals. (Reproduced from Jordan 1982c with permission from American Scientist)

8.3.1.2 Feedback Control: Tree-fall Gaps

The feedback cycle that controls stability of the system begins with nutrient flow from leaf litter to root hairs through mycorrhizae. Within the plant, nutrients are used in the synthesis of new leaves that when they fall, complete the feedback cycle. Stability occurs when these processes are in balance. Disruption in energy flow occurs when a canopy tree gets so large that it needs more nutrients than can be supplied through the mycorrhizal feedback. The tree dies and falls over, creating a light gap in the canopy. Under the canopy of large trees there occur stunted saplings that have survived for many years because their small biomass and slow growth demanded few nutrients. As a fallen tree decomposes, nutrients are released and become available to the saplings. The increased light, in combination with increased nutrients stimulates growth of the saplings and enables the forest to resume nutrient and energy flow. This endogenous feedback control is apparent in tropical rain forests where moisture and temperature are ideal for decomposition, and where a reservoir of soil nutrients is not available as a buffer against instability.

8.3.1.3 Feedback Control: Shifting Cultivation

Endogenous control of resilience was studied by an experiment that replicated indigenous shifting cultivation on nutrient poor soil. Two adjacent 1 ha plots (experimental and control) were established as part pf the San Carlos project on the mixed species tierra firme ecosystem in Fig. 2.1, Appendix 2. Before cutting, the soil's exchangeable calcium, potassium and magnesium contained only 8% of the ecosystem's stock but 80% of the total nitrogen which was bound in undecomposed organic matter. Standard extraction techniques to determine available phosphorus in the mineral soil revealed undetectable amounts. Iron and aluminum oxides, the main constituents of Oxisols (Cuevas 2001), bind phosphorus in insoluble iron and aluminum phosphates (Brady 1974).

The experimental plot was cut and burned according to traditional shifting cultivation techniques. The fire consumed only dried leaves, twigs, and the above-ground root mat. The bulk of tree biomass was left intact, but otherwise the soil surface was bare. The principal crop was manioc, and plantain was also included. Immediately after the burn, the quantity of calcium, magnesium, and potassium in the mineral soil increased due to input from ash. Nitrogen stayed constant. Over the next 3 years, the plot was continually weeded. Quantities of calcium, magnesium, potassium in the mineral soil changed little, as leaching losses were compensated for by input from the decaying tree trunks and branches. Nitrogen also remained high due to decomposition of primary forest roots. Nevertheless, productivity of manioc decreased by more than 50%.

8.3.1.4 Phosphorus Dynamics

Phosphorus is the most limiting element in Oxisols (Medina and Cuevas 1994). The only system input at San Carlos is through dust clouds from the Sahara region. These clouds carry phosphorus that is rained out over the Amazon (Bristow et al. 2010; Prospero et al. 1981; Prospero and Mayol-Bracero 2013). These exogenous inputs can be irregular (Jordan 1982a). Species have evolved mechanisms to tighten nutrient cycles so that recycling fills their nutritional needs. Although phosphorus availability is extremely low in the mineral soil, its efficient recycling in the living biomass compensates for low phosphorus availability. To examine the efficiency of phosphorus recycling on the Oxisol site, Stark and Jordan (1978) sprayed surface root mats with ^{32}P to simulate the addition of nutrient input from Saharan dust. Collections from lysimeters placed below the above-ground humus penetrated by roots showed that $<0.1\%$ of the radiotracer leached past the root-organic mat. Analysis of root mat samples showed that the remainder of the radioisotope was taken up and translocated by living roots. A study of the nutrient cycling index (Finn 1976, 1978) at the San Carlos site indicated that recycling of calcium, potassium, and magnesium was not as efficient as that of phosphorus (Jordan 1982b). Nevertheless recycling was sufficiently efficient so that the net primary productivity of the forest on Oxisols in the San Carlos project was 12.81 T/ha/year (Jordan 1989), within the range for tropical rain forests (10–35 T/ha/year, Lieth 1975).

Organic acids leached from decomposing organic matter keep phosphorus in the mineral soil available to plant roots (Powell et al. 1980) through the chelation of iron and aluminum (Lee et al. 1990), thereby sustaining and accelerating phosphorus recycling. This explains why best crop growth in the experimental plot was always adjacent to a decaying trunk. However, the good crop growth did not last. By year three, trunks became dry skeletons that produced few organic acids. We attributed the decline in crop yield to a decrease in phosphorus released from iron and aluminum in the soil, as remaining biomass in the plot disappeared. The indigenous farmers showed us how production could be stimulated again by piling leaves collected from the forest around the base of the manioc plants. This litter provided the organic acids necessary to resume phosphorus liberation in the soil (Jordan 1989). The critical feedback that maintains productivity in this system is the solubilization of bound phosphorus.

After cultivation for 3 years, the area was abandoned to secondary succession. Leaves and wind-blown trees falling into the plot were a source of nutrient replenishment. Birds and small mammals carried seeds from the forest into the plot. Because of uptake by successional vegetation, levels of cations and nitrogen in the soil dropped to levels that approximated those in the undisturbed forest. Nevertheless, recovery of the forest proceeded rapidly (Uhl 1982; Uhl and Jordan 1984). The key to recovery was that the disturbance was small in relation to the area of the forest. Input of organic matter into the damaged plot from the surrounding forest healed the local disruption to energy flow and phosphorus recycling. Vitousek and Reiners (1975) found similar resilience in a mature temperate watershed. They concluded that "If the patch size of vegetation turnover is small relative to the size of a watershed, then the watershed can integrate all of the patches within it to a steady state whole."

8.3.1.5 Tropical Agriculture on Richer Soils

In the moist tropics of Central America, soils are generally richer. Large amounts of nutrient input into agricultural ecosystems comes from mineral soil derived from volcanic ash, an exogenous source for the system (Guariguata and Ostertag 2001). Although weed competition is one reason for rapidly declining crop production on rich soils, Nye and Greenland (1960) showed that even on fertile soils, repeated cycles of slash and burn agriculture gradually deplete the soil of nutrients resulting in reduced productivity of crops and/or successional vegetation.

8.3.2 The Jarí Project: A Large-scale, High Intensity, Long Duration Disturbance

When organic debris on the soil surface is deliberately removed to make way for plantations, the recycling mechanism is destroyed, and limits of resilience and stability are exceeded. A striking example is the Jarí plantation of pulp species in the

Amazon rain forest of Brazil (Jordan 1985b). In the 1960s an area of several hundred square kilometers was cut down followed by "complete site preparation". A chain was fastened to two giant bulldozers about 100 m apart. As the bulldozers plowed through the forest in parallel lines, the chain pulled down all the trees in the path. The downed biomass was allowed to dry and then burned with fires so intense that they produced thunderclouds. The remaining debris was mechanically removed by bulldozers. The nutrient recycling capacity of the highly productive natural ecosystem was greatly reduced, (Russell 1983) and the monoculture of *Gmelina arborea* planted for pulp succumbed from low resistance to the fungus *Ceratocystis fimbriata* (Fearnside and Rankin 1985). In the official reports, difficulties were attributed to foreign ownership, difficult access, land tenure, high infrastructure costs, low pulp prices, and social costs (Palmer 1988). Low fertility of the organically deprived soil was never mentioned. The stubborn refusal of Daniel Ludwig, Jari's billionare financer to admit he was wrong about productivity of Amazon forests, and the laying of blame by project managers for failure on factors besides soil fertility shows how preconceived ideas conflict with scientific reality. In the year 2000, the entire plantation including the $340,000 pulp mill was sold for one dollar to a group of speculators (Lamirande 2000).

In contrast to the Jarí case, in an experiment that was part of the Biological Dynamics of Forest Fragments Project in the Amazon Rain Forest near Manaus Brazil (Bierregaard 2001), Mesquita (1995) found that when a forest was cut but the root mat and soil organic matter were not destroyed, recovery of the forest was through normal succession as would occur on nutrient rich soils. Recycling of nutrients by decomposers feeding on soil organic matter was the top-down ecosystem control that stabilized Mesquita's ecosystem, and facilitated recovery of the shifting cultivation plot.

8.4 Bottom-Up Control

Bottom-up control concerns the role of species populations in dampening the oscillations of energy flow through the ecosystem.

8.4.1 The El Verde Project

8.4.1.1 Perturbation = Ionizing Radiation

Following atmospheric tests of nuclear weapons after World War II, scientists became concerned with the effects of ionizing radiation that entered global ecosystems through atmospheric fallout. As a result, the U.S. Atomic Energy Commission in the 1960s initiated a series of projects to determine the effect of radiation on natural ecosystems (Golley 1993). One of these projects was in the rain forest near the

El Verde research station in the Luquillo forest of Puerto Rico. Its objective was to look at the effect of ionizing radiation on individual species, and how their response affected ecosystem dynamics. The site was in a "tabanuco" (*Dacryodes excelsa*) forest type that exists on well drained soils of moderate fertility (Edmisten 1970). From January to April of 1965, a portion of the forest was irradiated with a 1000 Ci source of ^{137}Cs (Odum and Drewry 1970). Another area was kept as a control.

Box 8.1
The radiation source used in the rain forest experiment was a cesium 137 (^{137}Cs) emitter. Unlike strontium-90 which emits beta particles that move through ecosystems, cesium 137 is a gamma emitter that emits energetic photons. It can be turned "on" and "off" in ecosystem experiments with the use of a removable lead shield over the cesium source.

Die-back of the canopy began before the end of irradiation. By July 1966, the canopy had disappeared to about 15 m from the source, and was intermittent to 30 m. A few months after the beginning of radiation, leaves of canopy and understory trees began to die, creating an energy pulse from the trees to the decomposers on the forest floor in the form of increased leaf fall. Due to this pulse, soil respiration increased as a result of greater activity of decomposers (Witkamp 1970). Gamma radiation extended horizontally out from the two-meter high source which was supported above the soil, so below-ground organisms were not killed by the radiation. However termites were exterminated close to the radiation source, and nests were abandoned out to 30 m (McMahan 1970). The protuberance of their nests above ground made them susceptible. Fungal populations increased on the soil surface after radiation ceased due to increased leaf and wood litter (Cowley 1970). The changed micro-climate around the radiation source caused an increase in mosquito populations, which may have caused an increased virus infection to rats in the area (Weinbren and Weinbren 1970). Populations of birds, lizards, and tree frogs were not affected because they were not confined to the affected area (Turner and Gist 1970; Recher 1970).

After the pulse of energy due to leaf fall had ceased, soil respiration not only decreased, it decreased below the level of that in the undisturbed forest, possibly due to exhaustion of energy reserves in the soil (Witkamp 1970). Evapotranspiration in the successional area was less than in the control forest, and as a result, soil was frequently saturated, a condition that could favor establishment of the "Colorado" forest type (*Cyrilla racemiflora*) that does better on wetter soils (Wadsworth 1970). Insect herbivory on the early successional plants was greater than herbivory in the closed forest because the successional plants devote more energy to growth than to maintenance and repair (Odum and Ruíz-Reyes 1970).

There was no recovery of the canopy, and stems of the radiation-killed trees fell during succeeding years. Secondary successional vegetation became established in the area shortly after the beginning of canopy dieback (Jordan 1971). Successional vegetation was able to restore the energy flow through the ecosystem since the nutrients in the mineral soil reservoir were scarcely diminished. The soil contained a reservoir that suppled vegetation with nutrients needed to re-stabilize the ecosystem. Because of the proximity of the undamaged forest, seeds from there stimulated recovery. As each component in the system was replaced, the feedback system gradually reorganized. Eighteen years after the end of the experiment, the species present in the irradiated area were the same as those present in the undisturbed forest (Jordan, personal observation, 1985).

8.4.1.2 Conclusion

Decomposer populations regulated the flow of energy through the ecosystem by adjusting their metabolic rates and population size. Most notable was the initial increase in decomposer activity when litterfall was heavy. This supplied the successional vegetation with an abundant nutrient supply. Once litterfall decreased, decomposer activity decreased and recycling became more efficient.

8.4.2 The Long-Term Ecological Research Project in Puerto Rico

The U.S. National Science Foundation created the Long-Term Ecological Research (LTER) Program in 1980 to understand the long-term dynamics of ecosystem processes. Puerto Rico's Luquillo site entered the program in 1988 with the objective of determining the effects of natural and human disturbances on tropical forests and streams in the Luquillo Mountains (Luquillo LTER 2019). The study site has been located on the slopes of El Yunque, the remnant of a super volcano during the Cretaceous period 145–66 million years ago. It rises 1075 m above the sea on the Eastern end of Puerto Rico. Although it is not nearly as high as the Andes, the zonation within that 1075 m is as distinct as the Andes that have an average height of 4000 m. The contrast is due to the Massenerhebung effect which means that mountains surrounded by large ranges will have life zones at much higher elevations than isolated island mountains, due to heat retention and wind shadowing.

8.4.2.1 Perturbation = Hurricanes

The earth's rotation causes daily and seasonal changes in the energy flow through ecosystems. Populations are adapted to such variations, and their tolerance is limited to the bounded limits prescribed by these variations. Hurricane damage effects energy flow through species (including decomposer) populations beyond their bounded limits. The question was, how do these changes in energy flow through species populations affect energy flow through ecosystems?

Hurricane Hugo hit Puerto Rico in 1989, and as it passed over the Luquillo mountains, it ripped the leaves off many canopy trees, and many trees were knocked down (Basnet et al. 1992), creating gaps in the forest that allowed increased solar radiation to enter the forest floor. However, regeneration of surviving seedlings and saplings was rapid (Hillinger 1990). Nutrient conserving and recycling mechanisms remained intact. The feedback loop between decomposers and primary producers remained intact. In 1992, Jordan personally observed that the forest began regenerating before there was significant loss of soil organic matter. Detritovores became abundant soon after the hurricane when leaf fall increased the amount of litter on the forest floor. As the forest recovered, sap-suckers became abundant in canopy gaps where new foliage growth was greatest (Schowalter et al. 2017). Herbivorous walking sticks in trees reduced the nutrient content of the leaves. The subsequent leaf litter fall was of low quality, and as result, leaf decomposition on the soil slowed. Due to canopy opening, lignin-degrading fungi also decreased (Shiels et al. 2015). Particularly reduced were organisms responsible for shredding of litter (Richardson et al. 2018). Consequently the diversity, and composition of communities in the litter and forest soil was reduced (Prather et al. 2018).

In a comparison of arthropod abundances taken between 1976 and 2012, dry weight of biomass of arthropods captured in sweep net samples declined 4–8 times, and in sticky traps 30–60 times compared to data from the 1970s. Because mean maximum temperatures have risen by 2.00 °C, Lister and Garcia (2018) attributed the decline to climate warming. Willig et al. (2019) attributed changes to changes in forest structure due to hurricane Hugo. In a survey of canopy arthropods taken 19 years after hurricane Hugo, Schowalter et al. (2017) found a legacy of the hurricane, despite droughts and other hurricanes of lesser intensities.

Recovery of the vegetation following hurricanes contrasted with that following ionizing radiation. In the irradiated area, recovery began with typical secondary successional annuals. Following the hurricane, adventitious shoots arising from stems of mature trees quickly restored the energy-supplying capacity of the ecosystem. As in the radiation study, the most important factor in recovery after hurricane damage was that the soil organic matter and the organisms that conserve nutrients were little damaged or diminished. Nutrient feedback enabled the damaged but still living trees to obtain nutrients and begin synthesizing new biomass. In both the radiation and the hurricane studies, the dynamics of the decomposing trophic web were similar but for different reasons. In the radiation study, decomposition first increased due to falling of dead leaves, then decreased as the trees died. In the hurricane study, decomposition increased following blowdown of the leaves, then decreased due to lower quality of the litter. In both cases, animal and plant populations responded to the change in ecosystem energy flow in different ways but both resulted in ecosystem recovery. Population interactions were an evident control of resilience in both systems.

The rain forest at El Verde had not fully recovered when hurricane Maria struck Puerto Rico in 2017. It was the most devastating storm to hit the island in recent times, and the effect on the rain forest was more severe than that of Hugo. Although leaves and branches were ripped off many of the trees, they did not die. Most have produced adventitious sprouts from the trunk. The hurricane altered but did not destroy the community of soil microbes that conserve and recycle nutrient elements because the organic matter in the upper soil horizons remained intact. A current project at the Luquillo site is to determine the energy reserve in the roots of these trees, to predict their ability to supply the energy source for the rest of the ecosystem's trophic webs (Amanodolare 2018). Despite the structural devastation to the forest, it appears that most species will recover, except perhaps for the Puerto Rican parrot (Associated Press 2018).

8.4.2.2 Conclusion

As in the radiation study, decomposer populations regulated the flow of energy through the ecosystem by adjusting their metabolic rates and population size. Most notable was the initial increase in decomposer activity soon after defoliation. Once the pulse ended, decomposition slowed, but after Hurricane Hugo there were sufficient nutrients stored in the soil organic matter to allow resumption of growth. Whether depletion of soil organic matter and its embedded nutrients after Hurricane Maria was severe enough to affect future productivity remains to be seen.

8.4.3 The Lago Guri Island Project

This project examined the effect of removing a top predator on energy dynamics of an ecosystem.

8.4.3.1 Perturbation = Elimination of Top Predators

Because species populations are continually expanding and contracting as energy flow through ecosystems varies, MacArthur and Wilson (1963) in their Island Biogeography Theory proposed that the number of species at any location depends on the balance between immigration and emigration or extinction. Consider a series of recently emergent volcanic islands, some large, some small, at various distances from the mainland, and that at the time of their emergence, they are completely barren of species. The theory predicts that islands closest to the mainland will increase their species richness faster than islands further away, and that larger islands will increase in richness faster than small islands. Equilibrium will be reached when each species has an equal likelihood of extinction and an equal likelihood of colonization or recolonization.

The theory was tested by on a series of islands formed by an impoundment on the Caroni river in Bolivar state, Venezuela (Terborgh et al. 1997; Shahabuddin et al. 2000). As the water rose following completion of the dam, populations of plants and herbivores on the islands were isolated. The absence of predators on these islands resulted in what the authors described as an "ecological meltdown in predator-free forest fragments" (Terborgh et al. 2001). Meltdown meant a severe reduction of seedlings and saplings of canopy trees due to herbivory by leaf-cutter ants, howler monkeys, iguanas and rodents whose population the predators controlled before island creation. After 7 years the founding avian communities had collapsed but was reconstituted (Terborgh et al. 2001). However, equilibrium had not been reached for other species. After 15 years, tree species were still in flux, (Terborgh and Feeley 2010), and they concluded that "Our results document transitory states in a process that we expect to continue for decades more." Thermodynamic theory predicts that equilibrium will never be reached until the herbivores are controlled. Can herbivores be controlled by means other than predators? Disease is often an alternative.

8.4.4 The Biological Dynamics of Tropical Rainforest Fragments Project

Since the origin of life, species have come into existence, populations have expanded, contracted, and become extinct. However in the twentieth century, rates of extinction have far exceeded historical rates. Increasing awareness of the potential cultural and economic value of species that are being lost has prompted a call for the establishment of reserves where it is hoped that endangered species could be conserved. While all agree that this is a good idea, there has been controversy over what size a reserve should be in order to save endangered species. Obviously, the larger the reserve, the better, but establishing reserves often conflicts with economic and development interests. In order to obtain a scientific answer to the question of optimum reserve size, a research project was established in 1979, near the center of the Amazon rain forest, about 90 km North of Manaus, Brazil. It was launched by the World Wildlife Fund and Brazil's National Institute for Research in Amazonia, and named the Biological Dynamics of Forest Fragments (BDFF) Project, also known as the Minimum Critical Size of Ecosystems Project. The project was designed specifically to investigate the relationship between the size of a forest fragment and both the stability and species carrying capacity of that forest fragment (Bierregaard et al. 1992).

8.4.4.1 Perturbation = Deforestation

Forest fragments ranging in size from one to 200 ha were delineated, and areas of comparable size in the undisturbed forest were marked for controls. Areas surrounding the fragments were clear-cut for at least 80 m, and then abandoned to successional vegetation. Population studies were made in the experimental areas before the fragments were isolated to enable quantification of changes specifically for each fragment. Results of the project have been summarized by Laurance et al. (2018), who cited almost 300 references to original publications.

8.4.4.2 Changes in Intact Forests

Over the past 2–3 decades, researchers at the BDFF site found that: tree mortality and recruitment have accelerated significantly over time; tree-community composition has shifted, generally in favor of faster-growing canopy trees and against shade-tolerant subcanopy trees; growth rates have increased for the large majority (84%) of tree genera in the study area; aboveground tree biomass has increased significantly over time (although tree-stem numbers have not changed significantly); lianas have increased in abundance. The trends appear consistent with those observed elsewhere in tropical forests. The most frequently invoked reason for rising tropical forest productivity is CO_2 fertilization. Atmospheric CO_2 levels have recently risen rapidly, and Rozema (1993) suggested that many plants show faster growth under enriched CO_2. This assumes that CO_2 is the limiting factor, a dubious assumption for Amazonian forests surviving on infertile soils.

8.4.4.3 Species Response to Fragmentation

Survival of species depends on the maintenance of their interactions with energy flow through the trophic webs of ecosystems. Species with small body mass, or species with limited range of movement are better able to survive in small reserves, where they are part of a complete trophic web. Examples include many small mammals, hummingbirds, frogs, and ants. Species that forage over large areas require larger reserves for the trophic web on which they depend to remain intact. Among birds, various species of understory insectivores, including army ant-followers, solitary species, terrestrial foragers and obligate mixed-flock members, are most susceptible to fragmentation. Many animal-eating bat species rarely persist in small fragments. Primates exhibit similarly predictable patterns of species loss, with wide-ranging frugivores being the most vulnerable, especially the black spider-monkey. A 100-ha reserve might be insufficient for wide ranging jaguars.

Another factor is the "edge effect" which strongly affects trophic stability. Energy flow through trophic webs near fragment edges is severely disrupted by fluxes of wind, heat and light. Establishment of nine 1-ha reserves is not equivalent to one 9 ha reserve because of the edge effects. A square nine ha plot has a perimeter of 1200 m. A one ha plot has an edge of 400 m. Nine one-ha plots have a combined perimeter of 3600 m. Although the area preserved is the same, when the area is divided into one ha plots, the perimeter increases by a factor of 3, meaning edge effects are much more pronounced in small preserves.

Elevated tree mortality can extend up to 250 m into the reserve. Many animal groups, such as bees, wasps, flies, beetles, ants, butterflies, understory birds and predatory bats decline in abundance near forest edges. Organisms that increase in abundance near edges, sometimes dramatically include many species of gap-favoring and frugivorous birds, hummingbirds, frugivorous bats that exploit early successional plants, light-loving butterflies and fast-growing lianas. As cleared areas fill with vegetation, trophic webs begin to heal. Where areas around fragments were not further disturbed after trees were cut, forest regrowth was rapid because the soil organic matter and the nutrient recycling decomposer food chains were left intact.

Both this study and the Lago Guri project showed, as predicted, that large reserves will conserve more species that require a large feeding range or highly specialized habitat, and that distance to reservoirs of replacement organisms strongly affects rate of recovery. Large reserves that have both old growth and successional forests may have higher species diversity than each forest type alone, but conservationists are more concerned with saving species in old growth forests, because they are more endangered than those in successional forests. Lindenmayer and Noss (2006) reviewed a series of papers on forest management, some of which advocated creation of secondary successional vegetation near old growth forests, because that would increase the species diversity of the area as a whole. The problem with that view is that the types of species that occur in successional forests are common, and in no danger of extinction. It is the species in the old growth that are threatened, and the larger the area covered by old growth, the greater the probability that these rare and endangered species will survive.

8.4.4.4 Conclusion

These cases of bottom-up control have shown the importance of species interactions with energy flow through the environment for control of stability and resilience. As species' populations adjust to changes in energy flow through ecosystems, the ecosystem responds by changing the rate at which energy is delivered to species. Deviation amplifying feedback controls the dynamic during recovery and negative feedback maintains stable equilibrium once recovery is complete.

8.5 What Have Case Studies Taught Us About Stability of Tropical Ecosystems?

8.5.1 Tropical Ecosystems Are Stable

Tropical rain forests were once thought to be "fragile" or unstable because of the many interactions (Farnworth and Golley 1973). The assumption was that if just one of the interactions breaks down, the whole system would collapse. Systems theory now predicts that it is *because* of their complex interactions and high species diversity, tropical rain forests should be the most stable of all terrestrial ecosystems (Hooper et al. 2005). The stability of a diverse ecosystem does not depend on the stability of particular populations that increase or decrease depending on environmental changes. If a minor perturbation such as disease outbreak or selective harvesting destroys one particular route of energy transfer through one species population, there are many others that compensate for the loss (Kardol et al. 2018).

8.5.2 Tropical Ecosystems Are Unstable

Ecosystem stability in the tropics depends more on the integrity of the nutrient cycling feedback system than ecosystem stability at higher latitudes. In the tropics, the feedback system is dependent on the energy in undecomposed biomass to feed the microbes that release the nutrients and transform them to soluble state in which they can be taken up by plants. When tropical forests are converted to monoculture plantations of coffee or of pulpwood as in Jarí and the nutrient cycling mechanisms are destroyed, the system will rapidly lose nutrients and become unstable. At higher latitudes where recent retreat of glaciers exposed fresh parent material high in nutrient content, systems are less dependent on endogenous recycling and more dependent on the exogeneous sources of nutrients.

Agricultural systems on soils derived from volcanic ash might be an exception to the trend of low fertility of tropical soils. These Andisols are relatively high in nutrients, but phosphorus may not be readily available because it is fixed in the amorphous structure of the clay. As in Oxisols, organic matter relieves the shortage of phosphorus. Another exception are the relatively fertile Inceptisols in the valleys of Andes mountains derived from erosion of relatively young geological material.

8.5.3 Energy Flow in Tropical Savannas and Rain Forests

Herbivory in savanna systems is dominated by large ruminant herbivores such as wildebeest that obtain their energy through consumption of grass. In tropical rain forests, vertebrate herbivores are much less important for energy flow and nutrient mobilization. Herbivores consume generally less than 5% of foliage area (Schowalter and Ganio 2003), probably because of the chemical defenses of leaves. Tapirs are common herbivores in tropical forests of the new world. Because tapirs are ruminants, they can digest leaves, but tapirs can't reach most of the leaves. Most of the other herbivores such as peccaries, sloths and monkeys subsist mainly on nuts, fruits, seeds and flowers. The dominant energy flow pathway is through leaf litter fall to soil organic matter where decomposers liberate nutrients.

8.5.4 Insects in Tropical Ecosystems

The difference in population cycles of herbivorous insects between tropical and high latitude systems is a cause of regional differences in stability. At high latitudes, life cycles are interrupted by freezing temperatures, and many populations have to recover from eggs that survived cold temperatures. In the tropics, population life cycles are not interrupted and so insect pressure on plants is higher. This may be the reason that tropical plants are the source of many hallucinogenic chemicals that provide a defense against herbivores.

8.6 Application of Lessons to Other Regions

To what extent can the conclusions from the tropics be applied to other regions of the world?

8.6.1 Relevance to Temperate Zones

The second portion of my career was spent in an agricultural ecosystem in the Piedmont region of Georgia. Soils there are Ultisols, very similar to Oxisols in the tropics, in that the underlying mineral soil is highly weathered and low in nutrient elements. Like tropical Oxisols, soil organic matter is critical for ecosystem stability, because it is where most of the nutrients are held for recycling (see Fig. 11.2a, Chap. 11). In regions farther north that have been glaciated during the Pleistocene, the situation is different. There the soils are relatively young, and higher in nutrient holding capacity. Nutrient recycling via organic matter in systems is not as critical. Systems are more resistant to perturbations because the mineral soil is an exogeneous source of nutrients.

8.6.2 Relevance to Aquatic Ecosystems

In terrestrial ecosystems, feedback theory deals with instability due to decrease in nutrient elements in the system. In fresh water ecosystems, instability results from an excess of nutrient elements.

8.6.3 The Experimental Lakes Project (Ecosystem Control of Species)

Since the 1940s, inorganic fertilizers have been commonly used in agricultural fields. Farmers found that maximum yield could be obtained by heavy applications of nitrogen and phosphorus, even though some of these chemicals could not be taken up by crops. The root systems of annual cropping systems do not efficiently exploit the soil volume in which the crops are growing. By the 1960s, it was evident that algal blooms and hypoxic conditions in lakes and waterways are caused by fertilizer runoff from excess fertilization (Environmental Protection Agency 2019). Unused nutrients are leached or eroded into nearby streams and lakes, where lack of dissolved oxygen caused by decomposing algal blooms disrupts energy flow in food webs. Environmental scientists wanted to know which nutrient caused the algal growth, and how it could be controlled. Laboratory experiments gave conflicting results (Stokstad 2008). Some scientists thought the culprit was phosphorus, principally in detergents and sewage; others thought it might be nitrogen from fertilizer and sewage, or carbon, or perhaps even trace metals.

The answer could not be determined by laboratory studies. To solve the problem, a group of scientists initiated a series of experiments in whole-lakes set in Precambrian Shield bedrock of northwestern Ontario. It was called the "Experimental Lakes Area" project (Schindler and Fee 1974).

Combinations of phosphorus, nitrogen, and carbon were added to several of the lakes at rates similar to those in lakes polluted by excess fertilizer runoff. Phosphate and nitrate caused rapid eutrophication. A similar result was obtained with phosphate, ammonia and sucrose, but recovery was almost immediate when phosphate additions only were discontinued. When two basins of one lake were fertilized with equal amounts of nitrate and sucrose, and phosphorus was also added to one of the basins, the phosphate-enriched basin quickly became highly eutrophic, while the basin receiving only nitrogen and carbon remained at pre-fertilization conditions. "These results, and the high affinity of sediments for phosphorus indicate that rapid abatement of eutrophication may be expected to follow phosphorus control measures" (Schindler 1974). The lake was resilient, and stability depended on the fact that phosphorus is tightly bound on the clay surface of mineral sediments (Brady 1974). Despite the evidence that chemical fertilizers cause pollution in streams, lakes, and oceans, control has been insufficient. Since the 1980s, there has been widespread global increase in intense lake phytoplankton blooms (Ho and Michalak 2019).

In another Experimental Lakes Area trial, the scientists tackled the issue of acid rain. Utility companies maintained that the observed increase in acid rain at that time could not harm the environment. The researchers added sulfuric and nitric acid to a lake at levels similar to that in acid rain, and found that plankton disappeared and fish reproduction was affected, but that lakes slowly recovered when the pH returned to normal. In yet another lake, researchers added synthetic hormones to study the effects of birth-control pills in wastewater. At hormone concentrations currently found in municipal wastewater, minnow populations crashed (Stokstad 2008).

8.6.4 Lake Mendota Studies (Species Control of Ecosystems)

Studies of lakes as ecosystems preceded studies of terrestrial ecosystems, perhaps because the dimensions of lakes are more clearly discernable. The studies of Edward Birge and Chauncy Juday in the late 1800s on the plankton Crustacea of Lake

Mendota in Wisconsin represent the first real beginning of limnology. The study on the seasonal distribution of the plankton in lakes led into an investigation of thermal stratification and chemical changes in the hypolimnion. This study led directly into quantitative studies of plankton standing crops and still later to an investigation of the dissolved organic content of lake waters as a means of studying the differences among lakes in their ability to produce organic matter (Frey 1963).

8.7 Case Studies as Tests of Thermodynamic Theory

Thermodynamic theory proposed that it is through feedback that each species adjusts to energy flow through ecosystems, and that ecosystems adjust to energy flow through species' populations. The case studies reviewed here validate these predictions. They have shown that during recovery from a disturbance, energy flow, nutrient recycling, and ecosystem biomass will increase when nutrient recycling mechanisms remain intact, but if feedback is damaged or destroyed, the ecosystem will lose biomass and decline in productivity. They have also shown the that the ability of species populations to adjust their biomass and metabolism to changes in ecosystem energy flow is a determining factor in ecosystem resilience. Thermodynamic theory further predicted that when nutrient cycling exceeds bounded limits in lakes, ecosystems become unstable and population dynamics are affected.

References

S. Amanodolare, Puerto Rico's catastrophic hurricane gave scientists a rare chance to study how tropical forests will fare in a stormier future. (2018). https://doi.org/10.1126/science.aav3739

Associated Press, Scientist's mission in post-hurricane Puerto Rico: saving endangered parrots. (2018), https://www.nbcnews.com/news/latino/scientists-mission-post-hurricanne-puerto-rico-saving-endangered-parrots-n938386. Accessed 7 Dec 2020

R.D. Bardgett, D.A. Wardle, *Aboveground – Belowground Linkages: Biotic Interactions, Ecosystem Processes, and Global Change* (Oxford University Press, Oxford, 2010)

K. Basnet, G.E. Likens, F.N. Scatena, A.E. Lugo, Hurricane Hugo: damage to a tropical rain forest in Puerto Rico. J. Trop. Ecol. **8**, 47–55 (1992)

R.O. Bierregaard, *Lessons from Amazonia: The Ecology and Conservation of a Fragmented Forest* (Yale University Press, New Haven, 2001)

R.O. Bierregaard, T.E. Lovejoy, V. Kapos, A. Augusto dos Santos, R.W. Hutchings, The biological dynamics of tropical rainforest fragments. Bioscience **42**(11), 859–866 (1992)

N.C. Brady, *The Nature and Properties of Soils* (Macmillan, New York, 1974)

C.S. Bristow, K.A. Hudson-Edwards, A. Chappell, Fertilizing the Amazon and equatorial Atlantic with West African dust. Geophys. Res. Lett. **37**, 3–7 (2010). https://doi.org/10.1029/2010GL043486

J.L.A. Campos, U.P. Albuquerque, N. Peroni, A.E. de Lima, Population structure and fruit availability of the babassu palm (Attalea speciosa Mart. ex Spreng) in human dominated landscapes

of the Northeast Region of Brazil. Acta Bot. Bras. **31**(2), 267–275 (2017). https://doi.org/10.1590/0102-33062016abb0255
G.T. Cowley, Fleshy fungi in relation to irradiation and cutting in the Luquillo experimental forest, in *A Tropical Rain Forest: A Study of Irradiation and Ecology at El Verde, Puerto Rico*, ed. by H. T. Odum, R. F. Pigeon, (Division of Technical Information, U.S. Atomic Energy Commission, Washington, DC, 1970), pp. F-9–F-13
E. Cuevas, Soil versus biological controls on nutrient cycling in terra firme forests, in *The Biogeochemistry of the Amazon Basin*, ed. by M. E. McClain, R. L. Victoria, J. E. Richey, (Oxford University Press, Oxford, 2001), pp. 52–67
J. Edmisten, Soil studies in the El Verde rain forest, in *A Tropical Rain Forest: A Study of Irradiation and Ecology at El Verde, Puerto Rico*, ed. by H. T. Odum, R. F. Pigeon, (Division of Technical Information, U.S. Atomic Energy Commission, Washington, DC, 1970), pp. H-79–H-87
C. Elton, *Animal Ecology. Sidgwick and Jackson, London*. Reprinted in 1939 (Macmillan, New York, 1927)
Environmental Protection Agency, Nutrient pollution: The sources and solutions: agriculture (2019), https://www.epa.gov/nutrientpollution/sources-and-solutions-agriculture. Accessed 17 Dec 2020
E. G. Farnworth, F. B. Golley (eds.), *Fragile Ecosystems: A Report of Research and Applications in the Neotropics* (Springer, New York, 1973)
P.M. Fearnside, J.M. Rankin, Jari revisited: changes and the outlook for sustainability in Amazonia's largest silvicultural estate. Interciencia **10**(3), 121–129 (1985)
J.T. Finn, Measures of ecosystem structure and function derived from analysis of flows. J. Theor. Biol. **56**, 363–380 (1976)
J.T. Finn, Cycling index: a general definition for cycling in compartment models, in *Environmental Chemistry and Cycling Processes. CONF-760429*, ed. by D. C. Adriano, I. L. Brisbin, (Technical Information Center, U.S. Department of Energy, Washington, DC, 1978), pp. 138–164
D.G. Frey, The Birge-Juday Era, an article adapted from Frey DG (1963), Limnology in North America. The University of Wisconsin Press, (1963), https://z0ku333mvy924cayk1kta4r1-wpengine.netdna-ssl.com/wp-content/uploads/2018/09/32-4-8.pdf. Accessed 19 Dec 2020
F.B. Golley, *A History of the Ecosystem Concept in Ecology* (Yale University Press, New Haven, 1993)
M.R. Guariguata, R. Ostertag, Neotropical secondary forest succession: changes in structural and functional characteristics. For. Ecol. Manage. **148**, 185–206 (2001)
R. Herrera, T. Merida, N. Stark, C.F. Jordan, Direct phosphorus transfer from leaf litter to roots. Naturwissenschaften **65**, S.208 (1978)
R. Herrera, C.F. Jordan, E. Medina, H. Klinge, How human activities disturb the nutrient cycles of a tropical rainforest in Amazonia. Ambio **10**(2/3, MAB: A Special Issue), 109–114 (1981)
C. Hillinger, Seeds of a storm: A Puerto Rican tropical rain forest sprouts new life in the wake of hurricane Hugo. Los Angeles Times (20 May 1990). Accessed 19 Dec 2020
J.C. Ho, A.M. Michalak, N. Pahlevan, Widespread global increase in intense lake phytoplankton blooms since the 1980s. Nature Online (2019). https://doi.org/10.1038/s41586-019-1648-7. Accessed 19 Dec 2020
D.U. Hooper, F.S. Chapin III, J.J. Ewel, et al., Effects of biodiversity on ecosystem functioning: a consensus of current knowledge. Ecol. Monogr. **75**(1), 3–35 (2005)
C.F. Jordan, Productivity of a tropical forest and its relation to a world pattern of energy storage. J. Ecol. **59**, 127–142 (1971)
C.F. Jordan, The nutrient balance of an Amazonian rain forest. Ecology **63**, 647–654 (1982a)
C.F. Jordan, Nutrient cycling index of an Amazonian rain forest. Acta Ecol. Gen. **3**(3), 393–400 (1982b)
C.F. Jordan, Amazon rain forests. Am. Sci. **70**, 394–401 (1982c)
C.F. Jordan, *Nutrient Cycling in Tropical Forest Ecosystems* (Wiley, New York, 1985a)
C.F. Jordan, Jari: a development project for pulp in the Brazilian Amazon. Environ. Prof. **7**, 135–142 (1985b)

C.F. Jordan, *Amazonian Rain Forests: Ecosystem Disturbance and Recovery* (Springer, New York, 1987)

C.F. Jordan, *An Amazonian Rain Forest. The Structure and Function of a Stressed Ecosystem and the Impact of Slash and Burn Agriculture*, Man and the Biosphere Series, vol 2 (UNESCO, Paris/Parthenon/Carnforth, 1989)

C.F. Jordan, G. Escalante, Root productivity in an Amazonian rain forest. Ecology **61**, 14–18 (1980)

P. Kardol, N. Fanin, D.A. Wardle, Long-term effects of species loss on community properties across contrasting ecosystems. Nature **557**, 710–713 (2018)

B.I. Kronberg, W.S. Fyfe, O.H. Leonardos, A.M. Santos, The chemistry of some Brazilian soils: element mobility during intense weathering. Chem. Geol. **24**, 211–229 (1979)

M. Lamirande., Pulppapernews.com. Headlines from 7 Jan 2000. Jordan CF files (2000)

W.F. Laurance, J.L.C. Camargo, P.M. Fearnside, T.E. Lovejoy, G.B. Williamson, R.C.G. Mesquita, C.F.J. Meyer, P.E.D. Bobrowiec, S.G.W. Laurance, An Amazonian rainforest and its fragments as a laboratory of global change. Biol. Rev. **93**(1), 223–247 (2018)

J.R. Leake, D.J. Read, Mycorrhizal fungi in terrestrial habitats, in *The Mycota IV. Environmental and Microbial Relationships*, ed. by D. T. Wicklow, B. Söderström, (Springer, Heidelberg, 1997), pp. 281–301

D. Lee, X.G. Han, C.F. Jordan, Soil phosphorus fractions, aluminum, and water retention as affected by microbial activity in an Ultisol. Plant Soil **121**, 125–136 (1990)

H. Lieth, Primary productivity in ecosystems: comparative analysis of global patterns, in *Unifying Concepts in Ecology*, ed. by W. H. van Dobben, R. Lowe-McConnell, (Springer, Dordrecht, 1975), pp. 67–88

R.L. Lindeman, The trophic-dynamic aspect of ecology. Ecology **23**, 399–418 (1942)

D.B. Lindenmayer, R.F. Noss, Salvage logging, ecosystem processes, and biodiversity conservation. Conserv. Biol. **20**(4), 949–958 (2006)

B.C. Lister, Garcia, Climate-driven declines in arthropod abundance restructure a rainforest food web. PNAS **115**(44), E10397–E10406 (2018)

A.J. Lotka, Contributions to the energetics of evolution. PNAS **8**(6), 147–151 (1922)

Luquillo LTER, (2019), https://luq.lter.network/home-luq. Accessed 26 Dec 2020

R.H. MacArthur, E.O. Wilson, An equilibrium theory of insular zoogeography. Evolution **17**(4), 373–387 (1963)

E.A. McMahan, Radiation and the termites at El Verde, in *A Tropical Rain Forest: A Study of Irradiation and Ecology at El Verde, Puerto Rico*, ed. by H. T. Odum, R. F. Pigeon, (Division of Technical Information, U.S. Atomic Energy Commission, Washington, DC, 1970), pp. E-105–E-122

E. Medina, E. Cuevas, Mineral nutrition: Humid tropical forests, in *Progress in Botany*, ed. by H. D. Behnke, U. Lüttge, K. Esser, J. W. Kadereit, M. Runge, vol. 55, (Springer, Berlin/Heidelberg, 1994), pp. 115–129. https://doi.org/10.1007/978-3-642-78568-9_6

R.C.G. Mesquita, The effect of different proportions of canopy opening on the carbon cycle of a central Amazonian secondary forest. Dissertation, University of Georgia, Athens, 1995

E.F. Moran, *Developing the Amazon* (Indiana University Press, Bloomington, 1981)

P.H. Nye, D.J. Greenland, *The Soil Under Shifting Cultivation*, Technical Communication No. 51 (Commonwealth Bureau of Soils, Commonwealth Agricultural Bureaux, Farnham Royal, Bucks, England, 1960)

H.T. Odum, G. Drewry, The cesium source at El Verde, in *A Tropical Rain Forest: A Study of Irradiation and Ecology at El Verde, Puerto Rico*, ed. by H. T. Odum, R. F. Pigeon, (Division of Technical Information, U.S. Atomic Energy Commission, Washington, DC, 1970), pp. C-23–C-36

H.T. Odum, J. Ruíz-Reyes, Holes in leaves and the grazing control mechanism, in *A Tropical Rain Forest: A Study of Irradiation and Ecology at El Verde, Puerto Rico*, ed. by H. T. Odum, R. F. Pigeon, (Division of Technical Information, U.S. Atomic Energy Commission, Washington, DC, 1970), pp. I-69–I-80

J.R. Palmer, Palmer reflects on Jari. Int. Soc. Trop. For. **9**(3), 1 (1988)

P.E. Powell, G.R. Cline, C.P.P. Reid, P.J. Szaniszlo, Occurrence of hydroxamate siderophore iron chelators in soils. Nature **287**, 833–834 (1980)

C.M. Prather, G.E. Belovsky, S.A. Cantrell, G. Gonazlez, Tropical herbivorous phasmids, but not litter snails, alter decomposition rates by modifying litter bacteria. Ecology **99**(4), 782–781 (2018)

J.M. Prospero, O.L. Mayol-Bracero, Understanding the transport and impact of African dust on the Caribbean Basin. Am. Meteorol. Soc. J. (online) (2013). https://doi.org/10.1175/BAMS-D-12-00142.1. Accessed 30 Dec 2020

J.M. Prospero, R.A. Glaccum, R.T. Nees, Atmospheric transport of soil dust from Africa to South America. Nature **289**(5798), 570–572 (1981)

H. Putzer, The geological evolution of the Amazon basin and its mineral resources, in *The Amazon, Limnology and Landscape Ecology of a Might Tropical River and Its Basin*, ed. by H. Sioli, (Junk, Dordrecht, 1984), pp. 15–46

H.F. Recher, Population density and seasonal changes of the avifauna in a tropical forest before and after gamma irradiation, in *A Tropical Rain Forest: A Study of Irradiation and Ecology at El Verde, Puerto Rico*, ed. by H. T. Odum, R. F. Pigeon, (Division of Technical Information, U.S. Atomic Energy Commission, Washington, DC, 1970), pp. E-69–E-93

S.C. Reed, A.R. Townsend, P.G. Taylor, C.C. Cleveland, Phosphorus cycling in tropical forests growing on highly weathered soils, in *Phosphorus in Action. Soil Biology 26*, ed. by E. Bünemann, A. Oberson, E. Frossard, (Springer, Berlin/Heidelberg, 2011), pp. 339–369

B. Richardson, M. Richardson, G. González, Responses of two litter-based invertebrate communities to changes in canopy cover in a forest subject to hurricanes. Forests **9**(6), 309 (2018). https://doi.org/10.3390/f9060309

J. Rozema, Plant responses to atmospheric carbon dioxide enrichment: interactions with some soil and atmospheric conditions. Vegatatio **104**, 173–190 (1993)

C.E. Russell, Nutrient cycling and productivity in native and plantation forests at Jarí Florestal, Para, Brazil. Dissertation, University of Georgia, Athens, Georgia, 1983

R. Sanford, J. Saldarriaga, K.E. Clark, C. Uhl, R. Herrera, Amazon rain-forest fires. Science **227**(4682), 53–55 (1985)

D.W. Schindler, Eutrophication and recovery in experimental lakes: implications for lake management. Science **184**(4139), 897–899 (1974)

D.W. Schindler, E.J. Fee, Experimental lakes area: whole-lake experiments in eutrophication. J. Fish Res. Board Can. **31**(5), 937–953 (1974). https://doi.org/10.1139/f74-110

T.D. Schowalter, L.M. Ganio, Diel, seasonal and disturbance induced variation in invertebrate assemblages, in *Arthropods of Tropical Forests*, ed. by Y. Basset, V. Novotny, S. E. Miller, R. L. Kitching, (Cambridge University Press, Cambridge, 2003), pp. 315–328

T.D. Schowalter, M.R. Willig, S.J. Presley, Post-hurricane successional dynamics in abundance and diversity of canopy arthropods in a tropical rainforest. Environ. Entomol. **46**(1), 11–20 (2017)

G. Shahabuddin, G.A. Herzner, C. Aponte, M.D.C. Gomez, Persistence of a frugivorous butterfly species in Venezuelan forest fragments: the role of movement and habitat quality. Biodivers. Conserv. **9**(12), 1623–1641 (2000)

A.B. Shiels, G. González, D.J. Lodge, M.R. Willing, J.K. Zimmerman, Cascading effect of canopy opening and debris deposition from a large-scale hurricane experiment in a tropical rain forest. Bioscience **65**(9), 871–881 (2015)

N.M. Stark, C.F. Jordan, Nutrient retention by the root mat of an Amazonian rain forest. Ecology **59**(3), 434–437 (1978)

E. Stokstad, Canada's experimental lakes. Science **322**(5906), 1316–1319 (2008)

J. Terborgh, K. Feeley, Propagation of trophic cascades via multiple pathways in tropical forests, in *Trophic Cascades: Predators, Prey, and the Changing Dynamics of Nature*, ed. by J. Terborgh, J. A. Estes, (Island Press, Washington, DC, 2010), pp. 125–140

J. Terborgh, L. Lopez, J. Tello, Bird communities in transition: the Lago Guri Islands. Ecology **78**(5), 1494–1501 (1997)

J. Terborgh, L. Lopez, V. Percy Nuñez, S.G. RaoM, G. Orihuela, M. Riveros, R. Acanio, G.H. Adler, T.D. Lambert, L. Balbas, Ecological meltdown in predator-free forest fragments. Science **294**(5548), 1923–1926 (2001)

F.B. Turner, C.S. Gist, Obervations on lizards and tree frogs in an irradiated Puerto Rican forest, in *A Tropical Rain Forest: A Study of Irradiation and Ecology at El Verde, Puerto Rico*, ed. by H. T. Odum, R. F. Pigeon, (Division of Technical Information, U.S. Atomic Energy Commission, 1970), pp. E-25–E-49

C. Uhl, Recovery following disturbances of different intensities in the Amazon rain forest of Venezuela. Interciencia **7**, 19–24 (1982)

C. Uhl, C.F. Jordan, Vegetation and nutrient dynamics during the first five years of succession following forest cutting and burning in the Rio Negro region of Amazonia. Ecology **65**, 1476–1490 (1984)

P.M. Vitousek, W.A. Reiners, Ecosystem succession and nutrient retention: a hypothesis. Bioscience **25**(6), 376–381 (1975)

R.K. Wadsworth, Point-Quarter sampling of forest type-site relations at El verde, in *A Tropical Rain Forest: A Study of Irradiation and Ecology at El Verde, Puerto Rico*, ed. by H. T. Odum, R. F. Pigeon, (Division of Technical Information, U.S. Atomic Energy Commission, Washington, DC, 1970), pp. B-97–B-103

M.P. Weinbren, B.M. Weinbren, Observations on the mosquito population in the irradiated forest at El Verde, in *A Tropical Rain Forest: A Study of Irradiation and Ecology at El Verde, Puerto Rico*, ed. by H. T. Odum, R. F. Pigeon, (Division of Technical Information, U.S. Atomic Energy Commission, Washington, DC, 1970), pp. E-159–E-167

F.W. Went, N. Stark, Mycorrhiza. Bioscience **18**, 1035–1039 (1968)

M.R. Willig, L. Woolbright, S.J. Presley, T.D. Schowalter, R.B. Waide, T.H. Scalley, J.K. Zimmerman, G. González, A.E. Lugo, Populations are not declining and food webs are not collapsing at the Luquillo Experimental Forest. PNAS **116**(25), 12143–12144 (2019)

M. Witkamp, Aspects of soil microflora in a gamma-irradiated rain forest, in *A Tropical Rain Forest: A Study of Irradiation and Ecology at El Verde, Puerto Rico*, ed. by H. T. Odum, R. F. Pigeon, (Division of Technical Information, U.S. Atomic Energy Commission, Washington, DC, 1970), pp. F-29–F 33

Chapter 9
Entropy and Maximum Power

9.1 Entropy

Entropy is the propensity of things to become disorganized. It is the reason machines wear out, bridges become unstable, teenager's bedrooms get messy, and people get old. The amount of entropy is a measure of the disorder or randomness of a system. In the case of the teenager's bedroom, the disorder is the randomness with which socks, books, pictures etc. are scattered on the floor. A jigsaw puzzle dumped on a table has maximum entropy. There is complete disorganization. As you start to piece the parts together, organization begins and randomness decreases. A completed puzzle has maximum structure and minimum disorganization. It has minimum entropy because it is completely organized. If the cat jumps on the table as you finish the puzzle and disorganizes it, entropy becomes maximum.

Box 9.1 Some Indicators of Entropy

- The color of a banana peel is an index of the disorganization of molecules inside the fruit. When the banana peel is green, most of the molecules are highly organized complex carbohydrates such as starches. When the banana peel turns yellow, it indicates the starches are breaking down into simple sugars. When the banana turns to brown mush, entropy is close to maximum.
- Forensic experts can determine the amount of time that has passed since a person has died by the decomposition of body parts. Corpses are refrigerated to retard entropy.

According to the physical definition, entropy is a measure of the amount of energy which is unavailable to do work. How can something be defined as the

C. F. Jordan, *Evolution from a Thermodynamic Perspective*,
https://doi.org/10.1007/978-3-030-85186-6_9

absence of something else? Physicists explain it this way: Entropy is maximum in a closed system such as a black box isolated from the environment. The molecules or atoms inside the box are moving randomly. No further disorganization inside the black box can occur. The box is said to be in thermodynamic equilibrium, because there is no energy transfer that could create order among the atoms. Work performed by a system is energy transferred by the system to its surroundings. Work can be measured; therefore, entropy can be measured and defined.

9.2 Entropy in a Steel Bar

A blacksmith places a steel rod on an anvil. As in all solid matter, the molecules within the steel rod have kinetic energy and are continuously vibrating. The random kinetic motion of the molecules bumping into each other keeps them subject to Newton's first law, that every object will remain at rest unless compelled to move by an outside force. The rod is close to thermodynamic equilibrium because the rod is at the same temperature as the anvil and the air around it, and disorganization of the molecules is maximum. Entropy is maximum because no further disorganization is possible.

The blacksmith puts energy into the system when he pounds the steel rod on his anvil with a hammer to flatten the rod. As his hammer hits the rod, molecules in the rod vibrate more vigorously causing the temperature of the rod to rise. As a result, movements of molecules are no longer random and entropy is no longer maximum. Molecules become organized to transfer heat away from the impact point of the hammer. Work is carried out as the kinetic energy is transformed into heat and the rod is transformed into a flat bar. Heat radiating from the bar can be measured and is an index of entropy. The blacksmith then stops, the bar cools and eventually regains thermodynamic equilibrium when the temperature becomes the same as its surroundings. Movement of molecules in the bar becomes random, and entropy is again maximum.

The process through which the blacksmith's energy has been transformed to heat is a one-way entropic transformation. Energy is "dissipated" (becomes disorganized) as it passes along an entropic gradient from a state that can perform "work" to a state that it cannot (Prigogine et al. 1972; Kleidon et al. 2010). The heat generated by the hammer blow cannot be used as an energy source to reshape the bar into a rod.

9.3 Thermodynamic Equilibrium

Thermodynamic equilibrium means there are no net macroscopic flows of energy, either within a system or between systems. As a result, no work can be performed. As long as the blacksmith continues hammering the steel bar on his anvil, the bar is not in thermodynamic equilibrium. The harder the blacksmith pounds the bar, the further it is from equilibrium, and the faster the work of flattening the bar is accomplished.

In contrast to a black box insulated from its environment, living systems are open systems with energy coming in in the form of solar radiation and exiting as heat while performing the work of metabolizing and growing. Organisms are highly organized systems of molecules, a state far from equilibrium. Ecosystems are highly organized systems of species. Disorganization due to entropy is prevented by the energy from solar radiation which works or keep things organized (Wolchover 2014).

Life exists in a far-from-equilibrium state. Molecules in an organism are not distributed randomly, they are exquisitely organized. Species in an ecosystem are not distributed randomly, they exist in a highly structured sequence. Energy from the sun keeps molecules and species organized while keeping entropy at bay. If energy flow through a living system were to stop, energy stored in the system would gradually be depleted. The system eventually would arrive at thermodynamic equilibrium and life would cease. Economic systems also are in a far-from-equilibrium state. If money flow through the system were to stop, the system would achieve equilibrium and collapse. To prevent collapse when the economy is slowing, the Federal Bank pumps money into the system to keep the economy going.

9.4 Entropic Gradients

An entropic energy gradient occurs when highly concentrated energy is dissipated into heat. An entropic gradient is produced when coal is burned. Carbon molecules are organized in the structure of a lump of coal in a form that stores energy. An entropic energy gradient is formed when coal is burned to produce heat. Molecules organized in the coal become disorganized as the coal burns. Heat is the last stage of energy degradation.

Box 9.2 Steam Engines
Doesn't heat drive a steam engine? To keep it working, coal has to be shoveled in. Heat under the boiler of a steam engine converts water into steam. Pressure builds inside the boiler, and the release of pressure as the piston valves open propels the engine. Pressure change, not heat is doing the work.

9.5 Capturing and Storing Entropy

An entropic gradient also exists between solar energy entering the Earth as concentrated energy and leaving the Earth as heat. When the sunlight strikes the Earth, the light energy causes molecules to vibrate faster, causing surface temperature to rise.

The ability of the sunlight to do work is diminished because light energy has been transformed into heat.

When sunlight strikes the leaf of a plant instead of soil, rocks or water, some of the energy from the global entropic gradient is captured by photosynthesis and is used in the chemical reactions that build living organisms and sustain life. Biologists describe these processes as "capturing or storing energy". Biophysicists prefer to call the process that takes energy from this gradient as "capturing entropy" or "storing entropy" because that is a more general term for energy-conserving processes such as evaporation of water (National Weather Service 2020), while photosynthesis is a process limited to living systems (Kleidon 2010; Kleidon et al. 2010; Wang and Bras 2011). When non-isolated systems like organisms and ecosystems capture entropy, the global entropic gradient decreases by the amount taken by photosynthesis. Part of the global entropy is entrapped in living biomass and does not reach the soil where it would be transformed to heat. Instead, it is held in biomass by organization of molecules. A living system just "borrows" some entropy that will be returned later when the organism dies.

Box 9.3 Getting Organized

"Getting organized" means storing entropy. Storing entropy means a system is building structure. "Decomposition" means releasing stored entropy when interactions between system components become random. War is release of entropy. Reconstruction represents re-storage of entropy.

9.5.1 Evapotranspiration and Entropy Reduction

In well-watered mature forests, latent heat of vaporization driving evapotranspiration of plants is more significant than increase in biomass for locally capturing entropy from the Earth's gradient (Wu et al. 2007; Surface energy balance 2019). Da Rocha et al. (2004) found evidence of this in the rain forests of Amazonia, where the reduction of entropy due to transpiration was much greater than entropy reduced through production of biomass. Evaporation of water through transpiration in rain forests also is more effective in reducing entropy than evaporation from lakes or oceans. Bodies of water have only one surface from which water is evaporated, while rain forests can have leaf areas that are seven or eight times as great as the soil area that they cover (Asner et al. 2003). Destruction of tropical rain forests increases global warming by eliminating evapotranspiration.

9.5.2 *Life Is a Balance Between Storing and Releasing Entropy*

Capturing entropy from the global gradient by a thermodynamically open system has also been called "entropy production" (Martyushev and Seleznev 2006). During entropy production, energy, instead of entropically dissipating as heat, it arranges the molecules as biomass (Collier 1988). Through the process of growing, living biomass "produces entropy" (Kleidon et al. 2010). This means that energy that would have dissipated to heat into the global gradient is stored instead as structurally arranged molecules in living biomass. It is really *potential* entropy, because entropy is minimum where organization is maximum. The larger and more complex the biomass, the more potential entropy that is produced. When the organisms dies and its structure decomposes, the borrowed entropy is released and returns to the global gradient. Life is a balance between storing and releasing entropy.

9.5.2.1 Potential Entropy

Physicists interested in the origin of life have confused us by saying that as organization increases during evolution, entropy is "stored" or "produced" (England 2020). We thought that entropy decreased as organization increased. When we completed the jigsaw puzzle, entropy was minimum. The two views of entropy seem to be contradictory. Does organization mean maximum or minimum entropy? What physicists mean is that "potential" entropy increases as organization increases. A completed puzzle on the table top has maximum "potential" or "stored" entropy, but the stored entropy is not evident until the system comes apart. When the puzzle breaks apart, its entropy is realized. Physicists may not like the terms potential entropy and realized entropy, but the terms help us understand the apparent contradiction.

9.5.2.2 Entropy and Life

When a child is born, a complex system appears, meaning entropy has been produced. More entropy is produced as the child grows. As an adult, he or she is not in thermodynamic equilibrium. They lose energy due to metabolic activity, but that energy is replaced whenever they have a meal or snacks. When they die, energy replacement stops. The structure of the system that once was a person is lost (except for the skeleton) and the entropy is released. Produced entropy becomes realized entropy. Ecosystems, including the living Earth ecosystem are not in thermodynamic equilibrium either. They are continually receiving energy from the sun, and storing entropy in highly organized complex food webs. The flow of energy in food webs is non-random, but is always in one direction only, towards dissipation. Carnivores eat herbivores, never the other way around.

9.5.3 The Law of Maximum Entropy Production

The "Law" of Maximum Entropy Production (LMEP, Kleidon et al. 2010) says a system will select the path or assemblage of paths out of available paths that maximizes entropy production at the fastest rate given the constraints (Swenson 1989; Martinez-Kahn and Matinez-Castilla 2010). This means that during evolution of a species, those metabolic pathways that are most efficient in facilitating energy flow are the ones most "fit", and so survive evolutionary selection. These pathways have the "steepest entropy gradient" (Beretta 2020). When these metabolic pathways become part of the ecosystem structure, these systems are said to "produce entropy".

9.5.4 Energy for Metabolism as Well as Growth

In addition to organisms and ecosystems storing entropy as they are built, they also must fight against degradation resulting from entropy inherent in living organisms themselves. There must be a continual input of energy to support metabolism that resists and repairs degradation of the system. Organisms must compensate for their own entropy, or increasingly random motions of molecules would degrade the system and it will decompose. When an organism no longer processes energy, it dies and decomposes. When a farmer grows a potato, he is decreasing global entropy. When you eat the potato, you are storing global entropy. Some of that stored entropy is released during metabolism (keeping your system organized). The rest is released when you die.

9.5.5 Unassisted Entropy Capture Is a Unique Characteristic of Life

While both batteries and living organisms store energy, batteries unassisted cannot capture energy as do plants. Batteries require engineered devices such as solar panels or wind mills to capture energy before it can be stored. The energy stored by water behind a dam is realized when the water turns a turbine, but here again, a man-made structure is required. The ability unassisted by physical structures to both capture and store energy is a unique characteristic of life. Since capturing and storing energy requires structure to do the storing, and structure is how energy is stored in the form of potential entropy, storage of entropy defines life.

9.6 Entropy Storage by Ecosystems

While organisms "borrow" entropy, ecosystems have stored it permanently, or at least they have for more than two billion years. Although the amount of entropy stored by ecosystems during geological times was reduced by at least five catastrophic events each of which eliminated the majority of living species, it was never eliminated, and life did not have to start anew after each event. The tendency for trophic chains and food webs to store entropy is a remarkable feature of nature, scarcely recognized by science.

9.6.1 What Causes Entropy to Be Stored?

What is the driving force that causes entropy to be produced and stored in living systems? Schneider and Sagan (2005) pointed out that the force results from the fact that nature abhors a gradient, and that the tendency for gradient reduction in solar energy transiting Earth is the force that produces stored entropy in living systems. The tendency for nature to reduce an entropic gradient is a very powerful force. If power lines are downed after a storm and you touch one of the lines, you will be reducing an energy gradient. It is the driving force behind hurricanes that reduce the pressure gradient between earth and the atmosphere. It is the gradient-reducing force exercised by lightning when it reduces the energy gradient between charged clouds and the earth. An expression of energy gradient reduction on earth between solar energy and heat is the increase in biomass and structure of life as ecosystems and their embedded species are formed.

9.6.2 Entropy Storage by Animals

Rate of entropy storage by systems is not constant. This is evident in animals, a type of living system. Early in the life of an animal, growth (entropy storage) is rapid and efficiency is low. Later in life, growth decreases as energy is used for metabolic repair and for hunting or scavenging. Depending on a species' position in the food chain, timing of entropy storage can be very important. In the Serengeti, wildebeest calves get big as fast as possible so they can follow the herd out of danger. Efficiency of energy use is low. For lion cubs, getting big fast is not so important. Efficiency is higher.

9.7 Capturing Pressure

To understand how the capturing and storing of entropy reduces global entropy and increases the entropy in the capturing system, it is helpful to look at capturing and storing pressure. River water flowing downslope has a current that produces pressure against anyone wading across the river. When meanders form that reduce the current (Fig. 9.1), pressure is captured by the river banks and pressure against someone wading across the river is reduced. The meanders are capturing and storing pressure. If the creek were to be straightened, pressure would suddenly be released from storage and current would increase. Wading across the stream would be more difficult. Capturing pressure is important to prevent erosion during storms. When life captures energy, it reduces the global entropic gradient. If life were to disappear, the global entropic gradient would increase.

Fig. 9.1 Meanders in a creek on Spring Valley Ecofarms in the Georgia Piedmont. With each meander, current is reduced and the potential pressure that would exist if the creek were straight is stored. Eroded sand is deposited on the inside of each meander where the current is slower and the pressure is less. (Photo by C.F. Jordan)

9.8 Entropy and Time

Time is an elusive concept. Most people would define time as something that is measured by a clock or a calendar. But what is that *something*? The events that measure our age are the spinning of the Earth on its axis, and circulation of the Earth around the sun. These rotations measure what humans call time, but the rotations in themselves are not time. Physicists say that time like entropy is an irreversible sequence of events. Because entropy of a system varies, the speed of time also varies. Imagine a child's bedroom as an example of how an increase in disorganization (entropy) is a definition of time. In the morning, the father organizes toy soldiers into battalions, companies, platoons, and squads. The child begins to play war. An hour later, the father comes back and finds that some of the squads and platoons are scattered throughout the room, but the battalions are still distinguishable. Two hours later he comes back again and finds that all the soldiers are randomly scattered. There is no evidence of military hierarchy. The movement of toy soldiers has proceeded in only one direction, towards increasing disorder. From the father's viewpoint, the amount of disorder depends on how long he has been away. The longer he was away, the more disorganization that occurred. From the child's viewpoint, the faster he has the battles play out, the greater the rate of disorganization. Time is entropy. With a greater rate of dissipation, less time is required to reach maximum entropy. What then determines the speed of entropy?

Box 9.4
Entropy also results in the loss of information. The structure of a military organization contains much information relevant to who reports to whom. As an army loses a battle, both organization and information are lost. No one is left to issue orders. When an ecosystem is destroyed, all the information embedded in the interactions necessary for stability is lost.

9.8.1 Time's Speed Regulator

Since entropy, like time, occurs only in one direction, it is called the arrow of time. It is a way of distinguishing the past from the future. Entropy has been called "Time's Speed Regulator". The number of energy-transforming events that occur determine the speed of time as well as entropy. H.T. Odum, a pioneer in systems ecology was curious if there is a general principle regarding what controls the speed of time (and therefore the speed of entropy) in ecological, economic, mechanical and social systems, as well as individual living organisms.

What prompted Odum to ask such a question? Perhaps he was imagining life on a space ship hurtling through space. Since the space ship is not orbiting around a sun and neither is there "day" and "night", there is no external measure of time. Time has to be measured by the length of time that it takes to complete a life cycle. A child would be one fifth life cycle old, a middle-aged person would-be one-half life cycle old. Humans on this space ship observed that dogs completed five life cycles while humans completed only one. That means time passes faster for dogs than for humans. The reason is that metabolic processes such as heart beat in dogs are faster than those in humans. For humans, rate of energy dissipation during each metabolic reaction is slower. The slower the rate of dissipation, the longer the time to perform a process (Falasco and Esposito 2020).

9.8.2 Efficiency of Energy Transformations

The basic metabolic rate in small animals is higher than that in large animals. This is because smaller animals have a greater surface area to volume ratio, so rate of heat loss is high. That means energy conversions in small mammals is faster, therefore they are more time efficient. Efficiency of energy conversions in ecosystems also varies. In ecosystems where energy transformations are efficient (prey is quickly caught by the predator), time is faster than where the prey is adept at eluding the predator. From the viewpoint of a scientist timing these systems from images on a trailside video, the speed of his clock is independent of the speed of the transactions. The scientist merely observes that rate of energy flow through an efficient system is greater than that through an inefficient system because more interactions are happening during a ten-minute interval on his clock. In ecological systems, time and entropy (an entropic sequence of events) are interchangeable; they are different forms of the same thing.

H.T. Odum's conclusion, theoretically determined in appendix 7, is that the efficiency of energy transformations regulates the speed of time and of entropy.

9.8.3 Passage of Time for Cats

A scientist wants to determine the difference in passage of time between two similar systems. She organizes two rooms of the same size. In each there is a table with a bowl filled with eggs, and a piece of cheese on top of the eggs. The baseboard of each room has a hole in it in which lives a mouse, and a door over the hole that the scientist can remotely open and close. There is a ramp from the floor to the table top by which the mouse can access the cheese. Also in each room there is a cat resting on a pillow. The only difference is that one cat has been recently fed, the other not.

The scientist starts a timer and opens the door to each mouse hole. The mice run out, climb the ramp and begin to eat the cheese. The hungry cat jumps onto the table and knocks the bowl off the table. The bowl and the eggs shatter, and egg yolk spreads on the floor. The mouse jumps down and scurries inside the pillowcase. The cat pursues and rips open the pillow case, and eats the mouse while spilling blood on the floor. In the other room, much less happens. The cat ignores the mouse who after nibbling part of the cheese, returns to its nest. In the hungry cat's room, many energy transformations have occurred, and at the end, entropy is very high, so from an ecosystem perspective, much time has passed. In the other room, entropy remains low, and little time has passed. Everything is still well organized. There has been a minimum of energy transformations – only the chemical energy of cheese to the kinetic energy of the mouse. For the scientist, exactly the same amount of time has passed as measured by the timer. For each cat, the number of energy transformations during the experiment determined the amount of time that passed.

9.9 The Maximum Power Principle

There are two types of efficiency. One is efficiency of time, where the systems that accomplishes the most work per unit time is the most efficient system, but in this system, resources are used inefficiently. That was demonstrated by the hungry cat. Human dominated systems on the frontier operate where time efficiency is greatest. Young or successional systems also operate to maximize time efficiency. Odum and Pinkerton (1955) called this the optimum efficiency for maximum power output. Derivation of the Maximum Power Principle is given in appendix 7. The other efficiency is efficiency of resource use, accomplished by fully exploiting resources and not wasting them. That was demonstrated by the lazy cat. She waited till the mouse ate the entire cheese before eating the mouse. Most mature systems operate where this efficiency is greatest.

Box 9.5
Workers that are paid by the job will work according to the maximum power principle. They will finish the job quickly at the expense of using resources efficiently. Sometimes that is desirable, because time is money. Conscientious workers that are paid by the hour will work carefully to ensure that resources are not wasted. A wasted resource is money lost.

In natural resource systems including farms, forests and fisheries, humans manage the system for speed of output, where they can get the most output in the shortest period of time. When a system matures, as in the case of old-growth forests, the system no longer is operating at the optimum efficiency for maximum power output. It operates at the optimum efficiency for stability. This is a higher efficiency than for maximum power output. As a result, light, nutrients and water are used more efficiently. Few resources escape being utilized by plants.

9.10 Optimum Efficiencies for a Truck and Its Driver

The two types of efficiencies can be understood by looking at a truck and its driver. First we talk about resource conversion efficiency, that is the efficiency with which a resource (in this case fossil fuel) can be used. Most people are familiar with the relationship between the speed at which a vehicle travels, and the efficiency with which the vehicle uses fuel. Let's say that the maximum fuel efficiency for a truck is 55 miles per hour. Once it has accelerated to 55 mph, its kinetic energy tends to keep it going at that speed, but that tendency is counteracted by wind drag and friction. Diesel fuel energy is needed to keep the truck going in the face of these resistances. In doing this, the engine converts high quality energy in the form of diesel fuel with high potential to perform work, to low quality energy in the form of heat. Heat that comes out as exhaust or goes into the radiator has been called wasted energy (Cullen and Allwood 2010), but it is an inevitable waste because of the second law of thermodynamics.

Figure 9.2 illustrates how the conversion efficiency is influenced by the speed of a truck. Speed of the truck is indicated on the Y axis, and fuel consumption per hour

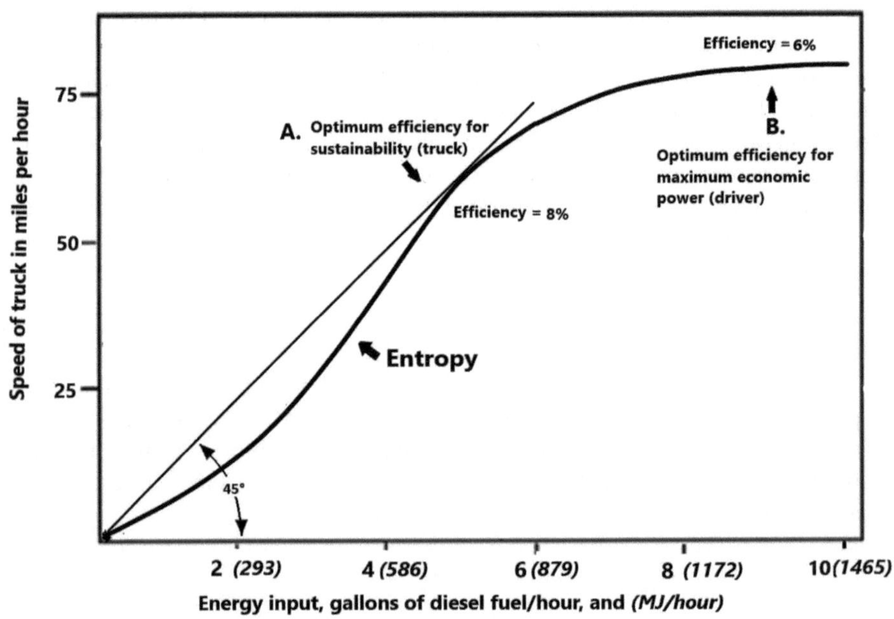

Fig. 9.2 Entropy as a function of the relation between speed and rate of energy use. Rate of entropy generation is indicated by the slope of the entropy curve. The more horizontal the line, the greater the speed of entropy increase. When the truck first starts, efficiency is low and entropy is high. At about 30 mph, rate of entropy increase begins to slow. At point A, rate of entropy increase changes from decreasing to increasing. The slope of the line changes from more vertical to more horizontal (less than 45° from horizontal). Beyond point A, rate of entropy increase is greater than the increase of speed per unit fuel input

on the X axis. The efficiency of energy use is calculated by dividing the speed of the truck by the rate at which energy is supplied to the engine. The speed for maximum conversion efficiency is indicated by point A. There the speed (55 mph) is divided by rate that energy is supplied to the engine (5 gallons per hour, 146.5 MJ/gallon = 732 MJ/h) for an efficiency of 8%. By traveling at 55 mph, the truck is minimizing entropy by getting maximum possible miles per gallon. Entropy is indicated by the slope of the curve. When the slope is less than 45° from horizontal, entropy is increasing faster than speed is increasing. When the slope is greater than 45°, speed is increasing faster than entropy. Point A is where the speed of entropy begins to change from decreasing to increasing.

Throughout the range of energy input rate, fuel uptake increases with each increase in speed. Below about 30 mph, rate of energy uptake increases more rapidly than rate of speed increase. That means entropy increases more rapidly than increase in speed. That is why driving in heavy traffic wastes fuel. Between about 30 and 55 mph, entropy increases more slowly than speed increases. Efficiency of fuel use increases. At point A, the slope of the curve changes to where entropy again increases more rapidly than speed increases. It marks the spot where energy use efficiency is maximum and entropy production is minimum. The ratio is 55 mph/732 MJ/h for an efficiency of 8% (efficiency at which fuel energy overcomes entropy). Fuel use of 5.0 gallons per hour results in a fuel use efficiency of 11 miles per gallon. As speed increases above this point, energy conversion efficiency decreases but power output increases until point B that marks the optimum efficiency for maximum power output.

So far we have been talking about energy conversion efficiency in the truck. Now we talk about time efficiency for the driver. Despite a lower fuel efficiency at 75 mph, the driver will operate at this rate because he can make more money. At 75 miles per hour, power output of the truck (rate that cargo is delivered) is greater, but fuel is consumed faster at 9 gallons per hour, resulting in 8.3 miles per gallon, and 1300 MJ/h. The ratio at point B is 75 mph/1300 MJ/h, for an efficiency of 6%. The driver operates at 75 mph because at that speed, power output is greater and he can make more deliveries per day and therefore make more money, even though efficiency of energy use (fuel use) is less. The increased cost of fuel is more than offset by the increased income gained from more deliveries. Six percent is the optimum time efficiency for maximum profit. It represents the optimum efficiency when maximum profit (as represented by cargo delivered) is the goal. The driver however will not operate above 75 mph. While at 85 mph power output of the truck would be even greater, efficiency would be so low that an increase in profit from an increase in rate of cargo delivery would no longer offset the greater cost of fuel. Point B marks the efficiency that gives the maximum power output for the driver. It is the point where the driver makes money at the fastest rate.

9.11 Sustainability

Operators of resource systems can manage their systems for high power output and low efficiency, or for lower power output and higher efficiency. The optimal rate for sustainability would be in a system with a lower power output than that for maximum power output. It would be optimal for trucks because they will not wear out as fast. It would be optimum for natural systems because production factors are used more efficiently.

Human dominated systems such as agriculture usually sacrifice efficiency for power output, because high power output is necessary for winning competition in the marketplace. This sacrifice also occurs in economic and military systems. In war, it is not high efficiency of the army's operations that wins the battle but greater power output. This means that soldiers and equipment must be sacrificed for a more intimidating attack. The general must sacrifice efficiency to gain power. New businesses do not use resources efficiently. They use them at a rate that will ensure maximum rate of growth, thereby defeating the competition and establishing a share of the marketplace. That rate is less than for the maximum efficiency. Resources are always used inefficiently and wasted on a frontier (bison extermination by pioneers in the old American West). Only after the war has been won, the frontier settled, or a monopoly secured does conversion efficiency become more important than time efficiency.

Box 9.6
The Viking era existed from around 793–1066. They were successful in battle because they maximized power output at the expense of discipline and organization. The era came to an end when opposing armies became better organized. It is better to have ten disciplined warriors in an organized formation than it is to have thirty powerful but rogue warriors. That's what eventually doomed the Vikings. Similarly, better organization is what causes mature ecosystems to replace successional ecosystems.

References

G.P. Asner, J.M.O. Scurlock, J.A. Hicke, Global synthesis of leaf area index observations: implications for ecological and remote sensing studies. Glob. Ecol. Biogeogr. **12**, 191–205 (2003)

G.P. Beretta, The fourth law of thermodynamics: steepest entropy ascent. Philos. Trans. R. Soc. A. (2020) 30 March 2020. https://doi.org/10.1098/rsta.2019.0168 Accessed 7 Dec 2020

J. Collier, The dynamics of biological order, in *Entropy, Information, and Evolution*, ed. by B. H. Weber, D. J. Depew, J. D. Smith, (MIT Press, Cambridge, 1988), pp. 228–242

J.M. Cullen, J.M. Allwood, Theoretical efficiency limits for energy conversion devices. Energy **35**(5), 2059–2069 (2010)

H.R. Da Rocha, M.L. Goulden, S.D. Miller, M.C. Menton, L.D.V.O. Pinto, H.C. de Freitas, A.M. Silva Figueira, Seasonality of water and heat fluxes over a tropical forest in Eastern Amazonia. Ecol. Appl. **14**(sp4), 22–32 (2004)

J.L. England, *Every Life is on Fire: How Thermodynamics Explains the Origins of Living Things* (Basic Books, New York, 2020)
G. Falasco, M. Esposito, Dissipation-time uncertainty relation. Phys. Rev. Lett. **125**, 120604 (2020).https://link.aps.org/doi/10.1103/PhysRevLett.125.120604. Accessed 2 Feb 2021
A. Kleidon, A basic introduction to the thermodynamics of the Earth system far from equilibrium and maximum entropy production. Philos. Trans. R. Soc. B. **365**, 1303–1315 (2010). https://doi.org/10.1098/rstb.2009.0310
A. Kleidon, Y. Malhi, P.M. Cox, Maximum entropy production in environmental and ecological systems. Philos. Trans. R. Soc. B. **365**, 1297–1302 (2010). https://doi.org/10.1098/rstb.2010.0018
M. Martinez-Kahn, L. Matinez-Castilla, The fourth law of thermodynamics: the law of maximum entropy (LMEP). An interview with Rod Swenson. Ecol. Psychol. **22**, 69–87 (2010). https://doi.org/10.1080/10407410903493160
L.M. Martyushev, V.D. Seleznev, Maximum entropy production principle in physics, chemistry and biology. Phys. Rep. **426**, 1–45 (2006). https://doi.org/10.1016/j.physrep.2005.12.001
National Weather Service, The earth-atmosphere energy balance (2020), https://www.weather.gov/jetstream/energy. Accessed 27 Dec 2020
H.T. Odum, R.C. Pinkerton, Time's speed regulator: the optimum efficiency for maximum power output in physical and biological systems. Am. Sci. **43**, 331–343 (1955)
I. Prigogine, G. Nicoles, A. Babloyantz, Thermodynamics and evolution. Phys. Today **25**(12), 138–141 (1972)
E.D. Schneider, D. Sagan, *Into the Cool* (University of Chicago Press, Chicago, 2005)
Surface Energy Balance, (2019), http://www.met.reading.ac.uk/~swrhgnrj/teaching/MT23E/mt23e_notes.pdf. Accessed 4 Jan 2021
Swenson R (1989) Emergence and the principle of maximum entropy production: Multi-level system Meeting of the International Society for General Systems Research, 32.theory, evolution, and non-equilibrium thermodynamics. Proceedings of the 32nd Annual Meeting of the International Society for General Systems Research, 32
J. Wang, R.L. Bras, A model of evapotranspiration based on the theory of maximum entropy production. Water Resour. Res. **47**, 1–10. WO3521. (2011). https://doi.org/10.1029/2010WR009392
N. Wolchover, A new physics theory of life. Quanta (2014), https://www.quantamagazine.org/a--new-thermodynamics-theory-of-the-origin-of-life-20140122/#comments. Accessed 6 Jan 2021
J.D. Wu, D. Guan, S. Han, T. Shi, C. Jin, T. Pei, G. Yu, Energy budget above a temperate mixed forest in northeastern China. Hydrol. Process. **21**(18), 2424–2434 (2007). https://doi.org/10.1002/hyp.6395

Chapter 10
A Thermodynamic View of Succession

10.1 The Population View

Although Clements (1916) pioneered the concept of plant succession, only later in the twentieth century did succession in terrestrial ecosystems become a leading-edge topic in ecology (Golley 1977; West et al. 1981). Many reviews argued that succession should be viewed as a population process, defined as a variation in rates of reproduction, establishment, growth, mortality, species life-history and physiological characteristics (Drury and Nisbet 1973; Pickett 1976; Peet and Christensen 1980; Bazzaz 1996). In the "Unified Neutral Theory of Biodiversity and Biogeography", Hubbell (2001, 2005) suggested that these differences between species in the same trophic level of an ecological community are irrelevant. The species present in a trophic level of a successional community depends solely on who arrives first. Wind is an important factor for rapid dispersion. Many early colonizers are grasses whose seeds are wind dispersed. Pines also invade, because their seeds can be wind-dispersed (Benkman 1995). Species present during succession also depends in part on the size of the disturbed area and the distance to seed source for the colonizing species (Denslow 1980).

10.2 The Thermodynamic View

The thermodynamic view is that succession is controlled by autocatalysis. Autocatalysis regulates the feedback system that controls biomass accumulation during succession in ecosystems, which in turn depends on the rate that nutrients are released from decaying organic matter. The autocatalytic reaction is between nutrients recycled from decomposed organic matter and glucose, the product of photosynthesis in autotrophs. The product is new energy in the form of biomass that

C. F. Jordan, *Evolution from a Thermodynamic Perspective*,
https://doi.org/10.1007/978-3-030-85186-6_10

moves back into the composer trophic chain of ecosystems and resets the energy balance in species interactions. Increased energy flow more than compensates for entropy that causes breakdown of biomass. For that reason, succession can be defined as entropy storage. Entropy storage reaches a maximum when structure of an ecosystem is a maximum. However power output continues to increase. The increase is used for stabilization and maintenance of the system.

10.2.1 Leaf Area Index and Succession

Leaf area index (LAI) denotes the layers of leaves above a point on the forest floor. It indicates the number of square meters of leaf area directly above a square meter of soil. When succession begins on abandoned agricultural fields, or land cleared by fire or other means, some solar energy reaches the earth without being intercepted by leaves, because LAI is less than 1.0. As long as there is less than a single layer of leaves, rate of community photosynthesis and rate of biomass (power output) accumulation per unit area is limited by the incomplete leaf coverage of the soil (Fig. 10.1). As soil coverage increases to the point of canopy closure, rate that

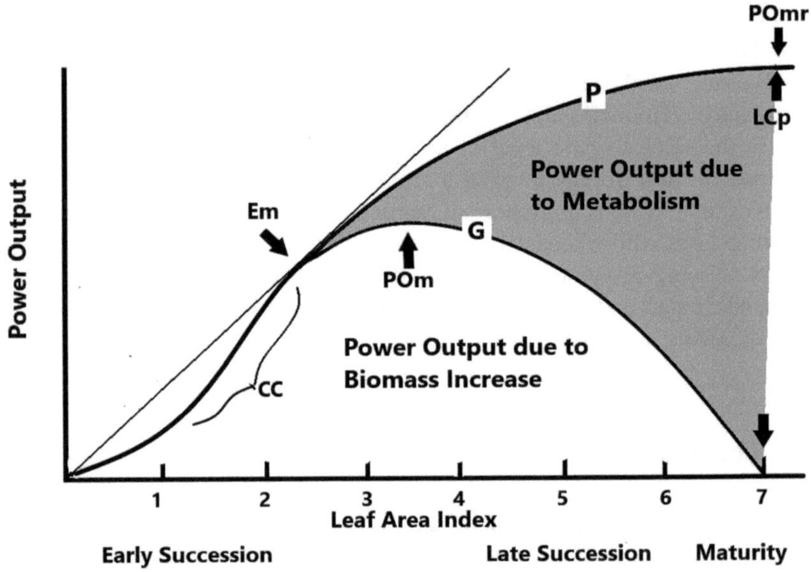

Fig. 10.1 Power output (P curve) of a forest ecosystem as a function of succession. The G curve at any value of LAI indicates the proportion of power output contributing to growth (biomass accumulation). The difference between the P and G curves indicates the proportion of power output contributing to metabolism. CC Canopy closure. Rate that the rate of power output of a forest ecosystem changes during succession is indicated by the slope of line P. Em the point where efficiency of energy conversion to biomass is maximum. POm point of maximum power output when power output is taken to mean only rate of biomass increment. POmr point of maximum power output when power output consists of biomass increment plus support for metabolism and trophic webs. LCp Light compensation point

biomass accumulation accelerates. If light entered the ecosystem only vertically and if the leaves were perfectly arranged, complete coverage (canopy closure) would occur at a leaf area index (LAI) of 1.0. Because of light diffusion and because in early mornings and late afternoons light enters the ecosystem in a partially horizontal direction, light is not completely absorbed by leaves before it reaches the soil until the leaf area index is higher than 1.0. Figure 10.1 assumes that canopy closure occurs between LAI of 1 and 2.5.

Box 10.1 The Rate that the Rate of Power Output Changes
The rate of biomass change is a first derivative function. The rate that that rate changes is a second derivative. It indicates the rate at which the first derivative changes. The second derivative at any point along line P is indicated by the slope of a line tangent to P. If the slope of the tangent is greater than 45° from horizontal, the rate that the rate changes is increasing. If it is less, the rate is decreasing.

10.2.2 Power Output as a Function of Leaf Area Index

Rate that biomass increases before canopy closure is limited by leaf coverage of soil. Rate that rate increases accelerates as leaves begin to cover bare spots in the soil. As the canopy closes between a LAI of 1.0–2.5, rate that rate accelerates begins to decrease. But efficiency of energy use continues to increase until point Em. Efficiency is power output divided by leaf area index. At point Em, efficiency of energy conversion to biomass is maximum. Beyond point Em, rate of biomass increase is no longer increasing and rate of increase in energy efficiency begins to decrease, but absolute biomass continues to increase and energy efficiency is still positive. The situation is analogous to that of the truck driver in chapter nine who continues to increase his speed even though the efficiency of the engine is decreasing, because by doing so, he will make more money.

Let's assume that the rate of biomass accumulation accelerates from zero to one ton per hectare during the development of the first leaf layer ($\Delta x/\Delta t = 1.0$). As LAI approaches 2.0, leaves in the second level are receiving less light on the average per day than those in the topmost layer. As a result, rate of biomass accumulation per level added is less than during development of level one, let's say 0.8 tons per hectare. While biomass of the ecosystem has increased to 1.8 tons at point Em, the rate that it increased decreased from 1.0 to 0.8 ($\Delta x/\Delta t < 1.0$). As succession proceeds to POm where LAI is about 3.5, rate of biomass increase per level added continues to decrease and efficiency decreases, but total biomass continues to increase. Power output in terms of biomass synthesis is greatest at POm. Efficiency there is Odum's *optimum efficiency for maximum power output*. Beyond that point, biomass continues to increase, but rate of power output increase decreases and efficiency continues to decrease. POm is where the truck driver in chapter nine will operate.

Point POm marks the point where during succession the rate of net primary productivity is greatest. Power output has increased since Em, but efficiency has decreased. As succession proceeds beyond point POm, the LAI increases but with each increase in LAI, the quantity of light reaching lower layers continues to decrease. As a result, the photosynthetic rate for each additional layer of leaves decreases. Each additional layer of leaves results in a marginally smaller increase in power output of the ecosystem until the trees reach the light compensation point (LCp), where no further layers of leaves are added. Before that point, the biomass of the entire ecosystem continues to increase, but the rate at which it increases decreases until point POmr where the rate of increase becomes zero. The photosynthesis and respiration rates of the forest become equal. In well-watered, fertile tropical ecosystems, this will occur when the LAI reaches about 7.0 (Asner et al. 2003). After this point, the system will eventually stagnate and deteriorate, unless there is an occasional moderate disturbance to the ecosystem that reinvigorates the system, gets rid of the older trees and maintains biomass production (Odum et al. 1979a, b). The disturbance re-directs energy from maintenance of old trees to replacement growth of young trees.

While H.T. Odum considered power output to be only biomass increase which maximized at point POm, we consider power output of the ecosystem to be the sum of the rate that biomass increases plus that rate at which photosynthetic energy is used to defend the system against entropy (point POmr). This power output includes defense against decomposers plus energy required to support the trophic web that stabilizes the ecosystem.

10.2.3 What Causes Changes in Leaf Area Index?

Bormann and Likens (1979) found that LAI changed because of species change during succession. Early fast-growing shade intolerant species such as pin cherry and trembling aspen produced few layers of leaves. They were gradually replaced by slower growing, shade tolerant species such as sugar maple and beech that could produce more layers of leaves because the leaves are more shade tolerant. Increase in LAI can also occur with species such as tulip poplar (*Liriodendron tulipifera*). This species occurs early in succession and grows very rapidly. It persists long after other trees reach the canopy. In the Joyce Kilmer Memorial Forest in the Southern Appalachians, some of them are more than 20 feet in circumference and over 400 years old (Joyce Kilmer Memorial Forest 2020).

Box 10.2 Ecosystem Relativity
In Chap. 9 we pointed out that that the rate of events that occur within a system determines the speed of time within that system. Figure 10.1 indicates that the speed of time within an ecosystem during its development is determined not by calendar time but by the rate at which new layers of leaves appear. Time scale can vary, depending on nutrient and moisture availability. In a dry forest, time moves slower than in a rain forest.

10.2.4 Maximum Entropy Production Principle

Early in succession, most of the power output is dedicated to building biomass. The process is sometimes called negative entropy, meaning that a system is becoming less disordered. We can also say the process is storing or producing entropy. As a system increases in organization during succession and evolution, entropy production is maximized. This is Kleidon et al.'s (2010) maximum entropy production (MEP) principle. It could also be called maximum information production, where information is entropy trapped in the intricate structure of organisms and ecosystems, just as entropy is trapped in the intricate structure of proteins in DNA. During succession and evolution, there are multiple combinations of species that will lead to a steady state ecosystem. The MEP predicts that the combination that actually occurs will be the one that has the highest entropy production rate (Meysman and Bruers 2010).

10.2.5 Successional Ecosystems Move Further from Thermodynamic Equilibrium

During succession, ecosystems are storing entropy at a rate faster than they are releasing it. They are moving further away from equilibrium. At maturity, distance from equilibrium no longer increases. If sunlight were suddenly shut off due to volcanic aerosols, the systems would not remain in constant biomass, but would decompose and stored entropy would be released. All the information stored in the species and in the ecosystems would disappear. The system would regress to thermodynamic equilibrium.

10.3 The Strategy of Ecosystem Development

"Ecosystem strategy" sounds teleological, but there is really no decision-maker within an ecosystem that decides if energy should be used for growth or defense. Teleology is used here to short-cut the explanation that it is really the interactions in nature that result in "strategy". There are no goals nor priorities among species. It is the interactions between species and individuals in all stages of succession that result in what E.P. Odum (1969) called "The strategy of ecosystem development". It is always toward maximization of power output which includes both biomass increment and stability support. As power output increases beyond point Em in Fig. 10.1, strategy of energy use efficiency changes. Efficiency of resource use replaces strategy to maximize time efficiency.

Power output does not increase indefinitely, but may gradually begin to decrease as forests become over-mature. Analyses by Gundersen et al. (2021) indicates that old-growth forest carbon sinks are overestimated. They no longer have a net uptake of CO_2. Decline occurs because photosynthesis in ecosystems becomes less than rate of respiration (Tang et al. 2014). Odum's view of ecosystem development during succession has been modified to show the beginning of biomass decline (Fig. 10.2). The tail end of the biomass curve drops off after 80 years. Also added in Fig. 10.2 to Odums original (1969) figure are: the point of canopy closure (CC); the point of maximum power output (POm); the light compensation point LCp); and Em, the point where the rate of biomass increase is most efficient.

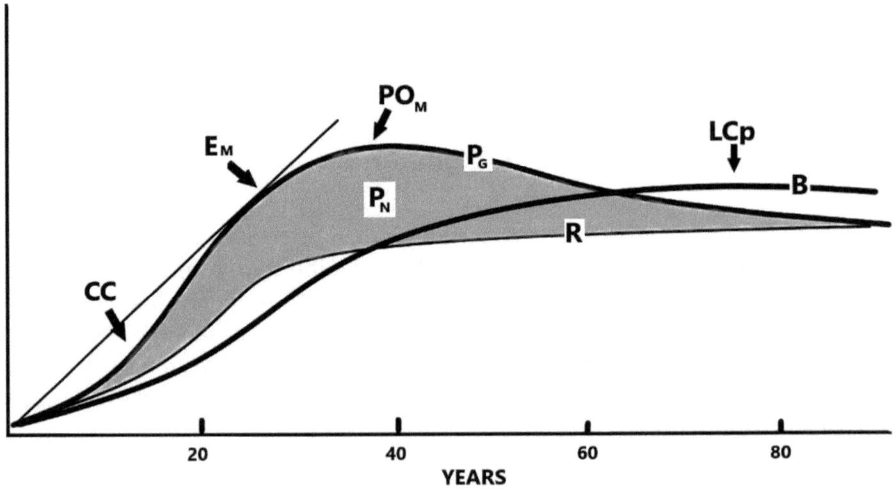

Fig. 10.2 Changes in ecosystem energy dynamics during secondary succession following abandonment from agriculture. Pg gross production, Pn net production, R total community respiration, B total biomass, POm point of maximum power output, CC canopy closure and the point where the rate of change of productivity change is increasing fastest, Em point where the rate of increase begins to decrease. LCp Light compensation point. (Figure is reprinted from Odum (1969) with permission from Science)

10.3.1 A Problem with Odum's Strategy

The decrease in power output (rate of biomass increase) after a maximum at point POm shown in Fig. 10.2 is a reflection of what field ecologists have long observed. Individual plant species as well as ecosystems grow rapidly during early succession, but slow down towards maturity at which point biomass no longer increases (Bond 2000). This has been taken to mean that the rate of photosynthesis is slowing down. Figure 10.1 indicates that power output does not decrease after POm as it does in Fig. 10.2, but continues to increase to POmr at LCp, the point of light saturation. As leaf area index increases, power output should also increase. Why does power output decline and respiration remain almost steady in Fig. 10.2 when the leaf area index continues to increase till point LCp? The answer is they do not. Power output continues to increase (Fig. 10.3). While rate of biomass increase declines, power output changes from making new biomass to supporting the mature ecosystem with a diverse array of defenses and a complex food web that stabilizes the system. Respiration also must increase because stabilization activities in mature systems burn the energy supplied by the increase in Pg.

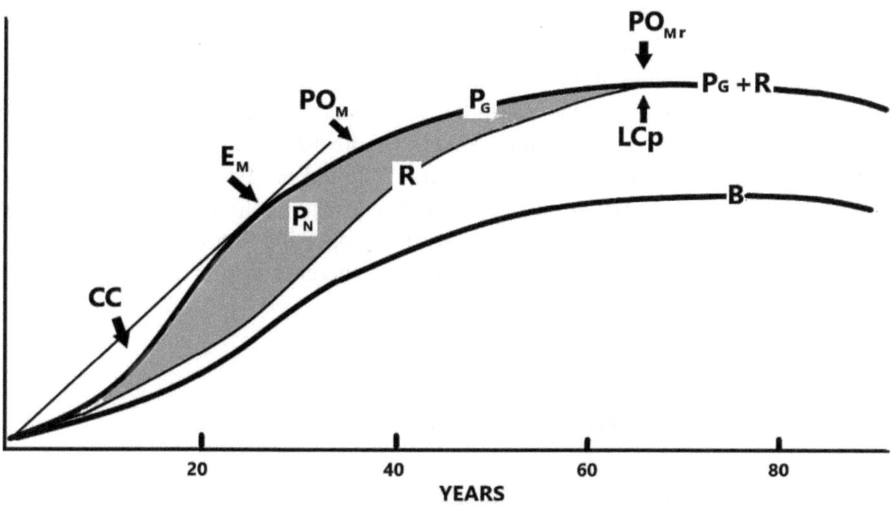

Fig. 10.3 Proposed changes in ecosystem energy dynamics during secondary succession. Pg gross production, Pn net production. Pn is greatest where the vertical distance between Pg and R is greatest, R total community respiration, B total biomass, CC canopy closure and the point where the rate of change of productivity change is increasing fastest, Em point where the rate of increase begins to decrease. It is the point where energy use is most efficient. POm point of maximum power output as estimated in Fig. 10.2, LCp Light compensation point, POmr revised point of maximum power output where power output includes energy for maintenance

10.3.2 Why Power Output Continues to Increase

Net photosynthesis and night-time respiration are measured by net increase or decrease of carbon dioxide in the air passing over the plant or the ecosystem. Gross photosynthesis is the rate that carbon is fixed as glucose. It can be measured directly in individual leaves by using carbon-14, but it cannot be measured directly in entire plants in the field nor in ecosystems because photosynthesis and respiration are occurring simultaneously. The two rates cannot be experimentally separated. Daytime respiration has been assumed to be equal to night time respiration adjusted for temperature differences. Gross photosynthesis has been assumed to be the sum of net photosynthesis and night time respiration. This is an underestimate of both. The true rate of daytime gross photosynthesis may be masked by a higher rate of respiration than occurs at night when carbohydrate stores in leaves become depleted (O'Leary et al. 2017). In Fig. 10.2, gross photosynthesis begins to decrease after 40 years. We propose that this is not the case. We suggest that gross photosynthesis in mature plants and ecosystems continues to increase after POm, which is really not the point of maximum power output. Gross photosynthesis continues to increase beyond point POm to point POmr, and rate of respiration increase is not almost steady as indicated in Fig. 10.2 but continues to increase as in Fig. 10.3.

10.4 Revised Definition of Maximum Power

According to H.T. Odum, the maximum power principle means that systems maximize power output and that the end point (in the case of succession) or evolution (during geologic time) is maximum biomass. Figure 10.3 summarizes the argument that the meaning of "maximum power output" should not be restricted to the biomass when ecosystems reach maturity, but should also include the energy used to stabilize a system once it reaches maturity by supporting a complex trophic web. Increase in maximum power should include the increase in power output as more complex food webs come to depend on that power.

10.4.1 Costs of Ecosystem Stabilization

As ecosystems mature, there is an increase in the complexity and diversity of the ecosystem. Trophic chains grow longer, and the increase in diversity within each trophic level results in increasingly complex food webs. Maintaining these interactions requires energy, but there is a benefit to the ecosystem in that interactions increase the stability of the ecosystem. In contrast to biomass synthesis costs, these costs result from respiration of individuals in the ecosystems, as well as from transactional costs between individuals. These costs are not included in the respiration curve of Fig. 10.2 but they are in Fig. 10.3.

10.4.2 Transactional Costs

Trophic webs incur "transactional" costs. In economics, transactional costs have been divided into three categories (Dahlman 1979). For each, we give an italicized ecological meaning.

- Search and information costs are costs involved in determining that the required good is available on the market, and which has the lowest price. *In ecosystems, predators must hunt for prey. Some prey are easier to catch than others.*
- Bargaining and decision costs are the costs required to come to an acceptable agreement with the other party to the transaction. *Prey make it difficult to reach an agreement with predators (transactional costs are high) but flowers make it easy for pollinators (transactional costs are low).*
- Policing and enforcement costs are the costs of making sure the other party sticks to the terms of the contract, and taking appropriate action if this turns out not to be the case. *In ecological systems there are no "contracts", but there are established patterns of behavior that if violated, upset the established flow of energy. For example, the common cuckoo lays eggs in the nests of other bird species and expects the next-builder to raise its young.*

Upon reaching maximum biomass (in ecosystems) or maximum infrastructure (in businesses and nations), energy and money flow do not stabilize but continue to increase. The increase just is used in a different way. Once a business becomes mature, more of its income is spent on defending itself from competitors or on fighting bureaucracy than is spent on growth. Once an ecosystem becomes mature, more of its energy is used to support trophic webs than is used for growth.

As the power output of a system grows linearly, the number of interactions within a system grows exponentially, and the cost of maintaining an infrastructure grows exponentially Each interaction has a cost and at some point, the cost of all the transactions in a system begins to exceed the ability of a system to increase power output. That is why there is a limit as to the size of any organization, from ecosystems to social systems (Odum 1971). When power output can no longer increase, systems enter a period of no further growth called the consolidation stage in Panarchy theory (Chap. 11).

10.5 Succession, Power Output, and Efficiency

The maximum power principle says that the efficiency at the optimum compromise between efficiency and power output is less than the maximum possible efficiency. During succession, efficiency is continually changing, but during early stages, it is always lower than maximum possible efficiency. Efficiency becomes maximum at point Em. As systems progress beyond this point, rate of biomass increment slows and efficiency decreases, but total biomass continues to increase until point POm.

POm is the point of maximum power output if power output consists only of growth. Efficiency here is Odum's "optimum efficiency for maximum power output". But power output continues to increase to point POmr. Power output beyond POm is increasingly used for stabilization.

Energy from photosynthesis for production of biomass is used at a relatively low efficiency. Cells get energy in the form of ATP through respiration, a series of chemical reactions oxidizing glucose. Growth occurs through cell division, a complex process with high energy demands. Cell division requires the coordinated generation of energy for multiple processes, including the synthesis of the machinery required for DNA replication and mitosis (Salazar and Malumbres 2017). This additional energy requirement lowers overall efficiency of photosynthesis to about 3.6–4.1% (Slattery and Ort 2015).

Mature systems are more energy efficient than successional systems because metabolic energy conversions that predominate after Em are more efficient than conversions for biomass production. Although total efficiency decreases beyond Em, efficiency used for respiration increases to point LCp. Species in mature ecosystems use almost all photosynthetic energy for metabolic life-sustaining processes and defense mechanisms which have a relatively cheaper cost that results in a higher efficiency of energy utilization (Coley et al. 1985; Poorter et al. 2004). Rate of increase of power output declines to zero at POmr, the light compensation point, and efficiency there is maximum.

Box 10.3

A strategy of some tree species to minimize metabolic costs as they mature is to develop heartwood which is resistant to rot. Redwood trees (*Sequoia sempervirens)* as they grow, they produce living sapwood which can rot quickly. However when they become old, they produce heartwood that renders the wood pest and rot resistant (Grey 2017). They have the advantage of being large (good competitor for energy) without the disadvantage of having to pay the cost for maintenance required by a large living system.

10.5.1 Kleiber's Law

Kleiber's law, the fact that for the vast majority of animals, the metabolic rate scales to the ¾ power of the animal's mass. The "law" has been used to argue that organisms have evolved to maximize their energy use efficiency (Banavar et al. 2014). In contrast, the maximum power principle says that natural systems tend to maximize power output, not efficiency. We now can see that the maximum power principle holds for plant species during growth stages of succession, and for animals when they are young. Kleiber's law holds when ecosystems reach maturity and when animals get old.

10.6 Are Ecosystems Spendthrifts?

H.T. Odum felt that because maximizing power output is not the most efficient use of energy, the maximum power principle is an ironic statement favoring the survival of the spendthrift. Odum clearly was thinking that increase in biomass was the only measure of power output. It is evident though that once an ecosystem reaches maturity, it no longer is a spendthrift. It uses its energy more efficiently for stabilization and sustainability. Succession and evolution produce spendthrifts only when systems are young. At maturity, systems become misers.

Ulanowicz's (2020) model of succession suggests that while species become more efficient as the ecosystem reaches maturity, their increase in efficiency limits their ability to adapt to changing environmental conditions. The ecosystem becomes less flexible and more prone to collapse following perturbation. That would account for the phenomena in "over-mature" ecosystems where biomass begins to decline (Tang et al. 2014). The high efficiency of species and their interactions comes at the expense of flexibility of ecosystems.

10.7 Interactions Between Species Facilitate Increase in Power Output

Power output and nutrient cycling during succession are influenced by three types of species interactions, facilitation, tolerance, and inhibition (Connell and Slatyer 1977).

10.7.1 Facilitation

10.7.1.1 Facilitation During Primary Succession

In primary succession on bare rock, early colonizers are algae, lichens, and mosses. Nutrient input is via atmospheric deposition. As the primitive plants die and decompose, they provide an organic bed in which higher plants can root. Increase of biomass depends on autocatalysis where nutrients added to the system interact with photosynthetic energy to produce organic matter that holds these nutrients until they are recycled. Rate of autocatalysis is often limited by phosphorus and nitrogen (Vitousek and White 1981). As soil organic matter increases during early succession, phosphorus availability increases due to solubilization of phosphorus in the substrate by organic acids leached from the soil organic matter (Rani et al. 2012). Phosphorus availability increases the ability of successional leguminous species to fix nitrogen (Walker and Syers 1976). Recycling of P back into the composing trophic web results in a more favorable N:P ratio that results in greater productivity by species later in the successional sere (Vitousek and White 1981).

10.7.1.2 Facilitation During Secondary Succession

Facilitation also occurs during secondary succession on abandoned agricultural land. As in primary succession, phosphorus is sometimes the limiting factor for secondary succession on highly weathered soils such as those of the Southeastern Piedmont region of the United States (Jordan 2013). While there is a large quantity of phosphorus in those soils, most of it is bound by iron and aluminum oxides and not available to plants (Brady 1974). As biomass increases during early succession, organic acids are leached into the mineral soil where they chelate the iron and aluminum that binds phosphorus, thereby increasing its availability to plants (Lee et al. 1990). The resulting increase in soil fertility facilitates the growth of later successional species.

Box 10.4 Facilitation by Animals
Around the year 2000, fire ants began invading old fields and pastures on my farm. I noticed that once a tree seedling or blackberry bush became established, it grew more vigorously if fire ants built a nest near its roots. The ants seem to loosen the soil and make it more physically amenable to penetration by plant roots. They seem to facilitate the growth of species that replace early successional herbs and grasses. The ants are pioneer species. They do not occur in my old growth stands of oak and hickory, presumably because they don't compete with native ant species.

10.7.2 Tolerance

The success of any particular plant species in a successional community depends on its shade tolerance. Early colonizers often are shade intolerant, that is they need full sunlight to survive. Shade tolerant species replace intolerant species by "shading them out". They produce multiple overlapping layers of leaves so that all of the incoming photosynthetically available solar radiation is intercepted before it reaches the ground. Quarterman and Keever (1962) described beech and magnolia as replacing oaks and hickories in the Southeast because of the very high shade tolerance of their leaves. These species are rarely mentioned in successional studies, but I have observed them coming in beneath the old growth oaks and hickories on my farm on the Georgia Piedmont. These shade tolerant trees are most efficient in capturing all the photosynthetically available sunlight, but they are not the fastest growing tree species. They apparently use relatively more of their energy for maintenance and suppression of competition because they live much longer than early successional species.

10.7.3 Inhibition

Because of the inability of other species to replace chaparral (Mediterranean type vegetation) and heath species in England, it has been suggested that these species produce an allelochemical compound that inhibits the establishment of other species, but evidence is controversial (Halsey 2004). *Lespedeza cuneata* that grows in the old fields of my farm and suppresses growth of other herbs and grasses may be allelopathic (Invasive.org 2019). More insidious in the Georgia Piedmont is Johnson grass (*Sorghum halepense*), a C4 species that maintains a competitive advantage through relatively rapid fixation of atmospheric carbon and extensive rhizomes. It inhibits succession by its very thick growth that crowds out seedlings of other species that may germinate in their midst. Similar C4 species that occur in the tropics include alang alang (*Imperata cylindrica*), and elephant grass (*Pennisetum purpureum*). It is difficult to convert fields dominated by these species to economic crops, or even other native species. Cutting or burning these grasses only stimulates more vigorous production.

10.8 Intermediate Disturbance Hypothesis

As an ecosystem reaches maturity, biomass no longer increases. All the products of photosynthesis go toward defense and maintenance of the system. In the intermediate disturbance hypothesis, Connell (1978) hypothesized that ecosystem productivity will be maintained if there is a light to moderate disturbance that sets succession back as suggested by (Odum et al. 1979a, b). Older trees that die or are removed are replaced with younger individuals or other species that are still growing. Selective logging of mature trees is an intermediate disturbance as long as the soil organic matter is not destroyed (Mesquita 1995). In contrast, clear cutting as a forest management technique sets back succession to time zero, and often destroys the soil organic matter that facilitates rapid growth. Light fires are another intermediate scale disturbance. In pre-colonial days of North America, occasional light fires would kill back trees which persisted in groves, surrounded by grassland that was maintained by fire (Jordan 1965). Periodic pre-colonial fires in the Piedmont region of Georgia maintained a "Piedmont Prairie" consisting of oak groves and native grasses, a successional stage well short of the potential climax (White and White 1996), but a stage which produced ample feed for the elk that grazed.

10.9 Nutrient Use Efficiency During Succession

There are two models of nutrient use efficiency during ecosystem succession, one for nutrient poor soils, the other for those rich in nutrients. Odum (1969) in his paper on the strategy of ecosystem development, suggested that during ecological succession, efficiency of nutrient cycling becomes higher (less leaky) as biomass of the forest increases because nutrient recycling mechanisms such as fine root growth become more developed. In mature systems, nutrient cycles become almost completely closed. This theory may have resulted from Odum's study of succession following abandonment of nutrient poor agricultural land on the Coastal Plain of South Carolina (Odum 1960; Quarterman and Keever 1962). The trend is similar to that found in the many studies of secondary forest succession on the North Carolina Piedmont (Christensen and Peet 1981; Peet 1981), and in the Piedmont of New Jersey (Bard 1952), where soil is derived from weathering of parent rock that is 480 million years old.

Vitousek and Reiners (1975) presented a different model of nutrient dynamics during succession. In contrast to Odum's model where nutrient use efficiency increases during succession, their model shows a decreasing nutrient use efficiency during succession. Their reasoning was that biomass is increasing during succession, and that an increase in biomass causes an efficient use of the nutrients stored in the soil, since they are taken up by the plants and are not available for leaching. Once a stand reaches maturity, there is no net uptake of nutrients by the trees, so there is more opportunity for leaching losses and the system becomes less efficient. This scenario would happen on a nutrient-rich soil. In a northern hardwoods forest on a comparatively nutrient-rich soil, they found that nitrogen losses through leaching increased steadily as succession passed beyond intermediate stages to maturity, apparently because nitrogen was abundant in the soil.

On nutrient-poor soils, rate of biomass increase for an ecosystem during succession is constrained by the rate that nutrients are added to the ecosystem via atmosphere deposition or leaching of parent soil material. Once these nutrients are incorporated into organic matter, the rate that organic matter in the soil decomposes controls the rate that nutrients are made available for autocatalysis. Decomposition of organic matter becomes the rate controlling factor for autocatalysis. During succession on nutrient rich soils, autocatalysis is not limited by nutrient availability through organic matter decomposition. The photosynthetic capacity of the plant community limits rate of biomass build-up.

Box 10.5. Nutrient Depletion

A number of years ago, a student wrote me about the nutrient content of the soil in a forest plantation that he was studying. He said he expected nutrient content to increase because he had read that trees increase the nutrient content of the soil. However, he found that the soil nutrients had actually decreased. I told him that the nutrients in the soil were being taken up by the plantation species as would happen in Odum's (1969) model of succession on a nutrient-poor soil. Trees however can prevent nutrient loss, so soils with mature trees will have higher nutrient contents than soils following agriculture or clear cut harvesting, despite replanting.

10.9.1 Succession Following Logging Versus Following Agriculture

Succession on land abandoned from agriculture may be different than succession following logging. Succession following agriculture on low fertility soils is slow because the autocatalytic cycle has few nutrients to work with. Succession following logging in Southern Appalachian Forests was much more rapid than following abandonment of agriculture. Hardwood stumps produced sprouts that shed leaves that increased organic matter input to the soil, thereby facilitating autocatalysis (Phillips and Shure 1990).

10.10 Thermodynamic View of Succession: Implications for Resource Management

The conventional view says that ecological succession should be viewed as a population process, defined as a variation in rates of reproduction, establishment, growth, mortality, species life-history and physiological characteristics. Thus in every location, the sequence of species that appear is likely to be different. Succession from a population view is stochastic. The thermodynamic view says that succession is deterministic. Regardless of the species involved, power output will always increase. Early in succession more energy is dedicated to biomass increase but later, more is dedicated to maintenance and stability. Energy use efficiency changes because synthesis of biomass early in succession uses energy less efficiently that energy for metabolism. Resource systems that are kept in the early successional stage to maximize rate of biomass production use energy inefficiently. Energy used inefficiently becomes pollution.

References

G.P. Asner, J.M.O. Scurlock, J.A. Hicke, Global synthesis of leaf area index observations: implications for ecological and remote sensing studies. Glob. Ecol. Biogeogr. **12**, 191–205 (2003)

J.R. Banavar, T.J. Cooke, A. Rinaldo, A. Maritan, Form, function, and evolution of living organisms. PNAS **111**(9), 3332–3337 (2014)

G. Bard, Secondary succession on the Piedmont of New Jersey. Ecol. Monogr. **22**(3), 195–215 (1952)

F.A. Bazzaz, *Plants in Changing Environments* (Cambridge University Press, New York, 1996)

C.W. Benkman, Wind dispersal capacity of pine seeds and the evolution of different seed dispersal modes in Pines. Oikos **73**, 221–224 (1995)

B.J. Bond, Age-related changes in photosynthesis of woody plants. Trends Plant Sci. **5**(8), 349–353 (2000)

F.H. Bormann, G.E. Likens, *Pattern and Process in a Forested Ecosystem* (Springer, New York, 1979)

N.C. Brady, *The Nature and Properties of Soils* (Macmillan, New York, 1974)

N.L. Christensen, R.K. Peet, Secondary forest succession on the North Carolina Piedmont, in *Forest Succession: Concepts and Application*, ed. by D. C. West, H. H. Shugart, D. B. Botkin, (Springer, New York, 1981), pp. 230–245
F.E. Clements, *Plant succession; an analysis of the development of vegetation* (Carnegie Institution of Washington, Washington, DC, 1916)
P.D. Coley, J.P. Bryant, F.S. Chapin, Resource availability and plant antiherbivore defense. Science **230**, 895–899 (1985)
J.H. Connell, Diversity in tropical rain forests and coral reefs. Science **199**(4335), 1302–1310 (1978)
J.H. Connell, R.O. Slatyer, Mechanisms of succession in natural communities and their role in community stability and organization. Am. Nat. **111**(982), 1119–1144 (1977)
C.J. Dahlman, The problem of externality. J. Law Econ. **22**(1), 141–162 (1979)
J.S. Denslow, Patterns of plant species diversity during succession under different disturbance regimes. Oecologia **46**, 18–21 (1980)
W.H. Drury, I.C.T. Nisbet, Succession. J Arnold Arbor **54**, 33–368 (1973)
F.B. Golley, *Ecological Succession: Benchmark Papers in Ecology*, vol V. 5 (Hutchinson & RossPublishing Co. Stroudsburg Pennsylvania, Dowden, 1977)
S. Grey, Weather resistance of cedar vs. redwood (2017), https://www.hunker.com/13401357/weather-resistance-of-cedar-vs-redwood. Accessed 19 Dec 2020
P. Gundersen, E.E. Thybring, T. Nord-Larsen, et al., Old-growth forest carbon sinks overestimated. Nature **591**, E21–E23 (2021)
R.W. Halsey, In search of allelopathy: an eco-historical view of the investigation of chemical inhibition in California coastal sage scrub and chamise chaparral. J. Torrey Bot. Soc. **131**(4), 343–367 (2004)
S.P. Hubbell, *The Unified Neutral Theory of Biodiversity and Biogeography* (Princeton University Press, Princeton, 2001)
S.P. Hubbell, Neutral theory in community ecology and the hypothesis of functional equivalence. Funct. Ecol. **19**, 166–172 (2005)
Invasive.org, (2019), https://www.invasive.org/browse/subinfo.cfm?sub=3033. Accessed 21 Dec 2020
C.F. Jordan, Fire influence in old fields of the New Jersey Piedmont. Bull. NJ. Acad. Sci. **10**, 7–12 (1965)
C.F. Jordan, *An Ecosystem Approach to Sustainable Agriculture: Energy Use Efficiency in the American South* (Springer, Heidelberg, 2013)
Joyce Kilmer Memorial Forest, (2020), https://www.blueridgeheritage.com/destinations/joyce-kilmer-memorial-forest/. Accessed 23 Dec 2020
A. Kleidon, Y. Malhi, P.M. Cox, Maximum entropy production in environmental and ecological systems. Phil. Trans. R. Soc. B **365**, 1297–1302 (2010). https://doi.org/10.1098/rstb.2010.0018
D. Lee, X.G. Han, C.F. Jordan, Soil phosphorus fractions, aluminum, and water retention as affected by microbial activity in an Ultisol. Plant Soil **121**, 125–136 (1990)
R.C.G. Mesquita, The effect of different proportions of canopy opening on the carbon cycle of a central Amazonian secondary forest. Dissertation, University of Georgia, Athens, Georgia, 1995
F.J.R. Meysman, S. Bruers, Ecosystem functioning and maximum entropy production: a quantitative test of hypotheses. Philos Trans R Soc Lond B Biol Sci **365**, 1405–1416 (2010). https://doi.org/10.1098/rstb.2009.,0300
E.P. Odum, Organic production and turnover in old field succession. Ecology **41**(1), 34–49 (1960)
E.P. Odum, The strategy of ecosystem development. Science **164**, 262–270 (1969)
H.T. Odum, *Environment, Power and Society* (Wiley, New York, 1971)
E.P. Odum, J.T. Finn, E.H. Franz, Perturbation theory and the subsidy-stress gradient. Bioscience **29**(3), 349–352 (1979a)
E.P. Odum, J.T. Finn, E.H. Franz, Perturbation theory and the subsidy-stress gradient. Bioscience **29**(3), 349–352 (1979b)

B.M. O'Leary, C.P. Lee, O.K. Atkin, R. Cheng, T.B. Brown, A.H. Millar, Variation in leaf respiration rates at night correlates with carbohydrate and amino acid supply. Plant Physiol. **174**(4), 2261–2273 (2017). https://doi.org/10.1104/pp.17.00610

R.K. Peet, Changes in biomass and production during secondary forest succession, in *Forest Succession: Concepts and Application*, ed. by D. C. West, H. H. Shugart, D. B. Botkin, (Springer, New York, 1981), pp. 324–338

R.K. Peet, N.L. Christensen, Succession: A population process. Vegetatio **43**, 131–140 (1980)

D.L. Phillips, D.J. Shure, Patch-size effects on early succession in southern Appalachian Forests. Ecology **71**(1), 204–212 (1990)

S.T.A. Pickett, Succession: an evolutionary interpretation. Am. Nat. **110**, 107–119 (1976)

L. Poorter, M. van de Plassche, S. Williams, R.G.A. Boot, Leaf traits and herbivory rates of tropical tree species differing in successional status. Plant Biol. **6**, 746–754 (2004)

E. Quarterman, C. Keever, Southern mixed hardwood forest: climax in the southeastern coastal plain. Ecol. Monogr. **32**(2), 167–185 (1962)

M.U. Rani, A. Reddy, G. Reddy, Screening of rhizobacteria containing plant growth rooting (PGPR) traits in rhizosphere soils and their role in enhancing growth of pigeon pea. Afr. J. Biotechnol. **11**(32), 8085–8091 (2012)

M. Salazar, M. Malumbres, Fueling the cell division cycle. Trends Cell Biol. **27**(1), 69–81 (2017)

R.A. Slattery, D.R. Ort, Photosynthetic energy conversion efficiency: setting a baseline for gauging future improvements in important food and biofuel crops. Plant Physiol. **168**(2), 383–392 (2015)

J. Tang, S. Luyssaert, A.D. Richardson, W. Kutsch, I.V. Janssens, Steeper declines in forest photosynthesis than respiration explain age-driven decreases in forest growth. PNAS **111**(24), 8856–8860 (2014). https://doi.org/10.1073/pnas.1320761111

R.A. Ulanowicz, Quantifying sustainable balance in ecosystem configurations. Curr. Res. Environ. Sustain. **1**, 1–6 (2020). https://doi.org/10.1016/j.crsust.2019.09.001

P.M. Vitousek, W.A. Reiners, Ecosystem succession and nutrient retention: a hypothesis. Bioscience **25**(6), 376–381 (1975)

P.M. Vitousek, P.S. White, Process studies in succession, in *Forest Succession: Concepts and Application*, ed. by D. C. West, H. H. Shugart, D. B. Botkin, (Springer, New York, 1981), pp. 266–276

T.W. Walker, J.K. Syers, The fate of phosphorus during pedogenesis. Geoderma **15**, 1–19 (1976)

D. West, H.H. Shugart, D.B. Botkin, *Forest Succession: Concepts and Application* (Springer, New York, 1981)

P.S. White, R.D. White, Old-growth oak and oak-hickory forests, in *Eastern Old Growth Forests*, ed. by M. B. Davis, (Island Press, Washington, DC, 1996), pp. 178–198

Chapter 11
Panarchy

11.1 The Universal Cycle of Systems

All systems including political, economic, social and ecological go through a four-stage cycle of growth, consolidation, collapse, and reorganization.

1. Growth, in which the system takes advantage of abundant resources to increase energy flow through the system.
2. Conservation or consolidation, when resources become scarce and when continually expanding controls hamper alternative ways that the system can adjust.
3. Collapse begins when costs to maintain the system are greater than system income.
4. Reorganization or renewal in which resources are restructured in a novel way by a new system.

Many ancient peoples were aware of this pattern as applied to civilizations. In Ancient Greek mythology, the sequence is symbolized by the phoenix, a long-lived bird that cyclically regenerates or is otherwise born again. A phoenix obtains new life by arising from the ashes of its predecessor. Some legends say it dies in a show of flames and combustion, others that it simply dies and decomposes before being born again. It symbolizes rebirth and eternity and is also representative of hope.

11.1.1 Panarchy

Holling and Gunderson (2002) in their book *Panarchy: Understanding Transformations in Systems of Humans and Nature* used the term "Panarchy" to characterize this cycle of accumulation, consolidation, degradation and renewal. The term Panarchy was coined as an antithesis to the top-down control of systems

C. F. Jordan, *Evolution from a Thermodynamic Perspective*,
https://doi.org/10.1007/978-3-030-85186-6_11

achieved through sacred rules. The name hints at the Greek god of nature – Pan – whose persona also evokes an image of unpredictable change (Resilience Alliance 2021). Panarchy envisions a series of smaller, faster subsystems nested hierarchically in larger, slower cycles. The subsystems adapt to the flows of energy controlled by the top-most level of a hierarchical system while simultaneously modifying the flows. That is a bottom-up view of ecosystem control. It contrasts with the rigidly structured deterministic pattern of rules controlled from above by thermodynamic laws and principles.

11.2 Thermodynamic Interpretation of the Sacred Rules

11.2.1 Growth and Consolidation

Succession as described by ecologists and evolution as described by evolutionary biologists represent the first two stages of the Panarchy cycle – growth and consolidation. The first stage is prescribed by the maximum power principle. The system increases in biomass as pioneer species move into virgin wilderness to exploit readily available resources. Stage two begins as resources become scarce. The system uses its energy to conserve the gains it has made by building a feedback system that preserves system stability.

11.2.2 Collapse

The collapse stage begins with the inability of populations in subsystems to adapt to changing conditions. There is an entrenched bureaucracy in the sense that species are reluctant (teleologically speaking) to exchange their established interactions for new ones. The old interactions gave them power, power they might lose by new interactions in the feedback system. As the species fail to adapt, entropy reigns and the system collapses. The classic example is the demise of giant reptiles that dominated the earth during the age of dinosaurs and their replacement by mammals better adapted to the conditions following the Cretaceous-Tertiary (K-T) Boundary Extinction sixty-five million years ago.

11.2.3 Renewal

The law of maximum entropy production ordains the establishment of a new cycle by taking energy from the global entropic gradient and building a new system by storing entropy in new structures and functions. A current example of system collapse and recovery has occurred in native Hawaiian ecosystems. Around 2010, ohia trees *Metrosideros polymorpha*, a "keystone of the native species" in Hawaii began dying off in groves due to a fungus and plant pathogen (Outside 2016). Stage four (reorganization) began with replacement of *Metrosideros* by "strawberry guava" (*Psidium cattleyanum*), a common invasive species native to Brazil (Lucchesi 2019).

11.3 Sub-systems

Within this deterministic cycle are numerous sub-cycles representing bottom-up forces that produce a stochastic yet adaptive response of participants within the system to the top-down rules. Veldhuis et al. (2018) gave an example: "*Imagine a grass species that grows using mineral nitrogen taken up from the soil. In doing so, it produces detritus that stimulates the growth of a population of a particular detrivorous bacteria that, in turn, makes nitrogen again available for the plants through organic matter decomposition and thus recycles it to the mineral pool.*" Such smaller, faster, adaptive levels invent, experiment and test, while the larger, slower levels stabilize and conserve accumulated memory of system dynamics. The slower, larger levels set the conditions within which the smaller faster ones function.

Sub cycles are controlled by the physiological characteristics and population dynamics of the involved species. The hare-lynx interaction in chapter six is a sub cycle within the larger flow of energy through the entire taiga ecosystem that includes dynamics of organic matter in the soil and the response of trees and shrubs to those dynamics (Krebs et al. 2001). Hare-lynx cycles are nested within the encompassing deterministic cycle, and the rate changes of these sub cycles through feedback helps the encompassing cycle maintain stability, albeit in this case stability becomes apparent only over a century.

11.4 Panarchy Over Two Billion Years of Evolution

The thermodynamic cycle has been strikingly apparent during evolution over the past 2.4 billion years. It occurred at least five times during geological history as evidenced by the five great extinction events (Donovan 1989) when energy availability was altered by climatic change or atmospheric aerosols. These events started the collapse stage of the cycle. The last event occurred 65 million years ago at the Cretaceous-Tertiary boundary stage. The dinosaurs and other large reptiles could not adapt to the decrease in solar energy caused by atmospheric aerosols when the Chicxulub asteroid smashed into Earth. The system based on giant reptiles collapsed. Mammals were more successful in adapting and began the recovery stage of the cycle. They reorganized the system and took over the functional niches formerly occupied by the giant reptiles. Ecosystems then proceeded to the growth stage, as they had after the four previous episodes of extinction.

11.5 Consolidation, Bureaucracy and System Collapse

The Consolidation stage begins when systems begin to devote more energy or more money to defense and maintenance of status quo than to expansion. In political and economic systems, this occurs with the rise of a bureaucracy that becomes more powerful as it becomes entrenched. The rise is inevitable. As number of stakeholders in a system increases, the number of interactions increases, but the potential for conflict also increases resulting in a bureaucracy necessary to adjudicate the conflicts. Once a bureaucracy is established, it is resistant to change and adapt to new circumstances because change means giving up power. Inability of a bureaucracy to adapt to changes in world geopolitics and/or economics results in system collapse. We can see the result of an inflexible bureaucracy in the fall of the great empires from the Roman through the Mayan to the overseas English and Spanish empires. The bureaucracies were unable to adapt to forces outside their system and beyond their control. Inflexible bureaucracy prevents systems from adapting to environmental, technical and social changes and unwittingly precipitates their downfall.

Box 11.1
Meadows et al. (1972) in their Report for the Club of Rome's Project on the Predicament of Mankind predicted limits to growth of the world's population based on limits of agricultural productivity. Within the world system of agriculture, bureaucracy is a more likely limit.

In ecological systems, a series of well-established mutualisms is the bureaucracy that inhibits adaptation as the energy environment changes. There is much current research regarding the ability of individual species to adapt to climate change. Less attention is given to change in species interactions, a more critical determinant for preservation of species.

11.5.1 Bureaucracy in Action (Case Studies)

A colleague of mine had a grant to study the pharmaceutical properties of plants from the rain forest in a tropical country. Soon after he and his wife began collecting samples to bring back to their lab in the U.S., a law was passed the required scientists to leave part of their collection with a local scientist. The law was passed because the local government felt that U.S. scientists were exploiting their countries resources without leaving any benefit for their country, especially since the results of their work could result in economic benefit for pharmaceutical companies. That was certainly a reasonable law, but the problem was that there were no scientists in the host country who had the knowledge or resources to do the necessary chemical analysis. As a result, the research program had to be canceled.

I experienced another example of bureaucracy impeding research, and while it did not have the implications for scientific advancement as the above example, I include it to show how regulations promulgated to benefit one sector of the population causes frustration in another. In 1988, I received a grant to study the energy balance of the taungya system of agroforestry in Thailand. In the taungya system, teak seedlings are widely spaced in a plantation so that later thinning of the stand is not required. Local farmers are allowed to grow upland rice between the seedlings for 3 years free of charge. Benefits for the forestry system is that weeds that would compete with the seedlings are removed by the farmers during cultivation of the rice (Jordan et al. 1992). Part of the grant was to buy a jeep to travel to the remote sites in Thailand, but it was stipulated that the jeep had to be made in America. The problem was that in Thailand, traffic flows on the left side of the road meaning that the steering wheel has to be on the right side of the vehicle. No such vehicles are made in America. The grant expired before bureaucrats could solve the problem.

11.6 Case Study: Panarchy in the Georgia Piedmont

In 1993, I bought a 100 acre farm on the Georgia Piedmont from the descendants of the founder, a Confederate soldier, and renamed the farm Spring Valley Ecofarms. We chose the Panarchy symbol as the icon for Spring Valley Ecofarms, because the logo (Fig. 11.1) represents the four stages of landscape dynamics that occurred in the Georgia Piedmont in the nineteenth and twentieth centuries.

The cycle in Georgia began in the early nineteenth century when colonists first settled the Georgia Piedmont, cleared the land, and began to cultivate cotton, tobacco and horticultural crops. Productivity was good for a few years due to the rich layer of soil organic matter that had formed under native forest. Soils in the Georgia Piedmont are Ultisols derived from weathering of the rocks that form the Appalachian Mountains. At Spring Valley Ecofarms, there are sites that represent how this soil changed during the last 200 years. When the first settlers arrived in the Georgia Piedmont in the early 1800s, they encountered soil that looked like that in pit A (Fig. 11.2) under old-growth stands of oak and hickory. The profile represents soil at ecosystem maturity. The top 10–12 inches (the "A" horizon) has a high content of blackish soil organic matter, rich in essential nutrients. Ecologists call this "natural capital" because of its potential for producing economic crops. At the bottom of the A horizon, there is a sharp transition to the underlying clay minerals, the B horizon. This clay mineral portion of the soil consists of reddish highly weathered iron and aluminum oxides, very poor in nutrients due to millions of years of extreme

Fig. 11.1 The Panarchy logo at Spring Valley Ecofarms. The green portion of the logo represents the growth and the consolidation stages of the Panarchy cycle. The gray is the collapse stage and the yellow is the stage where reorganization and emergence of a new system takes place. (Photo by C.F. Jordan)

weathering under a hot, humid climate. When dried, it served to cement together rock walls of sheds and smoke houses used by the settlers.

Original cultivation by the first settlers was the growth or exploitation stage in which the system takes advantage of plentiful soil organic matter to increase energy flow and high production. As farms expanded into plantations, the cotton empire grew but high production lasted only a few years. Plowing oxidized the soil and stimulated microbial decomposition of the soil organic matter. Fields were left barren during the winter, and erosion carried the topsoil into streams (Jordan 2013). Some farmers moved westward, but others tried to consolidate gains by spreading manure, guano, and greensand on the soil. These imported natural fertilizers were expensive to purchase and to spread, in contrast to the services of nature supplied by the original soil organic matter. The price of cotton could not support indefinite purchase of these inputs. As soils became void of organic matter (soil pit B), the consolidation stage gave way to collapse. In pit B, there are a few brown traces of organic matter originating from decomposed roots of grasses and annual herbs, but the soil mostly is red clay. Some farmers hung on until the early part of the twentieth Century, but low soil fertility so weakened plant resistance to the boll weevil that the system of cotton production crashed. The void in energy flow created when agriculture was abandoned was filled with pines that can survive on infertile soil because

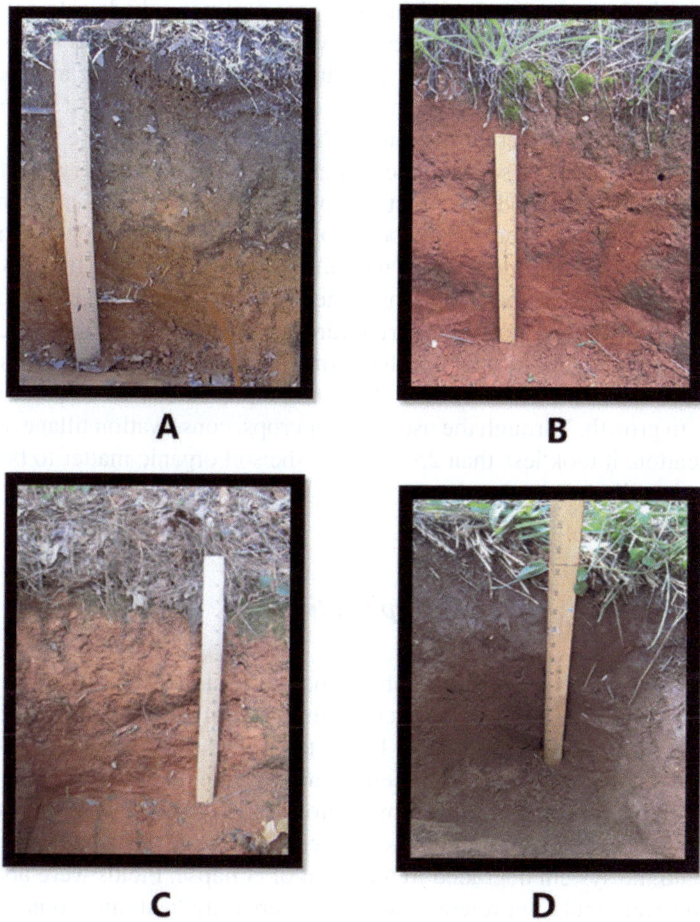

Fig. 11.2 Change in organic matter content during the cycle of Panarchy on an Ultisol in the Piedmont region of Georgia (The ruler is 18 inches.) Photo (**a**) represents the original untilled soil. (**b**) is the soil after a century of cotton cultivation. (**c**) is after abandonment when pines invaded the abandoned fields. (**d**) is the soil rehabilitation by organic agriculture. (Photo by C.F. Jordan)

their evergreen needles are a nutrient conserving mechanism (Monk 1966). Organic acids leached from the pine needle litter oxidized all the remaining organic matter in the soil leaving an impoverished soil profile (pit C).

Succession however rarely reached the mature stage of old growth hardwoods. In some places, the pines were clear-cut, hardwood saplings bulldozed, and pines established in plantations. Pulp mills sprung up to process the pines that now cover much of Georgia. After pines were harvested, some landowners tried to reestablish pines, but the system had entered an advanced stage of reorganization. Hardwoods, the natural successional stage following pines already had become established. The early successional "r" type pines were replaced by "K" types of species (MacArthur and Wilson 1967) such as oak and hickory that are more shade tolerant (Pianka 1970; Odum 1969, 1989), but they grew slowly on the impoverished soil.

In other places, energy intensive agriculture using fertilizers and pesticides revived productivity, but the soils remained void of organic matter. At Spring Valley Ecofarms, there was a different trajectory. Pit D is in an area of the farm used for experimental restoration of soil organic matter. In 1993, it looked very much like pit B. For 25 years it was used for vegetable production, using conservation tillage, cover cropping, green manuring, and occasional compost application. The re- organizational phase of Panarchy began as natural capital was restored. In 2018, the soil looked like that in pit D. About the same time, Georgia Organics, a non-profit organization, began promoting organic agriculture throughout Georgia and local farmer's markets sprang up to support the farms. Piedmont agriculture rose again, Phoenix-like, in a new format (Jordan 2013) that represented a transition from reorganization to growth. Through the use of cover crops, conservation tillage, and compost application, it took less than 25 years for the soil organic matter to build up to that shown in soil pit D.

11.6.1 Thermodynamic Interpretation

Panarchy is a succinct way of describing thermodynamic control of energy flow through a system, in this case the agricultural system of the South in the nineteenth and twentieth centuries. Entropy had been stored in the mature hardwood forests that covered the Piedmont. When the settlers arrived, the forests were in the consolidation phase. All the energy captured by photosynthesis was used to prevent entropy from being realized. As the colonists cleared the forest, the stored entropy was released, and the system degraded to the point of collapse. Fields were abandoned. Nature, however, could not tolerate the steep energy gradient, and so new types of systems rushed in (teleologically speaking) to ameliorate the gradient caused by the absence of entropy storage. Entropy was produced as structure of the system was rehabilitated, and the cycle became complete.

References

S.K. Donovan, *Mass Extinctions: Processes and Evidence* (Columbia University Press, New York, 1989)

C.S. Holling, L.H. Gunderson, Resilience and Adaptive Cycles, in *Panarchy: Understanding Transformations in Human and Natural Systems*, ed. by L. H. Gunderson, C. S. Holling, (Island Press, Washington, DC, 2002), pp. 25–62

C.F. Jordan, *An Ecosystem Approach to Sustainable Agriculture: Energy Use Efficiency in the American South* (Springer, Heidelberg, 2013)

C. F. Jordan, J. Gajaseni, H. Watanabe (eds.), *Taungya: Forest Plantations with Agriculture in S.E. Asia* (C.A.B. International, Wallingford, 1992)

C.J. Krebs, T. Boonstra, S. Boutin, A.R.E. Sinclair, What drives the 10-year cycle of snowshoe hares? Bioscience **51**, 25–35 (2001)
N. Lucchesi, The invader that grows in Hawaii is erasing a marker of natural history (2019), https://www.inverse.com/article/58204-hawaii-strawberry-guava. Accessed 26 Dec 2020
R.H. MacArthur, E.O. Wilson, *The Theory of Island Biogeography* (Princeton University Press, Princeton, 1967)
D.H. Meadows, D.L. Meadows, J. Randers, W.W. Behrens, *The Limits to Growth. A Report for the Club of Rome's Project on the Predicament of Mankind* (Potomac Associates, Falls Church, 1972)
C.D. Monk, An ecological significance of evergreeness. Ecology **47**, 504–505 (1966)
E.P. Odum, The strategy of ecosystem development. Science **164**, 262–270 (1969)
E.P. Odum, *Ecology and Our Endangered Life-Support Systems* (Sinauer, Sunderland, 1989)
Outside, What's killing Hawaii's trees? (2016), https://www.outsideonline.com/2406596/what's killing Hawaii's trees? Accessed 30 Dec 2020
E.R. Pianka, On r and K selection. Am. Nat. **104**(940), 592–597 (1970)
Resilience Alliance, Panarchy (2021), https://www.resalliance.org/panarchy. Accessed 16 May 2021
M.P. Veldhuis, M.P. Berg, M. Loreau, H. Olff, Ecological autocatalysis: a central principle in ecosystem organization? Ecol. Monogr. **88**(3), 304–319 (2018)

Chapter 12
A Thermodynamic View of Evolution

In 1973, the evolutionary biologist Theodosius Dobzhansky wrote an essay entitled "Nothing in Biology Makes Sense Except in the Light of Evolution." Thermodynamic theory says that "Nothing in evolution makes sense except in the light of energy flow".

12.1 Life: A Physicist's View

Erwin Schrödinger was a Nobel Prize-winning Austrian physicist who developed a number of fundamental results in the field of quantum theory. In his book *What Is Life?*, Schrödinger (1944) looked at the phenomenon of life from the viewpoint of physics. He focused on the need to explain life's apparent defiance of the second law of thermodynamics by maintaining order despite entropy that counteracts organization. Schrödinger's paradox was that despite the second law, living beings preserve and even elaborate exquisite atomic and molecular patterns (Schneider and Sagan 2005). The mystery for Schrödinger was how living systems keep from degrading. For biologists, it is obvious. It's the energy in food that we eat. Why was this considered so mysterious for physicists? A physicist's definition of life is not confined to the characteristics that distinguishes life from inorganic matter. Rather they use it in a much broader sense in which they look at the entropy in the universe. Schrödinger's puzzle was that living organisms and ecosystems remain organized despite the degradation going on all around them. We can see degradation in non-living systems when buildings decay and machines continuously wear out. For physicists, the defining characteristic of life is the ability to prevent decay, that is, to maintain for a significant period, a steady state in which the entropy (or disorder) within the living system is significantly lower than in its non-living surroundings (England 2013; MIT 2020). The system remains organized while its surroundings are becoming disorganized. As long as biological systems are alive, they are resisting decomposition.

C. F. Jordan, *Evolution from a Thermodynamic Perspective*,
https://doi.org/10.1007/978-3-030-85186-6_12

12.1.1 Life Is Produced by Capturing Entropy

A thermodynamic gradient exists between the solar radiation entering the Earth's atmosphere and the heat radiated back into the atmosphere from the Earth. Solar energy when captured by photosynthesis has the ability to perform "work", that is to power life and all the work done by living creatures. If solar energy interacts with the atoms and molecules that comprise non-living Earth, including soil, rocks, and water, the molecules vibrate more rapidly and give off heat. The ability of solar energy to perform "work", is lost, but if light strikes electrons in the chlorophyll of a living leaf, they become excited and jump to a higher, unstable energy level. Pairs of excited electrons pass through a series of electron donors. The energy they produce in the process is ultimately trapped in ATP and then in glucose used to synthesize new biomass or to fuel metabolic functions. The ability to do work is conserved.

Box 12.1
Entropy is "produced" in a system as its complexity increases – the more complexity, the more stored entropy. A completed jigsaw puzzle of 1000 pieces has more stored entropy than one of 10 pieces, and the person who put together the 1000-piece puzzle has produced a lot of entropy. The stored entropy is not evident however until the system collapses. There is an important difference though between a puzzle and a living system. The completed puzzle on a table top will not decompose on its own, whereas a living system will spontaneously decompose unless it obtains energy because the essential functions such as heartbeat and even just breathing require energy.

When plants photosynthesize, biologists say the plants are capturing energy. The plants use captured energy from sunlight to produce molecules that are structured in various ways that become the roots, stem, leaves, flowers and fruits and to fuel the metabolism within these structures. Physicists say that when plants photosynthesize, they are capturing entropy by reducing the global entropic gradient of energy. When entropy is captured by plants, there is less heating of the earth, i.e. less entropy. As long as the plant is growing, it is storing entropy in the structure of the plant. One could also say it producing entropy because with greater structure, there is more "potential" entropy to be released. When the plant dies and the plant structure decomposes or the wood is burned, the entropy stored by the plant is released and rejoins the global gradient as heat. We can say that living organisms "borrow" entropy from the global gradient and later return it when they die. The process of borrowing entropy and temporarily storing it slows down the global rate of entropy by using it to structure life. Biologists and physicists are talking about the same thing but with different frames of reference.

The answer then to Schrödinger's puzzle as to how life defies entropy is that entropy is being stored in the structure of living systems. But for structure to be maintained, there needs to be a continual energy input into the system to maintain order. If input stops, the system decomposes.

12.1.2 The Origin of Life

For England (2013) who models how clumps of carbon atoms first became alive, there is one essential difference between living things and inanimate clumps of carbon atoms: The former tends to be much better than inanimate aggregations of molecules at capturing energy from their environment and using that energy to maintain structure and reproduce (Wolchover 2014). The ability to capture energy depended on the ability of aggregate clumps of molecules to resonate to the energy source. The source of energy for the first aggregate of molecules that were "alive" must have been an entropic gradient, probably a chemical gradient around hydrothermal vents (Wicken 1988). A natural energy gradient exists in the gradients of chemical concentration between hydrothermal vents and sea water. Lane and Martin (2012) and Jackson (2016) proposed that the necessary aggregation of molecules that could capture energy to drive chemical reactions life first occurred in porous hydrothermal vents where heated, mineral-laden sea-water spews from cracks in the ocean crust. As the water from the vent mixed with sea water, there was created a gradient of positively charged protons that served as a "battery" to fuel the chemical reactions that were the basis of life in the aggregations of organic molecules and protocells. Other researchers have proposed that life originated in fresh-water ponds in terrestrial hydrothermal vents (Marshall 2020).

Horowitz and England (2017) sought to model the origin of life by analyzing the dynamics of a computer-generated chemical network that could capture the energy necessary for life. They concluded that "strong thermodynamic forcing is available only to rare combinations of chemical concentrations". Formation of aggregations that exhibit thermodynamic characteristics occurs because molecules that come together in a specific configuration "have a large influence on the accessibility of energy from external drives".

Once energy is captured by an aggregation of molecules, the next step in development of life is the ability to reproduce. About 3.8 billion years ago, aggregates of organic molecules acquired the ability to self-replicate. Replication may have arisen through the development of an initial molecular autocatalytic set. An autocatalytic set is a collection of entities, each of which can be created catalytically by other entities within the set, such that as a whole, the set is able to catalyze its own production (Mossel and Steel 2005; Lincoln and Joyce 2009). Mann (2021) has reviewed recent theories on the origin of life, but England's work stands out as a thermodynamic approach because the role of energy-powered autocatalytic feedback in the thermodynamic model is the basis for reproduction.

12.2 Two Approaches to Evolution

In the eco-evo-devo (ecological evolutionary developmental biology) approach to evolution, species are organized by taxonomic similarity (who is related to whom within trophic levels). In the thermodynamic approach, species are organized functionally (who eats whom as energy flows through trophic levels). The distinction is critical. In an auto parts store, parts are organized taxonomically. The mechanic has to know where to find them. In the automobile itself, parts are organized functionally. The mechanic has to know how they fit together. To repair a car and ensure that it operates properly, a mechanic must understand both classification (in which bin he finds the spare part needed) and function (where that part fits in the car's drive train). To understand evolution and to apply the understanding to conservation of species, the scientist has to understand how species are related to each other functionally as well as taxonomically within an ecosystem.

12.2.1 The Eco-Evo-Devo View

The goal of the eco-evo-devo approach is to uncover the rules that underlie the interactions between an organism's environment, genes, and development, and to incorporate these rules into evolutionary theory (Abouheif et al. 2014). Research is focused on using the fossil record in combination with molecular phylogenies to determine the mechanisms responsible for generating species diversity over geological timescales and the mechanisms that promote species formation and persistence at an ecological timescale including molecular genetic details that promote reproductive isolation (Editorial 2019). The Eco-Evo-Devo view is a bottom-up view of evolution. It explains how species respond to top-down control exerted by the physical laws and principles. Eco-evo-devo biology does not deal with the evolution of functional characteristics between species in different trophic levels, nor does it tell us why evolution has led to higher forms of life.

Box 12.2
The endangered northern white rhinoceros and the southern white rhinoceros in Africa differ mainly by size. Eco-evo-devo biologists are interested in the mechanisms through which they diverged (Naish 2017). Knowledge of whether these are different species or different races of the same species is important for conservation planning.

Butlin et al. (2008) criticized the Eco-Evo-Devo approach to speciation because its genetic emphasis ignores the spatial context in which divergence occurs. It does not consider the force driving differentiation, namely the disassociation of part of a population as food becomes limiting. Search for new sources of energy leads to genetic isolation as the sub-population finds new sources in distant landscapes (allopatric speciation), or because it has adapted more quickly to differences in ecosystem energy flow (sympatric speciation). In parapatric speciation the population is continuous but nonetheless individuals are more likely to mate with their geographic neighbors than with individuals in a different part of the population's range.

12.2.1.1 The Genetic View

The basic components of higher organisms, the linear polymers of DNA, RNA, protein and the structures they form were built up from simpler predecessors step by step. Most important in the evolution of complex organisms was division of labor (Cooper and West 2018; Pennisi 2018). Division was facilitated at the molecular level as protein chains were deformed by chemically fueled nanomachines called chaperones. Molecular chaperones are proteins that assist the conformational folding or unfolding and the assembly or disassembly of other macromolecular structures. With their help, chains can rearrange rapidly and irreversibly, acquiring new stable functional shapes they could never have found on their own (England 2020). In this process, single celled organisms adopted traits that entrenched them in a group lifestyle, thereby stabilizing the group and paving the way for the evolution of multicellular complexity (Libby and Ratcliff 2014). Multicellularity increased an organism's ability to compete for energy and consequently increased its power output.

12.2.2 The Thermodynamic View

The thermodynamic view of evolution deals with the laws and principles that control energy transformations as species and ecosystems increase in power output.

12.2.2.1 Evolution Is the Storage of Entropy

Physicists call the evolutionary process of building complex structures "storing entropy". "Entropy storage" means that solar energy, instead of dissipating as heat in the Earth, is stored in the structure of living organisms and ecosystems. During evolution, energy embodied in organisms increases as the complexity of proteins increases, and the structure of organisms becomes more complex. Energy is embodied in ecosystems as the configuration of food webs becomes more elaborate.

12.2.2.2 Embedded and Embodied Energy

Biomass consists of reduced carbon compounds. Energy is stored in reduced compounds and released when the compounds are oxidized or burned. It is sometimes called embedded energy. Embedded energy is the energy that would be given off as heat if an organism were oxidized (burned) in a calorimeter. Embodied energy is an expression of all the energy used in the work processes that generate a product or service in units of energy. Embodied energy is calculated by "life cycle analysis" (Murphy et al. 2016). For non-living systems, it would include all the energy needed to design and manufacture a product, for example an airplane. It includes the energy needed to extract aluminum from bauxite mined from the earth, processing the aluminum, and using it to fabricate the fuselage and wings. The sum of the pro-rated energy used to complete these transformations is the energy embodied in the final product (Patten 1995). Odum (1988) called it "Emergy" with an m instead of an n for embodied energy. The energy transformations are not random. They must be configured correctly to function properly. Selection of the proper pathways requires information. Information is "the configurations of the parts of a system" (Odum 1995). Design and testing of the design are included in the embodied energy.

Living systems contain an immense amount of embodied energy. The "Central Dogma" of molecular biology explains the flow of genetic information from DNA to RNA to protein. Embodied energy would include all the energy used to acquire information in the structure of proteins. For ecosystems, the embodied energy is energy consumed in a search for the combination of species interactions in food webs that enabled the species and the ecosystem to survive. It includes the energy expended in the development of the information in proteins, the energy used in storing the structure of cells in an organism, and the energy used in organizing those structure though time. Development of information consisted of a lot of trial-and-error efforts, most of which failed. The species and ecosystems that we see today are the few that survived. They embody a tremendous amount of energy that took billions of years to store in a way that resulted in information that produced fantastically elaborate species and ecosystems.

12.2.2.3 Entropy Storage Is a Powerful Tendency of Nature

Entropy storage is not permanent. When an organism dies or when a species becomes extinct, stored entropy is "released". For ecosystems, entropy storage has so far been permanent. It has lasted for over four billion years. It has built trophic chains and food webs, and preserved them albeit in a diminished form throughout at least five major global climate changes that drove most species in the corresponding epoch to extinction. It is the reason complexity and power output of higher species and ecosystems has increased during evolution. Entropy storage has driven evolution.

12.2.3 Fitness

Charles Darwin's 1859 theory on the origin of species was initially met with extreme skepticism, in part because of the absence of known mechanisms through which evolution could occur, and in part because it contradicted religious thought on

Creation (Bregliano 2019). While Creationists still reject the theory, contemporary understanding of genetics has secured among scientists the Neo-Darwinian view, sometimes called the "Modern Synthesis" that merged natural selection with genetics. Its premise is that evolution results from natural selection for characteristics that enable species' survival. A problem has been defining what those characteristics might be. One view has been that they facilitate adaptation to the environment, but what exactly are those facilitating characteristics? Spencer (1864) in his Principles of Biology suggested that they embody "fitness". "Survival of the fittest" then became a way of describing the mechanism of natural selection. However, "survival of the fittest" is a circular argument or tautology (Box 12.3), because those that are fit are the ones that survive, but to survive, a species must be "fit". Today, most evolutionary biologists circumvent the problem by saying that fitness involves the ability of species to propagate heritable characteristics which affect their reproductive success. Surviving through reproductive success is still a tautology. To survive, a species must reproduce. To reproduce, a species must survive. There is nothing inherent in reproductive success alone that would necessarily result in higher forms of life. Reproductive success alone leads only to more of the same. Organisms need no more than one cell to survive, so there is no evolutionary advantage to multicellularity if survival is the only characteristic for which organisms are selected. If reproductive success were the only criterion for selection, life on earth today would consist only of the one-celled protists such as algae that are well adapted to environments ranging from the tropics to the tundra. Nevertheless, Neo-Darwinian theory holds that higher forms of life have evolved from one celled organisms solely by the selection of offspring resulting from random mutation of genes.

Box 12.3
A classic tautology is as follows: What is God? God is Righteousness. What is Righteousness? Righteousness is God. That explains everything, so no further research is needed.

12.2.3.1 The Probability of Higher Forms of Life

Evolution has occurred much too quickly towards higher forms of life to have happened through adaptations resulting from selection from random mutations. Mutations that could lead to a higher form of life are very rare events. The number of mutations that increase adaptation to the environment per genome per replication of Escherichia coli ranges from 5.9×10^{-8} to 2×10^{-9} (André and Godelle 2006). Of these mutations, some are lost by drift (James and Jain 2016), and of those not lost, only a portion are permanently fixed within a population (Rozen et al. 2002). Because permanently fixed beneficial mutations occur so infrequently, the probability that selection from randomly occurring mutations could be responsible for the evolution of higher forms of life is similar to the probability that an "infinite number of monkeys banging on typewriters could produce King Lear" (Schneider and Sagan 2005). Only if evolution is selective for a characteristic that would lead to a higher form of life can organisms develop the complexity necessary for such an outcome. Reproductive success alone as the definition of fitness is inadequate.

12.2.3.2 Thermodynamic Fitness

The alternative to the definition of fitness as reproductive success was proposed by Ludwig Boltzmann, a nineteenth Century Austrian physicist and philosopher. "*The fundamental object of contention in the life-struggle, in the evolution of the organic world, is available energy. In accord with this observation is the principle that, in the struggle for existence, the advantage must go to those organisms whose energy-capturing devices are most efficient in directing available energy into channels favorable to the preservation of the species*" (Boltzmann cited in Lotka 1922).

Boltzmann recognized the importance of maximizing power output for survival. Maximization of power output is a non-tautological definition of fitness. Selection for power output means that organisms with a high energetic output relative to their size win out in the competitive struggle for existence. If a species can capture more energy than its competitors, it can better overcome limiting factors and become a winner. If water is limiting as in many seasonally dry tropical forests, those species that have enough energy can extend their roots deeply into the soil to obtain water (Nepstad et al. 1994). If nutrients are limiting as in some regions of the Amazon rain forest, those tree species that use their energy to interact with ectomycorrhiza to obtain nutrients directly from leaf litter on the forest floor are the better competitors (Went and Stark 1968; Herrera et al. 1978). The most important survival characteristic of the Galapagos finches in the Preface to this book was their ability to compete for and adapt to a new source of energy, i.e. a different size plant seed. Originally the finches' beaks may not have been well adapted to this new source of energy but through several generations of natural selection, those offspring with beaks best suited to this new energy source were the ones selected for survival because appropriate beak size gave the greatest power output. Power output was what defined their fitness. Adaptation to the Ecological Theater (climate, nesting sites, predators, and disease) was secondary. Robertson (1955, 1967, 1968) and Falconer (1981) pointed out that in evolution, adaptation to secondary factors occurs only after adaptation to the primary factor defining fitness.

Brown et al. (2018) have suggested that all species are equally fit. The question then arises, if all organisms are equally fit, why do animals of small or intermediate size exist in the same trophic level as large powerful organisms, since large powerful organisms would be the best competitors for energy and therefore the most "fit"? Small organisms make their living by using energy sources unavailable to larger ones because of functional constraints, using energy sources made available by larger and more complex organisms, or using energy left by scavengers and detritovores.

Box 12.4
Size of trees is not necessarily an indication of their power output. Size of living tissue in trees is the criterion, because that is what determines energy flow. Most mature trees have living sapwood and non-living heartwood. The giant redwoods and sequoias have achieved large size although their proportion of living sapwood to non-living heartwood is small. Their power output is low relative to their size.

Many early studies of species metabolism focused on efficiency of energy transformations (Wiegert 1976). An assumption seemed to be that fitness is characterized by efficiency of energy transformations. Brown (1995) disputed the assumption. "If this were true, endothermy could never have evolved." Endothermic birds and animals are extremely inefficient compared with ectothermic reptiles and amphibians. Endotherms expend energy at high rates in order to maintain a high, constant body temperature and associated capacity for high levels of activity, enabling them to live in environments where reptiles and amphibians cannot.

12.2.3.3 The Opposable Thumb, Intelligence, and Power Output

Evolution toward bigger organisms following the first four great extinctions in geological history (Donovan 1989) was based on power output, with physical strength as the principal characteristic that ensured survival of species in each clade. Power output in the phylogenetic branch that led to dinosaurs explains the success of *Tyrannosaurus rex*, but it cannot explain the increase in intelligence in primates that began only a few million years ago. That is explained by the evolutionary development of the opposable thumb that allowed pre-humans to manipulate tools and control energy flow. The first unquestionable stone tools were evidently made and used by early transitional humans in East Africa about 2.6 million years ago (Choi 2009). Making and using tools requires a degree of intelligence. Tools allow a species to control energy flow through their environment and work for the species advantage. The smartest individuals were the ones that used tools most effectively, so there was selection not only for physical power output (the only criterion until after the extinction of dinosaurs), but also for intelligence which gave humans greater power output than dinosaurs ever had. Evolutionary selection for power output once tool-using was accomplished meant selection for intelligence for the evolutionary line that gave rise to *Homo sapiens*. This has resulted in the ecosystem in which *Homo sapiens* lives (the planet Earth) to steadily increase its rate of energy flow.

12.2.3.4 Fitness and the Complexity of Higher Forms of Life

Complexity increases power output. Power output increases complexity. It is a circular relationship. It is like saying chickens are necessary to produce eggs, but eggs are necessary to produce chickens. What propels the chicken-egg cycle? It is the food that chickens eat. Food is the power that drives the cycle. It is an energy source from outside the system. Likewise, solar energy from outside a biological system drives the cycle leading to increased structure, and this complexity abets the systems' ability to acquire yet more energy. The ability to increase power defines fitness for living systems. Power output helps complex systems win the competition for energy needed for survival. Evolution is the survival of the powerful.

12.2.3.5 The Role of Competition

Dawkins (1976) maintained that evolution occurs because genes act in their own self-interest. Self-interest includes the ability to survive. But to survive, an organism that contains the genes must out-compete rivals for energy, so selection of species based on the ability to out-compete others for energy is the characteristic that ensures survival of genes in a competitive world. The genetic mechanism underlying selection for power output is the force exerted on protein-coding genes by organisms competing for energy (Shenvav and Zeevi 2020).

Competition among one-celled planktonic autotrophs for light and nutrients was not important in early evolutionary history because circulation by ocean currents exposed the plankton to all the light they needed, and nutrients were abundant in upwellings. Capability to absorb light and nutrients were the factors that ensured survival. There was no impetus for cells to evolve beyond the ability to reproduce. Reproductive success alone did not enable phytoplankton to evolve beyond a single-celled structure. Winning a competition for energy as a factor for survival was important only among the heterotrophs that fed upon the plankton. Selection for power output resulting from competition for energy is the factor that drove the evolution of bigger, more complex marine consumers.

The lack of multicellular complex autotrophs in the ocean could mean that organisms will not become complex without competition that drives selection for increased power output. While ocean-dwelling algae and kelp can be multicellular, they are relatively simple and their energy-capturing ability is simply the sum of the individual cells that make up the algae or kelp. Seagrasses are more complex, but they evolved from terrestrial plants which recolonized the ocean. Coral reefs are a complex high-powered mutualistic system between zooxanthellae, a photosynthetic algae and coral polyps, but the algae are simple one-celled organisms. Complexity results from the complex coral infrastructure, not from competition.

When life moved to terrestrial habitats, the situation changed for autotrophs. Autotrophs no longer moved freely to obtain nutrients and energy. Reproductive success was not enough to assure survival of a species. There was competition for energy at each and every spot of ground, so a high power output was necessary to win the competition and survive. As a result, competition resulted in selection for more complex plants that had a greater power output. By virtue of many layers of leaves, trees became much more complex and efficient than plankton or seaweed at capturing light per unit of the Earth's surface.

12.2.4 The "Goal" of Evolution

Critics have interpreted thermodynamic fitness as saying that increasing power output is a goal of evolution. Goal seeking is restricted to humans, therefore the theory is faulty. But species do not evolve with an objective of increasing power output. The only goal of species is to survive, and winning the competition for resources is the key to survival. It is just that the winners of the competition were those that got biggest fastest and therefore maximized power output. In a primitive competition, the strongest – those with the greatest muscle power – are the ones selected for survival.

Box 12.5
The "goal of evolution" can be expressed in a non-teleological way by calling evolution an "inevitable result of the influence of competition for energy".

12.3 The Relationship Between Species and Environment

The genetic mechanisms that underlie speciation – selection, drift, migration, and mutation (Vellend 2016) – are agreed upon. Theory regarding the relationship between species and environment is still developing.

- In the "classic" view, species adapt to the environment in which they find themselves. Although most evolutionary biologists today would probably agree that this view is an oversimplification of reality, for simplicity's sake it has been, and still is, widely used in theoretical evolutionary biology as an implicit conceptual framework (Loreau et al. 2004).
- The "modern" view recognizes that the environment is always changing. As the environment changes, species adapt to these changes (Pelletier et al. 2009).
- In the ecosystem view, organisms modify and interact with their environment. This generates an organism–environment feedback in the operation of natural selection (Ferriere and Legendre 2013).

- In the thermodynamic view, energy flow is the environment in which speciation occurs. Energy feedback between species in an ecosystem results in simultaneous and reciprocal evolution of both species and the ecosystem.

12.3.1 Evolution's "Theater"

Just as in Neo-Darwinian theory, the mechanism of species evolution in thermodynamic theory occurs through selection of mutations that are beneficial. It just assumes a different venue in which mutations take place and a different meaning for "beneficial". The difference is that in the Ecological Theater (the Neo-Darwinian view), beneficial means accommodation to local ecological conditions (Turcotte et al. 2012). In the Thermodynamic theater, beneficial means accommodation to an energy flow that is changing day to day, season to season, century to century and during evolutionary time. Due to the variations in energy availability in space and time, there is opportunity for different species to win competitions at various times and at various locations. Losers do not necessarily perish if they can evolve in a way to use some of the energy that escaped the winners. Trees in the forest understory use energy that slipped by those in the canopy. Losers in one ecosystem can also become winners in another to which they migrate.

12.3.2 Is Evolution Stochastic or Deterministic?

Population theory cannot predict what type of species will appear next during evolution. In the ecological theater, the fate of the players is stochastic. It depends on the ever-changing dynamics of species interactions in food webs. In contrast, evolution of ecosystems in which the species reside is deterministic. In the thermodynamic theater, the outcome of the play is predetermined. Thermodynamic theory can predict very well what will happen following ecosystem disturbance – increase in power output that is reflected in an increase in biomass and diversity of species. The only question is, who will play what role.

Box 12.6
Stochastic evolution could not have led to higher forms of life. There has to be a control that directs evolution deterministically. In a thermodynamic sense, species evolution in the aggregate is deterministic. Aggregate evolution toward maximum power is determined by thermodynamic laws and principles. We don't know what will happen to a particular species, but we know what will happen to all of them together. Although evolution increases their power output, they have no idea that is occurring. All they "know" is that they are competing.

12.4 Ecosystem Evolution

Alfred Lotka, a mathematician and physical chemist, theorized that competition of species for energy results in a power increase not only of species, but also of the system of which species are a part. *"It appears, however, at least a priori probable that, among the certainly very large (if not infinite) variety of types presented for selection, sooner or later those will occur which give the opportunity for selection to operate in the direction indicated, namely so as to increase the total mass of the system, the rate of circulation of mass through the system and the total energy flux through the system"* (Lotka 1922).

Odum and Pinkerton (1955) in their formulation of the Maximum Power Principle put it this way: "Under competing conditions, the primary producers, the plants, which are best adapted (most "fit") may be the types that can as a group give the greatest power output (flow of useful energy) in the form of growth". "Maximization of Power Output" is a non-tautological definition of fitness of species, and it is this characteristic that drives maximum power output of ecosystems. Not only do ecosystems evolve, they evolve towards an increase in their power output (growth, or energy output per unit time) as a result of competition between species. Maximization of power output is a definition of ecosystem fitness as well as of species fitness.

12.4.1 Succession Was the Clue

Studies of ecosystem succession helped ecologists develop the thermodynamic perspective that could explain the evolution of complex ecosystems. Succession resulted from an increase in energy flow from early stages consisting of a few small species to mature stages with many large species. Increase in energy flow was the key, and competition between species is what drove species and ecosystems to maximize power output. The same pattern occurs in evolution. Power output of ecosystems increases in both succession and evolution.

12.4.2 Ecosystems Moved Away from Equilibrium

Thermodynamic equilibrium occurs when there are no energy transformations within the system. Since energy transformations are the basis of life, living systems are never in thermodynamic equilibrium. If the sun ceased to shine, ecosystems would regress to thermodynamic equilibrium, organization of species would be random, energy would not flow between them, and they all soon would be dead. But since the beginning of life, ecosystems moved further from thermodynamic equilibrium. The theory has been advanced in recent decades by ecosystem modelers and system analysts. *Evolution occurs as ecosystems use surplus energy to move further away from thermodynamic equilibrium* through *development of ever more complexity and diversity that improve resource flow utilization* (Nielsen et al. 2020).

Ecosystems are always using solar energy to keep organized, and the distance from thermodynamic equilibrium depends on the complexity of the system. A eukaryote and the bacteria that decomposed it existed in a state not far from thermodynamic equilibrium. Diversity and complexity of the system were low, and little work was accomplished. As evolution progressed, the systems become more complex and in the process, rates of energy intake and heat discharge both increased. Information accumulated in the system, and more and more entropy was stored. There was evolutionary selection for increased complexity because ecosystems that increased in complexity used energy more effectively, increased their power output, and obtained an evolutionary advantage. As evolution proceeded, systems moved further away from a state of thermodynamic equilibrium.

12.4.3 Thermodynamic Mechanisms

Thermodynamic mechanisms that contribute to the evolution of ecosystems away from thermodynamic equilibrium and towards increased power output include entropy storage, gradient reduction, and stabilization.

12.4.3.1 Entropy Storage

In Chap. 3, ecosystems were defined as self-contained, self-regulating, self-organizing, interconnected, open thermodynamic networks of component parts which evolve as a complex regime that combines stability and adaptability in the phase transition between order and chaos, through activities of plants, animals, fungi, and microbes. A key phrase is "the phase transition between order and chaos". Entropy continually tries to pull ecosystems down into chaos, while the ecosystem struggles to remain organized by pumping solar energy into the food chain. By keeping food webs organized and the species in them alive, ecosystems are (teleologically speaking) storing entropy.

Ecosystems would become disorganized, but they get energy to counter entropy (sometimes called negative entropy) from the larger entropic gradient in which they are embedded. By way of photosynthesis, plants siphon off some of this energy from the global entropic gradient before it becomes heat, and again speaking teleologically, pass this energy through trophic chains to keep the interactions intact and organized, since interactions must be fueled to be maintained. Because structure will deconstruct if energy input ceases, a stable ecosystem can be viewed as "storing entropy". It is more intuitive to call entropy stored in a system "potential entropy", and if the system deconstructs, to call it "realized entropy", or perhaps "kinetic entropy". As ecosystems evolve, they increase in potential entropy, or as some authors have said, entropy is produced. The process can also be characterized as information storage as it is embodied in the structure and organization of the food web. Structure-enhancing configurations defines ecosystem information (Ulanowicz 2009). As evolution proceeds, information in ecosystems increases.

12.4.3.2 Gradient Reduction

As systems grow in size and complexity, they reduce the Earth's entropy gradient over the patch of earth on which the ecosystem sits by capturing some of that entropy. As they reduce the earth's entropy gradient by incorporating some of the energy that otherwise would be degraded, they store the entropy that was removed. The move is reflected in an increase in system complexity through increasing biomass of existing species and through increasing the number of species and their transactions in trophic webs. The system becomes better organized, and the utilization of energy becomes more efficient. The system has moved further from thermodynamic equilibrium.

12.4.3.3 Thermodynamic "Obligations"

Nature abhors a gradient. When you produce a vacuum in a straw inserted in a glass of water, the pressure gradient is eliminated. Life originated because nature abhors an energy gradient. The gradient existed since the formation of the Earth, and billions of years later when a group of molecules happened to be arranged in a formation that absorbed energy, they acquired the ability to reproduce, a criterium for living systems. Organisms and ecosystems have maintained their structure through more than two billion years because of nature's obligation to reduce the Earth's energy gradient. Nature of course does not have any obligation to reduce the Earth's energy gradient any more than Newton's apple had an obligation to fall so as to reduce the gravitational gradient between the apple and the Earth. These "obligations" are very powerful thermodynamic forces. The tendency to reduce the global entropic gradient through storage of entropy in living biomass is so powerful a force that energy flows through trophic levels have been resilient despite species decimation through five major cataclysmic extinction events in geologic history. Although species populations were unstable during the extinction events in geological history, trophic webs maintained their structure and in that sense were stable. The species involved changed, but the basic thermodynamic structure – producer-consumer-decomposer – was constant. Following each extinction event, trophic webs and ecosystems recovered, each time with strikingly different species but with the conserved thermodynamic structure. While ecosystems lost species throughout geological history, their basic structure remained intact. They were performing their obligations.

12.4.4 Biological Mechanisms

Biological mechanisms that contribute to the evolution of ecosystems include autocatalysis, feedback and interactions including predator-prey and competition.

12.4.4.1 Autocatalysis

Autocatalysis in ecosystems occurs when interacting populations of different species form a loop whereby each species produces the resources needed by the next species in the loop, in such a configuration that the whole set of species is self-promoting and self-sustaining given sufficient input of energy and essential carbon and nutrients (Veldhuis et al. 2018). The microbial loop in marine ecosystems (Fig. 5.4) is an autocatalytic system. Autocatalysis speeds up the rate of energy flow through ecosystems when nutrients interact with products of photosynthesis to produce compounds in excess of those needed for ecosystem maintenance. The process increases the biomass of the ecosystem. Productivity of a marine microbial loop is enhanced when nutrients carried by upwellings becomes available to autotrophic plankton. Productivity in excess of that needed to maintain the loop becomes available to higher trophic levels. The evolution of consumer species (mainly fish and filter-feeding whales) toward an increase in power output results from the evolution of power output in the microbial loop.

In terrestrial systems, autocatalysis starts with energy capture by plants. Energy capture increases as a result of competition. The plants are acting individually by competing for energy, but the individual actions when taken together result in increasing energy flow through the ecosystem. As collective power output of the plants increase, the trophic webs dependent on plants increase. Greater rate of energy flow through ecosystems permits more specialized types of consumers that in turn increases the efficiency of energy and nutrient flow through the trophic webs of ecosystems (Novotny et al. 2006). The trophic webs increase energy and nutrient flow to decomposers. That allows an increase in nutrient recycling that in turn increases autocatalysis and an increase in primary productivity and power output of the system.

12.4.4.2 Increased Rate of Feedback

Survival of an ecosystem depends on the maintenance of energy flow and nutrient cycling by species within the ecosystem. This network consists of energy and nutrient flow in one direction and control in the other, much as humans live in an economic system consisting of goods and services in one direction and flow of money (a type of control) in the other. Nutrient cycles and energy flow in nature are maintained by feedback cycles through the species that are embedded and interconnected in the ecosystem. As feedback increases, ecosystems evolve.

12.4.4.3 Power-Enhancing Interactions

Ecologists have recognized the relationship between a pollinator and a flower as a co-evolved mutualism. Such mutualisms increase the power output of the interacting species over and above the level achieved when they act alone. Other commonly recognized mutualisms include oak-hickory ecosystems that include squirrels that plant acorns at a distance from the parent tree. This increases the probability that the resulting seedlings will survive because of reduced competition where they are planted. Interaction between ruminant species and the bacteria in their rumen increases the power output of the ruminant. Interactions between leguminous trees and nitrogen fixing bacteria give a power output advantage to their ecosystems. Ecosystems that contain many mutualistic interactions will have an increased power output and be more likely to be dominant in a natural landscape.

While some interactions such as predator-prey are detrimental to one participant, all interactions are mutualistic at the ecosystem level. From a thermodynamic perspective, predator-prey interactions are co-evolved mutualisms. That statement may seem to contradict conventional wisdom that says evolution occurs because genes act in their own self-interest (Dawkins 1976). If evolution occurs only at the self-interest level, what self-interest is there for a wildebeest to sacrifice itself to a lion? How can that be a mutualism? The mutualistic benefit to the prey is difficult to perceive except in the context of ecosystem stability. The Sewall Wright Effect (Fisher and Ford 1950) says it is in the self-interest of the herd, not the individual. The benefit of lions to wildebeests at the herd level is that lions keep the wildebeest population under control so their herd does not expand to the point where the grasslands cannot support the herd (Abrams 2000). If mutualisms such as predator-prey were not beneficial to ecosystems, they would not have persisted, just as characteristics that do not benefit a species are often lost.

It is evident that grasslands benefit from lions in that lions prevent wildebeests from overgrazing that would result in invasion of the landscape by low-productivity woody shrubs. Grasslands also benefit from wildebeests when wildebeests eat mature leaves so new grass productivity is increased through increased access to sunlight and also through increased nutrient recycling by wildebeest digestion of decay-resistant leaves. All this benefits the lions in that it keeps constant their supply of meat.

Box 12.7

The maximum power principle predicts that it is the efficiency of energy transfer between species that determines the power output of the interaction. Power output of systems where these interactions are efficient (prey is quickly caught by the predator), is greater than where the prey is adept at eluding the predator. Taken all together in an ecosystem context, ecosystems increase in power output as the efficiency of energy transformations within it increase.

Although parasitic-host interactions are detrimental to the host being parasitized, parasite-host interactions like predator-prey interactions are beneficial to the ecosystem when populations of the host organism grow to the extent that the ecosystem is destabilized. In the long-term, even competitive interactions are beneficial to ecosystems because they increase power output of ecosystems through evolution of increased species diversity.

12.4.5 Ecosystem Fitness

Because of entropy storage, gradient reduction, thermodynamic obligations, autocatalysis, feedback, and power-enhancing interactions, energy flow through ecosystems increased, resulting in increased power output of ecosystems. With increased fitness, systems were able to survive perturbations that would have degraded or destroyed less fit systems. Maximization of power output is a definition of ecosystem fitness.

12.4.6 Ecosystems Evolve One Step at a Time

Opponents of the thermodynamic view could argue that ecosystem evolution is unlikely to happen since simultaneous evolution of autotrophs, herbivores and predators at the same rate and in the right direction would be required. They would be right, but that is not how ecosystems evolve. Ecosystems do not evolve as an integrated unit, and they do not compete with each other head-to-head as do two similar species. Evolution occurs as less fit species in the system are replaced by more fit species, or when a new species evolves that increases energy flow and power output of the system. Evolution of ecosystems occurs step by step, species by species. The emergence of new or unique ecosystems occurs because with time, better adapted species replaced lesser adapted ones within each trophic level of an already existing ecosystem. For this to happen, a new species must be better synchronized than the species it replaced. That could happen through sympatric speciation where the new species evolves from an existing species and eventually replaces it. It could happen when a sub- population of a species moves to another ecosystem where a species with the same functional niche is in decline. Due to global warming, trout could be in decline in northern lakes and streams where they are the top predator. They could be replaced by bass that are top predators in southern lakes. Maple trees could decline in northern ecosystem as temperatures warmed, and be replaced by oaks from temperate zones, but in no case could the oaks do well unless there was a squirrel or similar creature to bury oak's acorns. Independent evolution of species simply cannot happen in the natural world outside of the context of ecosystem structure and function. But ecosystems depend on species to increase their fitness, because it is at the species level that selection occurs.

12.4.6.1 Ecosystem Migration

When aquatic ecosystems first migrated onto land some 500 million years ago, it proceeded in a step-by-step process in which each species as it moved landward integrated itself into the terrestrial flow of energy and nutrients. The first land plants consisted only of simple moss-like plants and their decomposers, but these plants created the habitats to which herbivores and predators from aquatic environments gradually adapted and occupied. Adaptation occurred relatively quickly because the necessary genetic information already inherent in the species was flexible enough to accommodate the change (Meyer et al. 2021).

Box 12.8
We are familiar with the phylogenetic tree of life that shows the evolution of taxonomic relationships between species. Understanding of ecosystem evolution would benefit from a thermodynamic tree of ecosystem evolution that shows how the diverse ecosystems of today evolved from the primal interaction of autotrophic eukaryotes and decomposers along a line of increasing power output with all the branches (ecosystem types) that radiate from the main line. This would differ greatly from the phylogenetic tree because it would show the importance of interactions in evolution. Interactions often are between organisms in different taxonomic units, sometimes between organisms in different phyla. The tree would show all the ecosystems of the world and how they are related functionally.

Brown (1995) has pointed out that as climate warms, species will migrate up the slopes of mountains or to higher latitudes. How fast species migrate compared to how fast their ecosystems evolve is significant for species conservation (Li et al. 2020). The "thermal mismatch" hypothesis suggests that mammalian and amphibian hosts adapted to cooler conditions will be at greater risk of infection under warming conditions because small-bodied parasites are more likely to maintain performance over a wider range to temperatures than larger -bodied hosts (Cohen et al. 2020). The thermal mismatch hypothesis also has implications for nutrient cycling. As temperature rises from global warming, decomposition increases faster than photosynthesis, meaning there will be more nutrients available for leaching loss and an increased impoverishment of ecosystems.

12.4.6.2 How Food Webs Were Built

Species in ecosystems do not move randomly like atoms in an insulated black box but move in a highly structured way that enables flow of energy and recycling of nutrients. Food webs were built as potential species were selected to fit into the web in a way that facilitated energy flow through the web and enhanced stability. The process of "fitting in" took evolutionary experimentation. The energy used for all species to "fit in" is energy embodied in the food web of an ecosystem. It is all the energy used during the self-organization of a food web. Just as economies self-organize as participants in the economy search for buyers and sellers (an energy consuming process as real estate brokers can testify), ecosystems self-organize as species search for both consumers and providers of energy.

12.5 The Origin of Ecosystems

Interactions between trophic levels such as between lions and wildebeests did not result from independent phylogenetic evolution of traits that just happened to fit together in an interaction. It is too improbable that lions and wildebeests evolved independently and it is just a lucky coincidence that the interaction works to stabilize the Serengeti ecosystem. Except for Thoreau when he lived at Walden, there seems to have been little interest in the question of how all the species in an ecosystem evolved to the point where they became functionally well-coordinated ("living in harmony", as Thoreau (1854) put it). Species in each of the trophic levels level had to evolve in tandem with their life support system – the interdependent autotrophs, herbivores, predators and decomposers – ever since trophic levels originated as different eukaryotic types. The question is, how did ecosystems with their circular trophic loops begin?

12.5.1 Origin of Feedback Loops

There are several competing hypotheses about how eukaryotic cells, which contain a nucleus and other membrane-bound organelles, evolved from their prokaryotic ancestors, whose cells lack a nucleus (Watson 2019). One theory is that about 2.7 billion years ago following some 1–1.5 billion years of prokaryotic evolution, a prokaryotic bacterium belonging to an archaeal phylum (Embley and Williams 2015) merged with another prokaryotic bacterium to form a symbiotic union (Martin and Sousa 2015; Javaux 2019). Eukaryotic cells appear to be the descendants of this symbiosis (Understanding Evolution 2018; Watson 2019). It is not clear if these microbes were autotrophic or heterotrophic (Brennan 2019). These cells took up nutrients from the environment, but they were not immortal. Eventually they died. There had to have been some type of organism that consumed and decomposed these eukaryotes (Levin 1999). Otherwise there would have been an infinite build-up of primitive cells.

No individual species is capable of performing all the ecological roles necessary for life. To survive, a species must be part of an ecosystem in which different species perform different functional roles to keep energy flowing and nutrients cycling. The problem of organic matter and nutrients bound up in eukaryotic cells was solved by the decomposing ability of prokaryotic bacteria. The first feedback loop emerged from the circulation of nutrients released by prokaryotic bacteria and nutrient uptake by eukaryotic cells. This became the microbial loop that during the course of evolution developed as part of the ocean's food web (Pomeroy 1974). The bacterio plankton in the microbial loop in marine systems of today as illustrated in Fig. 5.4 of Chap. 5 may represent early decomposers.

12.5.2 Origin of Trophic Levels

The original feedback loop consisted only of Protists (eukaryotic organisms). As the feedback loop formed, each node (proto-species) became a stake-holder in an ecosystem's life support system. Since then, species have increased in size and complexity, as the autocatalytic interaction between recycled nutrients and photosynthetic energy increased the ability of systems to capture energy. With increased energy there was the opportunity for evolution of primary and secondary consumers that later became herbivores and carnivores.

12.5.3 Why Are There Trophic Levels?

The existence of trophic levels in ecosystems is a well-established concept. A question that seems to never have been asked is, why are there trophic levels in the first place? Why was there a need to expand the original two-component model of producer and decomposer? What was the irresistible force that created the need? Life in ecosystems could very well exist without animals. It is just that animals increase the entropy-storing capability of ecosystems, and reduction of global entropy is so universal a phenomenon that it could be considered a law of nature. The actions of herbivores and predators increase the decomposition rate of plant material, thereby stimulating autocatalysis and a more rapid build-up of ecosystem structure that reduces global entropy.

12.5.3.1 Phylogenetic Structure of Trophic Levels

In some trophic loops, the herbivores and predators still are in the same phylum. In the Chordata phylum, lions eat wildebeests. However, evolution of cross-phyla trophic interactions has been a common occurrence. Feedback loops in some ecosystems have components that occur in two or more phyla. In the food chain made up of birds-insects-plants, two phyla of consumer species are represented. The integration of different phyla into one ecosystem may have occurred as the environment changed or as ecosystems migrated to a new environment. A predator from a different phylum may replace a predator from the same phylum as the herbivore because the replacement is more efficient in transforming energy supplied by herbivores. As plants moved from aquatic environments to terrestrial environments, the optimal combination of species changed, optimal meaning the best maximizer of energy flow. An optimal herbivore of plankton is not an optimal herbivore of trees.

Box 12.9
When the Chordata branched off from the phylogenetic tree, there must have been an omnivorous progenitor that gave rise to both herbivores and predators, but optimal predator-prey combinations for an ecosystem do not always occur in the same phylum. Some insects may be better herbivores than herbivorous mammals on certain plants. Incorporating a species from a different phylum, class, order or family in another ecosystem is a more economical way to improve an ecosystem function than waiting for a better functioning species to evolve within the current taxonomic group.

12.5.3.2 The Importance of Animals

The amount of ecosystem energy flow through animals is small compared to the direct flow of energy from autotrophs to decomposers. As you walk through a forest, dead leaves cover the soil, but rarely do you see a dead animal. They are important even though energy flow is small because they control much larger energy flows through ecosystems, much as a small amount of energy to flip an electric switch controls a large amount of energy moving through a circuit.

12.6 The "Goal" of Ecosystem Evolution

Evolution of species results from random mutations and selection of mutations based on species' characteristics that affect their ability to capture energy. As species increase their power output, the ecosystems in which they are embedded also increase in power output. As energy flow through the system increases, species evolve and adapt to the newly increased energy environment. Species and ecosystems simultaneously increase power output through reciprocal feedback in a deviation amplifying feedback system.

The evolution of ecosystems along a pathway toward maximum power output through increased energy fluxes, storages, transformations, and nutrient cycling does not require goal functions for ecosystems or natural selection operating at levels beyond the individual (Phillips 2008). That is a key point against those who argue that anthropomorphic goals such as maximization of power output are not possible in non-human systems. Maximization is not a goal. It is the inevitable result of competition among members of the system. The only goal of species is to survive, and to do that they must win the competition for resources. It is just that the winners of the competition were those that got biggest fastest and therefore maximized power output for themselves and their ecosystem.

12.6.1 *Conflicting Goals?*

The maximum power principle says that ecosystems are selected for maximum power output. Power output of ecosystems is expressed in two goals: maximum growth rate and maximum stability. Chapter 10 showed that early in succession, maximum power output is expressed as maximum growth, but at maturity, it is expressed as stability. For ecosystem stability, power output at maturity has to remain high to support the longer food chains and more complex food webs that assures the diversity that maintains ecosystem stability (Hooper et al. 2005). At maturity, power output of ecosystems for growth is less important for survival than power output for stability. Ecosystem goals are not conflicted. The goal in evolution as well as in succession is always maximum power output. It is just that maximum power output is expressed differently in immature systems than in mature ones. Energy flow and species diversity increases until the energetic cost of maintenance began to exceed cost of further growth, at which time power output is dedicated to stability.

12.6.2 *"Motivations" of Species*

Species become more powerful through species interactions, as energy feedback through the ecosystem increases their power output. But increase in their own power output does not come from a desire to help their ecosystem have maximum power. The only "motivation" for a species is to out-compete other species for a functional role on the ecosystem team. The motivation is survival through winning a place on the team. "Motivation" of species here is analogous to the motivation of a potential football player competing for a spot on the team. His primary motivation is not based on the desire for the team to win the championship. It is for him to beat out a competitor for a particular position, but that competition inevitably results in a more powerful team.

12.6.3 *The Earth Ecosystem*

Increase in ecosystem power output due to autocatalysis and deviation amplifying feedback makes sense when the boundaries of the ecosystem are defined. The problem is that ecosystems exist in a nested series. Figure 2.1 of Appendix 2 illustrates seven ecosystem types within a square kilometer of a tropical rain forest. The system in which jaguars exists encompasses all seven ecosystems. As the scale of the system being studied enlarges from ecosystem to landscape to biome to the whole Earth, the ecosystem concept merges with the Gaia concept. Gaia Earth is an ecosystem on a global scale. Critics have interpreted Gaia as being teleological because it seems to say that evolution deliberately shaped the planet in a way that made it an optimal environment for species, but it was reciprocal feedback that made the result close to, but never optimal for both the species and the Gaian ecosystem (Appendix 8). Species adapted to energy flow through the environment, but the adaptation by species changed the environmental energy flow, and that required further adaptation by species. At no point was energy flow optimum for either species or ecosystems, but there was always evolution toward an optimal state. An optimal state is never reached because movement towards optimality in one part of the system changed what would have been optimal for other parts. Presumably optimization could have occurred for both species and environment if energy input to ecosystems had remained absolutely steady and unwavering throughout geological ages, but we know that past changes in climate resulted in changes of energy input. Change is the only constant in evolution and all we can do is to work towards an optimal state, but we will never achieve it. The energy flow will change before we get there.

12.6.4 Why Is There Resistance to the Idea of Ecosystem Evolution

The idea that ecosystems evolve has not been generally accepted because it seems to contradict the established view that evolution occurs only at the level of species. Critics of the idea that ecosystems evolve point out that DNA is necessary for reproduction and since only species have DNA, ecosystems don't evolve. True, ecosystems don't have DNA, but species within them evolve and species have DNA. Ecosystems change as the species within them evolve. Then why not say simply that ecosystems change instead of evolve? Because evolution implies change in a direction, and ecosystems change in the direction of maximizing power output.

Another issue is that some ecologists have a problem with the idea that ecosystems evolve toward maximum power output as a result of the evolution of species "that have no idea" what their interactions mean for maximizing power output of their ecosystem. They reject any idea that seems to say species deliberately evolve to increase ecosystem power. It is not necessary for species to have any idea about where their evolution is heading for them to affect an increase in ecosystem power. In the more sophisticated field of economics, economists have no problem with the idea that economic systems have evolved from primitive barter systems to computerized trading, and that evolution toward increasing Gross Domestic Product depends on individual decisions of buyers and sellers who have no idea about how their activities affect a Gross Domestic Product (Friedman 1998). Maximum power of systems just happens to be what happens when participants in a system are competitors.

12.6.5 *Evolution of Economic Systems*

Ecosystems and economic systems have much in common, especially with regard to power output. A micro-economic study of the motivations of corporations buying and selling in their own self-interest does not tell us why national economies grow. Only by looking at the macroeconomic picture can we see that it is the interactions between actors in the economic system – interactions that form the invisible hand of the marketplace – that leads to increased power output of the system as a whole. It is the actors competing to make the most money that propels the national economic cycle to grow.

In the economic world, the passing of money from one person or business to another keeps the economy going. When money flow stops, the economy stops. Flow of money keeps people interdependent and the economy going. The motivation of people to seek money is not to keep the economic system functioning, but to improve their own life. It just so happens that people seeking money keeps the economy functioning (economists already know this). Although biologists understand the need for energy to fuel metabolism of individual organisms (Parent 2019), some may not be aware of the need for energy flow to keep ecosystems organized and functioning. The need is because species depend on each other to obtain the energy they need for life. It is the passing of energy from one species to another in food chains that keeps the system organized and functioning. The motivation of species to seek energy is not to keep their ecosystem functioning, but to improve their own life. It just so happens that species seeking energy keeps energy flowing, and that flow of energy keeps species interdependent. That is the mechanism that keeps the ecosystem functioning.

12.7 A Thermodynamic Model of Ecosystem Evolution

Models synthesize disparate information in a way that has meaning. So far, we have discussed competition, power output, autocatalysis, and feedback when working together result in simultaneous evolution of species and ecosystems. Now we present a model that shows how these factors working together influence evolution in a direction that resulted in the complex food webs we observe today. The model does not present details of how particular species and ecosystems evolved. It is a thermodynamic model that shows how and why energy transformations increased during evolution of ecosystems.

12.7.1 Network Models

The thermodynamic model is a network model (Fath et al. 2007). Network theory is concerned with complex networks such as telecommunication networks, computer networks, biological networks, cognitive and semantic networks, and social networks. Networks consist of basic elements called nodes, and the connections between nodes called links. In ecological models, nodes are species or proto-species (aggregates of organic chemicals).

Many ecosystem models are static in the evolutionary sense. They show flows of energy and nutrients only at one point in time. The model presented here shows how ecosystems changed during evolutionary time. The sequence shows how food chains lengthen, food webs increase in diversity, and ecosystems acquire greater power output.

12.7.1.1 The First Ecosystem

The first ecosystem consisted of a eukaryotic producer and a bacterial decomposer. In these first, two-component ecosystems, an autotrophic eukaryotic population produced biomass, and when the older individuals died, they were eaten by decomposers, the prokaryotic bacteria (Fig. 12.1a).

12.7.1.2 Evolution of Trophic Chains and Trophic Levels

Consumer eukaryotes evolved because it was more efficient for them to seek out and eat autotrophic cells as a source of energy than to capture solar energy themselves. But for a new species to be a permanent part of a trophic chain, DNA fixation in the species must be permanent (Seeman 2005).

12.7.1.3 Fixation of Links in Food Chains

While beneficial mutations are the basis for adaptation, not all beneficial mutations are of consequence to the long-term evolutionary outcome. Only those beneficial mutations with large fitness effects achieved fixation (Rozen et al. 2002). Population geneticists have proposed mechanisms that could accelerate fixation of beneficial mutations, but none have proved capable of explaining the fixation of rare beneficial mutations that would enable a new trophic level (Barton 2010; Wylie et al. 2009; James and Jain 2016). That's because the benefit occurred at the ecosystem level, not the species level. The benefit to ecosystems of a longer food chain was increased power output and greater stability (Johnson 1988), a process facilitated by autocatalytic reactions (Hordjik et al. 2010; Hordijk 2013). Longer trophic chains and more complex food webs contributed to increased power output that enabled ecosystems

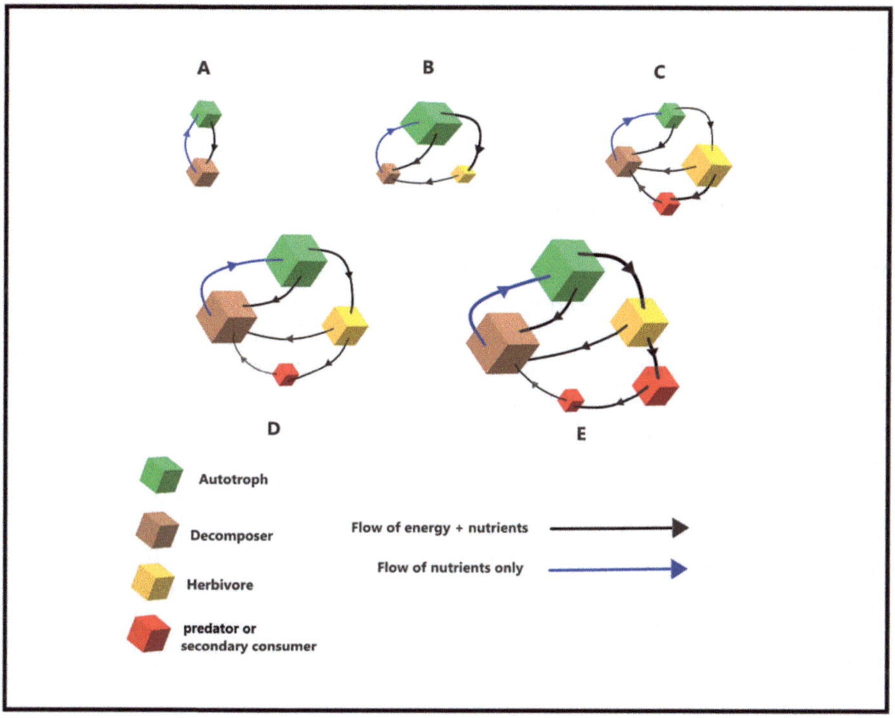

Fig. 12.1 Growth of food chains during evolution. From each trophic level there is flow of energy directly to decomposers. For autotrophs there is energy flow both to the next trophic level and to decomposers. For top predators, energy flow goes directly to decomposers. In marine systems some dissolved organic carbon moves from decomposer bacterio plankton to autotrophs

to develop even longer chains and more complex webs. A system with increased power output increased the chances of capturing resources necessary for survival of all who were part of the system. This selective advantage led to the fixation of a mutation in a new consumer that increased ecosystem power output. The new consumer was like a new player that benefitted the team and so the team protected the new arrival and made sure its position was permanent.

12.7.1.4 Addition of Herbivores

Once a eukaryotic cell joined the ecosystem as a consumer, it added a new route through which nutrients and energy could flow (Fig. 12.1b). The ecosystem functioned more efficiently. The descendants of this consumer evolved into animals. Teleologically speaking, the reason animals came into existence was to facilitate nutrient movement to the decomposers that would increase recycling and increase ecosystem productivity. Their advantage to the ecosystem was greater power output. Because power output increased, ecosystems of type B were better competitors than type A. The combination of autotroph, consumer, and decomposer in this enhanced energy flow enjoyed a selective advantage over cells that were unattached and free-floating or systems with only two nodes. The combination was a more effective way of capturing energy.

12.7.1.5 Addition of Predators

We know from studies of energy flow through present day ecosystems that an uncontrolled herbivore population may eventually destroy their own food base, but if predators are introduced, moderated energy flow will be restored (Ritchie et al. 2012). The same dynamic may have resulted in the appearance of secondary consumers. As herbivore populations increased, they exceeded the ability of autotrophs to sustain them. Herbivore biomass increased, while autotroph biomass decreased (Fig.12.1c) illustrated by the relative mass of the compartments. At this point, systems may have begun to destabilize, because of the imbalance between energy need by herbivores and energy supply by autotrophs. However the increase in herbivore biomass provided an energy source for the first predator, thereby allowing a readjustment between autotroph and herbivore to a more sustainable relationship (Fig.12.1d). As system feedback continued to stimulate increase in biomass, there was enough excess energy to permit the addition of another trophic level (Fig. 12.1e).

12.7.1.6 Minimum Energy Necessary for Food Chain Elongation

Food chains elongated one step at a time. There is some minimum amount of energy required in an ecosystem before a new trophic level can be added (Odum 1971). Once there appeared an autotrophic eukaryote fixing energy at a rate greater than that needed for its own maintenance and replication, there was an opportunity for an herbivore to begin a trophic chain. There was and still is however, a photosynthetic limit to the rate that autotrophs can transform solar energy to biomass. Because each trophic level uses part of the energy flow for metabolism and growth, there is less available energy remaining in each successive transformation. The number of steps in an ecological food chain is limited by the amount of energy supplied by autotrophs at the base (Kay 1989). Trophic chains will elongate as long as there are sufficient nutrients or sufficient energy to interact during autocatalysis to increase power output of the system. Where nutrients are scarce as in oligotrophic lakes, or where energy is limited as in high latitude environs, trophic chains remain stunted.

12.7.2 Increase in Complexity of Trophic Webs

Simultaneous and reciprocal evolution between species and ecosystems resulted in the increase in power output of both, and in the increased complexity of food webs. As the amount of energy passing through a food chain increased, there was greater opportunity for more multicellularity, and greater diversity of organisms within each trophic level (Schirrmeister et al. 2013). With more energy moving through a system, it was divided and used more ways.

12.7.2.1 The "Great Oxidation Event"

In the Precambrian era, conditions on earth were anerobic. Methane was a product of anerobic metabolism, an inefficient pathway. Ecosystems evolved slowly. The "great oxidation event" occurred some 2.4 billion years ago. Cells emerged that could take in sunlight and carbon dioxide and obtain energy, releasing oxygen as a by-product (Knoll and Nowak 2017). In these phototrophic eukaryotes, photosynthesis took place inside a chloroplast, an organelle that arose in eukaryotes by endosymbiosis with a photosynthetic bacterium. Oxygen produced by photosynthetic cyanobacteria began to build up in the atmosphere and poisoned most of the anaerobic bacteria (Marshall 2009). Aerobic decomposition became more important than anaerobic decomposition. Decomposition rate became faster, recycling rate of nutrients increased, autocatalysis at the decomposer-autotroph interface increased, and energy flow through ecosystems increased. This was a major step towards maximizing power in species and ecosystems and increasing the rate of evolution. Species changed the environment, while the environment changed the species. Through feedback, species and their environment (the global ecosystem) evolved simultaneously and reciprocally (Braakman et al. 2017).

12.7.2.2 The Braakman Model

Braakman et al.'s (2017) mechanistic evolutionary model of the globally abundant oceanic phytoplankter *Prochlorococcus* supports evolutionary self-organization of ecosystems. Their model indicates that "*Prochlorococcus* and the dominant co-occurring heterotrophic bacterium SAR11 form a coevolved mutualism that maximizes their collective metabolic rate. This led them to propose a general theory relating metabolic evolution to the self-amplification and self-organization of the biosphere. In the author's words, *"This leads to an evolutionary dynamic that increases total ecosystem biomass and paves the way for self-amplifying feedback loops that recycle organic carbon and reinforce the maximization of cellular metabolic rate. It has been argued that the hierarchical organization of metabolic pathways reflects the outgrowth of self-amplifying feedbacks that increased the free energy consumption of the emerging biosphere. Our framework extends those arguments into the world of phenotypically differentiated cells and microbial ecosystems, and is consistent with the theorem that the flow of energy through the biosphere promotes its self-organization into chemical cycles".*

12.7.3 Evolution of Trophic Webs

Lindeman (1942) proposed a trophic-dynamic view of ecology where nodes (species) are linked sequentially in a line between trophic levels that represent food chains. That was the first but simplistic step in understanding energy flow through ecosystems. Klaise and Johnson (2017) found a tendency of species to organize in interactions called motifs. Ecosystem evolution begins with small groups of nodes coming together to form "network motifs" (Mangan and Alon 2003). A basic and recurrent motif in all networks consists of three nodes, an input node, an intermediate node and an output node. When motifs come together in clusters, the redundant pathways between producers and consumers can buffer perturbations thereby rendering the system resilient against external damage (Gilarranz et al. 2017). Three motifs that might occur frequently in food webs are A, B, and C in Fig. 12.2 (top). In motifs A and B, the input node is an autotroph, the intermediate node is an herbivore and the output node is an omnivore. In motif C, the input is a small omnivore, the intermediate is a large omnivore, and the output is a predator. These "feed forward" motifs come together in various clusters (Fig. 12.2 bottom), a term used to characterize and understand the architecture of living and artificial systems (Vespignani 2018). Note that the path of energy to decomposers is not indicated. It exists in a third dimension where energy flows from each node to the decomposer, and nutrients flow from the decomposer node to the autotroph as in Fig. 12.1.

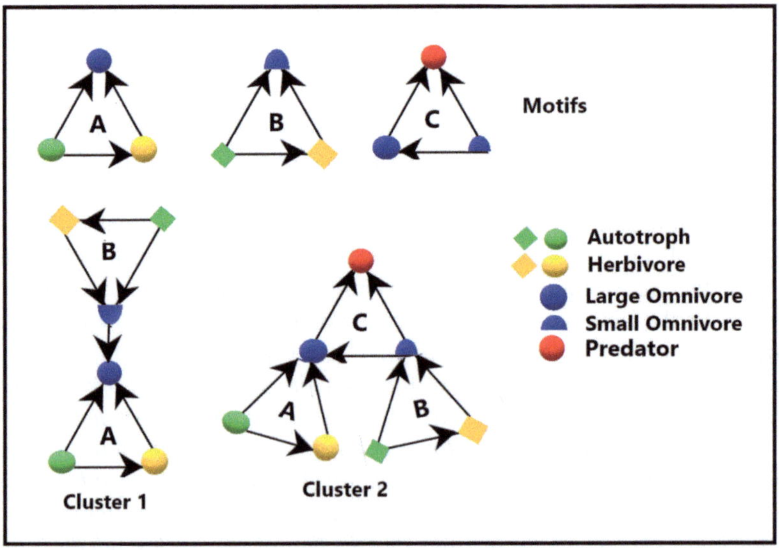

Fig. 12.2 Organization of trophic groups in trophic webs. Clusters typical of ecological food webs consist of basic motifs where energy is transferred from an autotroph through an herbivore to an omnivore and predator. Arrows indicate direction of energy flow

Gorchowski et al. (2018) analyzed 12 different types of networks: *E. coli* metabolism, A. *fulgidus* metabolism, *S. cerevisiae* transcription, *C. elegans* neural, Wikipedia vote, Gnutella file-sharing, Little Rock Lake food web, Air traffic control, European Union email, and two random networks. They found that in the lake food web, cluster, types 1 and 2 (Fig. 12.2 bottom) were most frequent.

Each cluster is made up of several different motifs. In cluster one, the longest chain contains three links: small autotroph to small herbivore; small herbivore to small omnivore; small omnivore to large omnivore. In cluster two, the longest chain contains four links: small autotroph to small herbivore; small herbivore to small omnivore; small omnivore to large omnivore; large omnivore to predator. Longer trophic chains require more energy, but chains with more energy are able to outcompete chains with less. The more you have, the more you get. The most important thing for the survival of cluster types would be a large energetic output in the form of growth, reproduction, and maintenance. Cluster types with a high output relative to their size should win out in the competitive struggle for energy. Cluster 2 is more likely to survive than cluster 1, and therefore be more available for further evolutionary selection. Important in these food webs is the presence of omnivory. Omnivory provides alternative paths of energy flow through trophic webs, should a key species in a pathway become extinct (Stouffer and Bascompte (2010).

As species in the clusters of Fig. 12.2 evolved into multicellular organisms, the cluster itself evolved into a microbial food loop (Fig. 5.4 in Chap. 5). As the microbial food web grew in an aquatic environment, consumers evolved to harvest abundant mesoplankton. That pathway gave rise to higher trophic levels based on filter feeders and carnivorous predators, as in Chap. 4, Fig. 4.4. The result was a bigger, more powerful system because each newcomer acted as a link that added a new route through which nutrients and energy could flow. Ecosystems evolved as the species that comprise them evolved. Increase in ecosystem power output resulted from species with greater power output replacing those with lesser power output, just like football teams become more powerful as stronger players replace weaker players.

12.7.3.1 Reinforcing Design Motifs

Reinforcing design motifs and clusters do not come together simultaneously, but modularly, one motif at a time, one cluster at a time, one aggregation at time. The thermodynamic theater was created like a builder building a house. A modern builder does not have to invent all the electrical, heating and plumbing fixtures every time a house is built, but merely puts together already existing modules that work best. Likewise, once the first interaction in a primitive ecosystem is stabilized, it is used in combination with another interaction to build a more powerful and more competitive system. Once a pattern is established, it is built upon by reinforcing designs. Odum (1995) explained it this way: *"In this process, units at one level of organization are chosen in large part by the reinforcement action of some other unit from a higher level of organization, one with longer turnover time and larger territory*." In this sense, a motif selects the species that increases its power output, a cluster selects the motif that increases its power output, likewise for aggregates and for ecosystems. A priority for component species to be selected is their contribution to power output of the system. In the football analogy, the left tackle that best increases the power output of an offensive line will be chosen by the offensive coordinator, and the offensive line that best increases the power output of the team will be chosen by the head coach. A winning percentage will ensure that the respective teams, lines, and players will have their contracts renewed.

12.7.3.2 A Test of the Model

Jørgensen and Fath (2006) proposed that network changes yielding the highest exergy (free energy) storage and throughflow will have a selective advantage and would be more favorable to the overall system organization. They tested their hypothesis using a network analysis that consisted of component parts of complex systems connected in ways that reflect different interdependencies and interrelationships. Their network analysis failed to reject the network evolution hypothesis, so the theory remains useful (sensu Popper 1959). Ecosystems with species interactions that are most favorable for high power output are the ones most likely to persist.

12.7.4 Life Moves Ashore

Approximately 465 million years ago, plants began colonizing the land. By that time, the ancestors of modern animals were already developed (Marshall 2009; Vandamme 2019). After plant colonization of land, the first four legged animals or tetrapods emerged, probably in shallow freshwater habitats. The popular image is that animals were the first to emerge from the water and colonize land (Fig. 12.3). But what prompted them to emerge? The thermodynamic answer is, they were following the energy. Plants were the first to colonize land, and plants were the energy source that drew up tetrapods from the depths. Evidence for herbivory is found in the fossil record simultaneous with the first land plants (Labandeira 1998, 2007).

12.7.4.1 Evolution of Terrestrial Ecosystems

In aquatic systems, the autocatalytic cycle that prompted evolution occurred in locations where nutrients were not limiting to the phytoplankton at the base of the food chain. Nutrient cycles were loose and not confined to a restricted space as were

Fig. 12.3 "Evolution in a Nutshell". (Reprinted from Bored Panda (2016) with permission from Bizarro comics)

ecosystems depicted in Fig. 4.2, Chap. 4. In terrestrial ecosystems, nutrients are frequently limiting, and consequently the nutrient cycles are "tight" (few leaks from the system). Development of food chains was different from chains in aquatic systems.

For consumer species to be fixed as part of the ecosystem, there had to be an evolutionary advantage for the system. In terrestrial ecosystems, the advantage was an increase in nutrient recycling that increased ecosystem power output. The incorporation of an herbivore occurred because there often is an impediment to the breakdown of plant biomass and recycling of nutrients. That impediment is defensive chemicals produced by the plant to protect them from decomposers that attack them while they are still alive. To defend themselves, autotrophs presumably produced secondary metabolites similar to those in modern day plants (Freeman and Beattie 2008) as a defense. As a result, autotrophic biomass increased (larger cube in Fig. 12.1b). That presented an energy source for consumer organisms that could overcome plant defenses and digest plant material. Herbivory also benefitted the ecosystem because the compounds that inhibited decomposition in living plants also inhibited decomposition in the soil and deterred nutrient recycling. Herbivory, especially by animals possessing a rumen, overcame this inhibition. Ruminate animals break down plant material more completely than non-ruminate animals. When herbivores egest fecal material or when they die, they supply soil bacteria with a rich source of energy that lack the defenses of plants. By increasing the rate of nutrient recycling, herbivores increased the power output of the ecosystem.

12.8 Biodiversity and the Five Great Extinctions

There have been five major spasms of biodiversity loss since the Great Oxygenation Event 2.4 billion years ago. The first extinction was in the Ordovician Period of the Paleozoic Era, 440 million years ago possibly due to an ice age. The second was during the Devonian Period of the Paleozoic Era, about 375 million years ago, caused by a lack of oxygen in the oceans, quick cooling of air temperatures, and volcanic eruptions and meteor strikes. The end-Permian mass extinction, 251 million years ago, brought about by temperature increase and ocean acidification (Shen et al. 2016), was the most devastating ecological event of all time. It was exacerbated by two earlier events, 270 and 260 million years ago (Sahney and Benton 2008). After the end-Permian, there was a major recovery evidenced by new floral species across global landscape (Gastaldo 2019). At the end of the Triassic Period of the Mesozoic Era, about 200 million years ago, the fourth major mass extinction wiped out more than half of all known species living at the time. The cause was major volcanic activity and global climate change.

12.8.1 The Cretaceous-Tertiary (K-T) Boundary Extinction

Sixty-six million years ago, a city-sized asteroid smashed into planet Earth near what is now Chicxulub in the Yucatan Peninsula of present-day Mexico. Atmospheric aerosols resulted in the demise of most of the dinosaurs that had reigned for almost 140 million years, and caused the extinction of 75% of plant and animal species including all non-avian dinosaurs. However, the extinction was not immediate. Fassett (2009) analyzed dinosaur bones found in the Ojo Alamo Sandstone in the San Juan Basin of New Mexico. Based on the age of the rocks in which the bones were found, it appears that some dinosaurs outlived the crash. It took five million years of the Tertiary for the dinosaurs to become extinct (Archibald and Clemens 1982). Meanwhile, mammals began playing important roles in trophic food webs. Herbivorous species of dinosaurs were replaced by herbivorous species of mammals (Sahney and Benton 2008). Within approximately 100,000 years, mammalian taxonomic richness doubled, and their maximum body size increased (Lyson et al. 2019). The survivors included some small, squirrel-like animals that were to give rise to the dominant life forms of the next era of fur-bearing, warm-blooded mammals.

12.8.2 The Amazing Sustainability of Trophic Chains

The entropy-capturing force of living systems is truly amazing. It enabled trophic chains to survive asteroid impacts and other catastrophes that eliminated species that dominated Earth for millions of years. The major extinction events in the

Earth's geological history (Donovan 1989) partially degraded many trophic chains, but they were not powerful enough to destroy them (Song et al. 2018). Ecosystems were reduced to states with lower diversity and consequently lower flows of energy, but the basic thermodynamic structure remained intact. As species became extinct, other functionally similar species evolved and replaced them. The interactions critical for maintaining energy flow and nutrient recycling persisted. However recovery to complexity and high-power output took "1.8 million years of exceptional volatility after the Chicxulub bolide impact before a more stable equilibrium community emerged that displayed hallmarks of resilience" (Alvarez et al. 2019). In some cases it required at least ten million years of evolutionary change to reestablish the lost diversity (Kirchner and Weil 2000).

Blanco et al. (2021) analyzed a series of mammalian faunas spanning the past 21 million years of dynamic environmental changes on the Iberian Peninsula and demonstrated increasing persistence of functional systems over time and the decoupling of those systems from species composition and duration. These findings are evidence of the persistence of ecosystem functioning despite disturbances that cause species loss, and that this persistence makes it easier for new species to evolve. Trophic interactions are still present, and all a new species has to do is fit in. Because of organizational structure that was conserved during perturbations, development of trophic chains did not have to revert to the basic eukaryote – bacterium system. Ecosystems were able to resume development from a more complex state, but each time with mostly different species. The fact that trophic structures not only remained despite the degradation going on all around them but also resumed increasing diversity and productivity is evidence of the power of this entropy-producing, gradient-reducing force that maintains life. Kauffman (2000) argued that this behavior of the biosphere is so law-like and pervasive that it could be considered a fourth law of thermodynamics.

12.8.3 A Test of Thermodynamic Theory

To test the idea that thermodynamic laws and principles underlie evolution, we use the increase in species diversity during evolution as a proxy for increase of power output. High diversity of participants means that there is a high diversity of interactions in the system. High diversity of interactions results in more powerful systems because high diversity provides more ways that energy can flow through the system. The relationship is reciprocal. An increase in energy flow results in an increase in species populations and/or an increase in the rate of growth of individuals within the population (Chap. 6). Species diversity in ecosystems then is an indicator of power output. Exceptions are in the autotrophic level of marine systems and tidal marshlands. There, autotrophic diversity is often low, despite high primary productivity. The positive correlation between energy flow and species diversity exists in these systems at the consumer trophic levels (Wu et al. 2009).

12.8.3.1 The Study

To study the influence of environmental change on biodiversity during evolution, Fan et al. (2020) used stratigraphic correlations to construct composite biodiversity curves for Cambrian to Triassic marine invertebrate genera and species. They found three distinct cycles of growth, consolidation, collapse and reorganization over 250 million years (Fig. 12.4). The first cycle began with a sharp increase in diversity

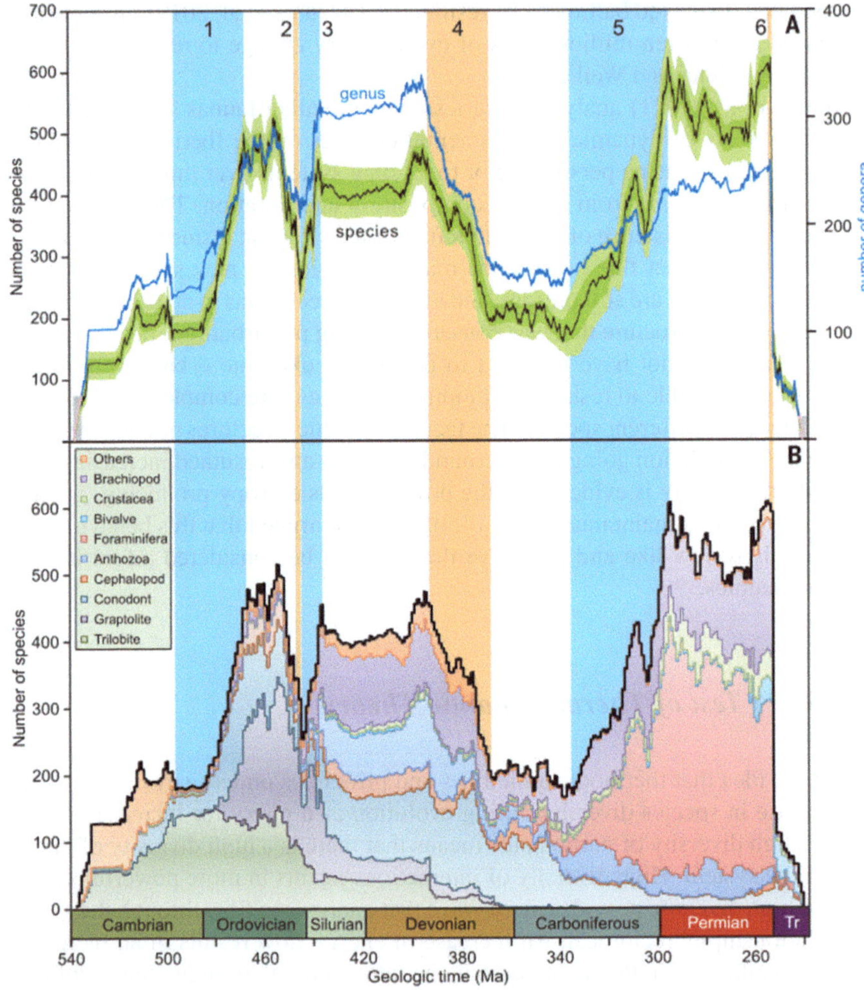

Fig. 12.4 General trajectories of Paleozoic genus and species diversity of marine fossil species (**A**) and species diversity for 10 major fossil groups (**B**). The light and dark green shading in **a** represent 1 and 2 standard deviations. Vertical blue bars indicate periods of increasing diversity, tan bars indicate extinction events. 500–480 = Great Ordovician Biodiversification Event; 460–450 = End-Ordovician mass extinction; 440 = Earl Silurian radiation; 390–370 = Middle to Late Devonian diversity decline; 340–300 = Late Carboniferous-early Permian biodiversification event; 255 = End-Permian mass extinction. (Reprinted from Fan et al. (2020) with permission from Science)

associated with the Cambrian explosion (growth phase), followed by a plateau (consolidation phase) during the Ordovician. This is where the evolutionary "goal" switched from growth to stability. Next was a rapid diversity decline during the end-Ordovician mass extinction (collapse phase), and a subsequent reorganization and beginning of the next growth phase at the beginning of the Silurian. Consolidation and subsequent collapse occurred in the Devonian and then a period of reorganization during the Carboniferous. The third growth phase was during the Late Carboniferous-early Permian biodiversification event, consolidation occurred during the Permian, and collapse happened during the End-Permian mass extinction event.

If we use high species diversity as an indicator of high energy flow through ecosystems, of high biomass, and of maturity of a system, and as an indicator that changes in diversity reflect changes in these parameters, the patterns reflect the thermodynamic laws and principles that underlie dynamics of all systems:

- The maximum power principle which says that systems evolve in the direction of increasing growth and power output.
- The second law of thermodynamics that says when energy leaving a system equals energy entering the system, growth ends and consolidation begins. When energy leaving the system is greater, the system degrades.
- The principle that the entropy-producing, gradient-reducing force that maintains life is behind the tendency for any system that is disassembled to reform itself and arise from the remains of the previous system.

12.9 Panarchy and Evolution

Figure 12.4 shows cycles of species diversity during ecosystem evolution that reflect cycles of energy flow as predicted by Panarchy theory. At the beginning of a cycle, diversity and energy flow evolved in a feedback loop, an increase in one facilitating an increase in the other. An increase in diversity was contingent upon an increase in energy flow, and an increase in energy flow prompted an increase in diversity. An increase in species diversity resulted in more complex food webs that in turn enabled more energy flow through ecosystems. Greater flow enabled the energy to be divided more ways. This enabled more complex food webs and development of new species. This deviation-amplifying system came about as a result of the reciprocal evolution of species and the ecosystem in which they resided.

Once ecosystems reached a maximum marked by the right-hand edge of the blue vertical bars in Fig. 12.4, consolidation replaced growth as the evolutionary strategy. After each phase of increasing diversity, there followed periods of tens of million years in which ecosystems were relatively stable, and diversity changed little. The increase in species diversity at the beginning of the Ordovician indicates that there was an increase in energy flow and biomass during that period. A period of energy stability during mid Ordovician is indicated by a plateau in species diversity, and a collapse at the end of the period is signaled by a sharp drop in species diversity. The cycle was repeated during the Silurian- Devonian and Carboniferous-Permian periods. Five times in the last 2.5 billion years, after ecosystems reached a maximum biomass, a catastrophic global event changed the energy environment to which most species were unable to adapt, resulting in ecosystem collapse followed by reorganization with new species (Donovan 1989; Jablonski 1995; Scoville 2018) and eventual return to consolidation.

12.10 Thermodynamic Requirements for Living Systems on Other Planets

In his paper entitled The First Three Billion Years of Community Evolution, Simpson (1969) defined the “relay and replacement” of continuing change in Earth’s species as “ecosystem evolution”. For example, on page 172 he states that “In land communities, the evolution of ecosystems, including the universal phenomenon of relay and replacement, is reflected in the changing systematics and ecological makeup of Cenozoic mammalian faunas.” His use of the term “ecosystem” simply meant the agglomeration of species that happened to exist at a particular time in evolution, quite different from how the term is used here for the feedback system of nutrients and energy that sustains the lives of participating species. However, during the discussion at the end of the talk, Bernie Patten, a pioneer in ecosystem ecology, asked Simpson: “Do you think it would be possible to go to another planet and find a system such as I described, an autotrophic and decomposing system?” Simpson answered, “The decomposers would almost certainly be primary heterotrophs that stayed on when the autotrophs arose and become integrated into the system as the decomposers.”

References

E. Abouheif, M.J. Favé, A.S. Ibarrarán-Viniegra, M.P. Lesoway, A.M. Rafiqi, R. Rajakumar, Eco-evo-devo: The time has come. Adv. Exp. Med. Biol. **781**, 107–125 (2014). https://doi.org/10.1007/978-94-007-7347-9_6

P.A. Abrams, The evolution of predator-prey interactions: Theory and evidence. Annu. Rev. Ecol. Syst. **31**, 79–105 (2000)

S.A. Alvarez, S.J. Gibbs, P.R. Brown, H. Kim, R.M. Sheward, A. Ridgwell, Diversity decoupled from ecosystem function and resilience during mass extinction recovery. Nature **574**, 242–245 (2019)

J. André, B. Godelle, The evolution of mutation rate in finite asexual populations. Genetics **172**(1), 611–626 (2006)

J.D. Archibald, W.A. Clemens, Late cretaceous extinctions: The fossil evidence currently available suggests that changes across the cretaceous-tertiary boundary were the result of gradual processes and not of a single catastrophic event. Am. Sci. **70**(4), 377–385 (1982). www.jstor.org/stable/27851545

N.H. Barton, Mutation and the evolution of recombination. Philos. Trans. R. Soc. B **365**, 1281–1294 (2010). https://doi.org/10.1098/rstb.2009.0320

F. Blanco, J. Calatayud, D. Martin-Perea, et al., Punctuated ecological equilibrium in mammal communities over evolutionary time scales. Science **372**(6539), 300–303 (2021)

Bored Panda, (2016) https://www.boredpanda.com/funny-satirical-evolution-charles-darwin-day. Accessed 7 Dec 2020

R. Braakman, M.J. Follows, S.W. Chishold, Metabolic evolution and the self-organization of ecosystems (2017), pnas.org/cgi/doi/10.1073/pnas.1619573114

J.C. Bregliano, *Theory of Evolution: Misunderstandings and Resistance. Encyclopedia of the Environment* (Université Grenoble Alpes, Gières, 2019). www.univ-grenoble-alpes.fr. https://www.encyclopedie-environnement.org/en/life/theory-of-evolution-misunderstandings-and-resistance/. Accessed 12 Dec 2020

J. Brennan, Did heterotrophs evolve from autotrophs? (2019), sciencing.com. https://sciencing.com/did-heterotrophs-evolve-autotrophs-23027.html. Accessed 15 Dec 2020

J.H. Brown, *Macroecology*. Univ. of Chicago Press. Chicago, (1995)

J.H. Brown, C.A.S. Hall, R.M. Sibly, Equal fitness paradigm explained by a trade-off between generation time and energy production rate. Nat. Ecol. Evol. **2**, 262–268 (2018)

R.K. Butlin, J. Galindo, J. Grahame, Sympatric, parapatric, or allopatric: the most important way to classify speciation? Philos. Trans. R Soc. B. (2008). https://doi.org/10.1098/rstb.2008.0076

C.Q. Choi, Human evolution: the origin of tool use. Live Science (2009), https://www.livescience.com/7968-human-evolution-origin-tool.html. Accessed 12 Dec 2020

J.M. Cohen, E.L. Sauer, O. Santiago, S. Spencer, J.R. Rohr, Divergent impacts of warming weather on wildlife disease across climates. Science **370**(6519), 933 (2020)

G.A. Cooper, S.A. West, Division of labour and the evolution of extreme specialization. Nat. Ecol. Evol. **2**, 1161–1167 (2018)

R. Dawkins, *The Selfish Gene* (Oxford University Press, Oxford, 1976)

S.K. Donovan, *Mass Extinctions: Processes and Evidence* (Columbia University Press, New York, 1989)

Editorial, Linking evolution across scales. Nat. Ecol. Evol. **3**, 1369 (2019)

T.M. Embley, T.A. Williams, Steps on the road to eukaryotes. Nature **521**(7551), 169–170 (2015). https://doi.org/10.1038/nature14522

J.L. England, Statistical physics of self-replication. J. Chem. Phys. **139**, 121923 (2013). https://doi.org/10.1063/1.4818538. Accessed 17 Dec 2020

J.L. England, *Every Life Is on Fire: How Thermodynamics Explains the Origins of Living Things* (Basic Books, New York, 2020)

D.S. Falconer, *Introduction to Quantitative Genetics*, vol 2 (Longman House, Essex, 1981)

J. Fan, S. Shu-zhong, D.H. Erwin, P.M. Sadler, et al., A high-resolution summary of Cambrian to Early Triassic marine invertebrate biodiversity. Science **367**(6475), 272–277 (2020)

J.E. Fassett, New geochronologic and stratigraphic evidence confirms the Paleocene age of the dinosaur-bearing Ojo Alamo sandstone and Animas formation in the San Juan basin, New Mexico and Colorado (2009), https://palaeo-electronica.org/2009_1/149/index.html. Accessed 17 Dec 2020

B.D. Fath, U.M. Scharler, R.E. Ulanowicz, B. Hannon, Ecological network analysis: Network construction. Ecol. Model. **208**, 49–55 (2007)

R. Ferriere, S. Legendre, Eco-evolutionary feedbacks, adaptive dynamics and evolutionary rescue theory. Philos Trans. R Soc. Lon B. **368**(1610) (2013). https://doi.org/10.1098/rstb.2012.0081

R.A. Fisher, E.B. Ford, The Sewall Wright effect. Heredity **4**, 117–119 (1950)

B.C. Freeman, G. Beattie, *An Overview of Plant Defenses against Pathogens and Herbivores* (The Plant Health Instructor. Iowa State University Digital Repository, 2008). https://doi.org/10.1094/PHI-I-2008-0226-01. Accessed 19 Dec 2020

D. Friedman, Evolutionary economics goes mainstream. J. Evol. Econ. **8**(4), 423–432 (1998)

R.A. Gastaldo, Plants escaped an ancient mass extinction. Nature **567**, 37–39 (2019)

L.J. Gilarranz, B. Rayfield, G. Liñán-Cembrano, J. Bascompte, A. Gonzalez, Effects of network modularity on the spread of perturbation impact in experimental metapopulations. Science **357**(6347), 199–201 (2017)

T.E. Gorchowski, C.S. Grierson, M. di Bernardo, Organization of feed-forward loop motifs reveals architectural principles in natural and engineered networks. Sci. Adv. **4**(3), eaap9751 (2018). https://doi.org/10.1126/sciadv9751. Accessed 19 Dec 2020

R. Herrera, T. Merida, N. Stark, C.F. Jordan, Direct phosphorus transfer from leaf litter to roots. Naturwissenschaften **65**, S.208 (1978)

D.U. Hooper, F.S. Chapin III, J.J. Ewel, et al., Effects of biodiversity on ecosystem functioning: A consensus of current knowledge. Ecol. Monogr. **75**(1), 3–35 (2005)

W. Hordijk, Autocatalytic sets: From the origin of life to the economy. Bioscience **63**(11), 877–881 (2013). https://doi.org/10.1525/bio.2013.63.11.6

W. Hordjik, J. Hein, M. Steel, Autocatalytic sets and the origin of life. Entropy **12**(7), 1733–1742 (2010). https://doi.org/10.3390/e12071733

J.M. Horowitz, J.L. England, Spontaneous fine-tuning to environment in many-species chemical reaction networks. PNAS **114**(29), 7565–7570 (2017)

D. Jablonski, Extinction in the fossil record, in *Extinction Rates*, ed. by J. H. Lawton, R. M. May, (Oxford University Press, Oxford, 1995), pp. 25–44

J.B. Jackson, Natural pH gradients in hydrothermal alkali vents were unlikely to have played a role in the origin of life. J. Mol. Evol. **83**(1), 1–11 (2016)

A. James, K. Jain, Fixation probability of rare nonmutator and evolution of mutation rates. Ecol. Evol. **6**(3), 755–764 (2016). https://doi.org/10.1002/ece3.1932

E.J. Javaux, Challenges in evidencing the earliest traces of life. Nature **572**(77700), 451–460 (2019)

L. Johnson, The thermodynamic origin of ecosystems: A tale of broken symmetry, in *Entropy, Information, and Evolution*, ed. by B. H. Weber, D. J. Depew, J. D. Smith, (MIT Press, Cambridge, 1988), pp. 75–105

S.E. Jørgensen, B.D. Fath, Examination of ecological networks. Ecol. Model. **196**(3,4), 283–288 (2006). https://doi.org/10.1016/jecolmodel.2006.02.029

S.A. Kauffman, *Investigations* (Oxford University Press, New York, 2000)

S. Kay, in *A Thermodynamic Perspective of the Self-Organization of Living Systems*, ed. by P.W.J. Ledington, Proceedings of the international Society for the Systems Sciences, 33rd Annual Meeting, 2–7 July 1989. p 24–30

J.W. Kirchner, A. Weil, Delayed biological recovery from extinctions throughout the fossil record. Nature **404**, 177–180 (2000)

J. Klaise, S. Johnson, The origin of motif families in food webs. Sci. Rep. **7**, 16197 (2017). https://doi.org/10.1038/s41598-017-15496-1

A.H. Knoll, M.A. Nowak, The timetable of evolution. Sci. Adv. **3**(5), e1603076 (2017). https://doi.org/10.1126/sciadv.1603076

C.C. Labandeira, Early history of arthropod and vascular plant associations 1. Annu. Rev. Earth Planet. Sci. **26**(1), 329–377 (1998). https://doi.org/10.1146/annurev.earth.26.1.329

C. Labandeira, The origin of herbivory on land: Initial patterns of plant tissue consumption by arthropods. Insect Sci. **4**(4), 259–275 (2007). https://doi.org/10.1111/j.1744-7917.2007.00152.x

N. Lane, W.F. Martin, The origin of membrane bioenergetics. Cell **151**(7), 1406–1416 (2012). https://doi.org/10.1016/j.cell.2012.11.050

S.A. Levin, *Fragile Domain: Complexity and the Commons* (Persus Publishing, Cambridge, 1999)

Y. Li, J. Roman, D.S. Wilcove, T. Male, H. Doremus, Species protection will take more than rule reversal: Key improvements are needed for implementation of the Endangered Species Act. Science **370**(6517), 665–666 (2020)

E. Libby, W.C. Ratcliff, Ratcheting the evolution of multicellularity. Science **346**(6208), 426–427 (2014). https://doi.org/10.1126/science.1262053

T.A. Lincoln, G.F. Joyce, Self-sustained replication of an RNA enzyme. Science **323**(5918), 1229–1232 (2009). https://doi.org/10.1126/science.1167856

R.L. Lindeman, The trophic-dynamic aspect of ecology. Ecology **23**, 399–418 (1942). https://doi.org/10.2307/1930126

M. Loreau, C. de Mazancourt, R.D. Holt, Ecosystem evolution and conservation, in *Evolutionary Conservation Biology*, ed. by R. Ferrière, U. Dieckmann, D. Couvet, (Cambridge University Press, 2004), pp. 327–343

A.J. Lotka, Contributions to the energetics of evolution. PNAS **8**(6), 147–151 (1922)

T.R. Lyson, I.M. Miller, A.D. Bercovici, et al., Exceptional continental record of biotic recovery after the Cretaceous-Paleogene mass extinction. Science **366**(6468), 966–971 (2019)

S. Mangan, U. Alon, Structure and function of the feed-forward loop network motif. PNAS **100**(21), 11980–11985 (2003). https://doi.org/10.1073/pnas.2133841100

A. Mann, Inner workings: Making headway with the mysteries of life's origins. PNAS **118**(16) (2021) https://www.pnas.org/content/118/16/e2105383118. Accessed 23 April 2021

M. Marshall, Timeline: The evolution of life (2009), https://www.newscientist.com/article/dn17453-timeline-the-evolution-of-life/. Accessed 26 Dec 2020

M. Marshall, The water paradox and the origins of life. Nature **588**(7837), 210–213 (2020) https://www.nature.com/articles/d41586-020-03461-4

W.F. Martin, F.L. Sousa, Early microbial evolution: The age of anaerobes. Cold Spring Harb. Perspect. Biol. **8**(2), a018127 (2015). https://doi.org/10.1101/cshperspect.a018127. Accessed 26 Dec 2020

A. Meyer, A. Schloissnig, P. Franchini, et al., Giant lungfish genome elucidates the conquest of land by vertebrates. Nature **590**, 284–291 (2021)

MIT, Physics of Living Systems at MIT (2020), https://cpls.scripts.mit.edu/wordpress/. Accessed 26 Dec 2020

E. Mossel, M. Steel, Random biochemical networks and the probability of self-sustaining autocatalysis. J. Theor. Biol. **233**(3), 327–336 (2005)

D.J. Murphy, M. Carbajales-Dale, D. Moeller, Comparing apples to apples: Why the net energy analysis community needs to adopt the life-cycle analysis framework. Energies **9**, 917 (2016). https://doi.org/10.3390/en9110917

D. Naish, How many white rhino species are there? The conversation continues. Sci. Am. (2017) 6 Nov 2017. https://blogs.scientificamerican.com/tetrapod-zoology/how-many-white-rhino-species-are-there-the-conversation-continues/

D.C. Nepstad, C.R. de Carvalho, E.A. Davidson, P.H. Jipp, P.A. Lefebvre, G.H. Negreiros, E.D. da Silva, T.A. Stone, S.E. Trumbore, S. Vieira, The role of deep roots in the hydrological and carbon cycles of Amazonian forests and pastures. Nature **372**, 666–669 (1994)

N.N. Nielsen, B.D. Fath, J.C. Marques, F. Müller, B. Patten, R.E. Ulanowicz, S.E. Jørgensen, E. Tiezzi, *A New Ecology Systems Perspective* (Elsevier, Amsterdam, 2020)

V. Novotny, P. Drozd, S.E. Miller, M. Kulfan, M. Janda, Y. Basset, G.D. Weiblen, Why are there so many species of herbivorous insects in tropical rainforests? Science **313**(5790), 1115–1118 (2006)

H.T. Odum, *Environment, Power and Society* (Wiley, New York, 1971)

H.T. Odum, Self-organization, transformity, and information. Science **242**, 1131–1139 (1988)

H.T. Odum, Self-organization and maximum power, in *Maximum Power*, ed. by C. A. S. Hall, (University Press of Colorado, Niwot, 1995), pp. 311–330

H.T. Odum, R.C. Pinkerton, Time's speed regulator: The optimum efficiency for maximum power output in physical and biological systems. Am. Sci. **43**, 331–343 (1955)

E. Parent, The Living Systems Theory of James Grier Miller (2019), http://www.isss.org/primer/asem14ep.html. Accessed 30 Dec 2020

B.C. Patten, Environs, emergy, transformity, and energy value, in *Maximum Power: The Ideas and Applications of H.T. Odum*, ed. by C. S. Hall, (University Press of Colorado, Niwot, 1995), pp. 255–278

F. Pelletier, D. Garant, A.P. Hendry, Eco-evolutionary dynamics. Philos. Trans. R. Soc. Lond. Ser. B Biol. Sci. **364**(1523), 1483–1489 (2009). https://doi.org/10.1098/rstb.2009.0027

E. Pennisi, The power of many. Science **360**(6396), 1388–1391 (2018). https://doi.org/10.1126/science.360.6396.1388
J. Phillips, Goal functions in ecosystem and biosphere evolution. Prog. Phys. Geogr. **32**(1), 51–64 (2008)
L.R. Pomeroy, The ocean's food web, a changing paradigm. Bioscience **24**(9), 499–504 (1974)
K. Popper, *The Logic of Scientific Discovery* (Routledge, London/New York, 1959)
E.G. Ritchie, B. Elmhagen, A.S. Glen, M. Letnic, G. Ludwig, R.A. McDonald, Ecosystem restoration with teeth: What role for predators? Trends Ecol. Evol. **27**(5), 265–271 (2012). https://doi.org/10.1016/j.tree.2012.01.001
A. Robertson, Selection in animals: Synthesis. Cold Spring Harb. Symp. Quant. Genet. **20**, 225–229 (1955)
A. Robertson, The nature of quantitative genetic variation, in *Heritage from Mendel*, ed. by A. Brink, (The University of Wisconsin Press, Madison, 1967), pp. 265–280
A. Robertson, The spectrum of genetic variation, in *Population Biology and Evolution Syracuse University Press*, ed. by R. C. Lewontin, (Syracuse, New York, 1968), pp. 5–16
D.E. Rozen, J. Arjan, G. de Visser, Fitness effects of fixed beneficial mutations in microbial populations. Curr. Biol. **12**, 1040–1045 (2002)
S. Sahney, M.J. Benton, Recovery from the most profound extinction of all time. Proc. Biol. Sci. **275**(1636), 759–765 (2008). https://doi.org/10.1098/rspb.2007.270
B.E. Schirrmeister, J.M. de Vos, A. Antonelli, H.C. Bagheri, Evolution of multicellularity coincided with increased diversification of cyanobacteria and the Great Oxidation Event. PNAS **110**, 1791–1796 (2013). https://doi.org/10.1073/pnas.1209927110
E.D. Schneider, D. Sagan, *Into the Cool* (University of Chicago Press, Chicago, 2005)
E. Schrödinger, *What Is Life? The Physical Aspect of the Living Cell* (Cambridge University Press, Cambridge, 1944)
H. Scoville, The 5 major mass extinctions (2018), https://www.thoughtco.com/the-5-major-mass-extinctions-4018102. Accessed 4 Jan 2021
N.C. Seeman, Structural DNA nanotechnology: An overview. Method Mol. Biol. **303**, 143–166 (2005)
S. Shen, J. Jin, G.R. Shi, Ecosystem evolution in deep time. Evidence from the rich Paleozoic fossil records of China. Palaeogeogr. Palaeoclimatol. Palaeoecol. **448**, 1–3 (2016)
L. Shenvav, D. Zeevi, Resource conservation manifests in the genetic code. Science **370**(6517), 683–687 (2020)
G.G. Simpson, The first three billion years of community evolution, in *Diversity and Stability in Ecological Systems. Report of a Symposium Held May 26–28, 1969*, ed. by G. M. Woodwell, H. H. Smith, (Biology Department, Brookhaven National Laboratory, Upton, 1969), pp. 162–177
H. Song, P.B. Wignall, A.M. Dunhill, Decoupled taxonomic and ecological recoveries from the Permo-Triassic extinction. Sci. Adv. **4**(10), eeat5091 (2018). https://doi.org/10.1126/sciadv.aat5091
H. Spencer, *Principles of Biology* (Appleton, New York, 1864)
D.B. Stouffer, J. Bascompte, Understanding food-web persistence from local to global scales. Ecol. Lett. **13** (2010). https://doi.org/10.1111/ele.2010.13.issue-2,154-161
H.D. Thoreau, *Walden* (Ticknor and Fields, Boston, 1854)
M.M. Turcotte, M.S.C. Corrin, M.T.J. Johnson, Adaptive evolution in ecological communities. PLoS Biol. **10**(5), e1001332 (2012). https://doi.org/10.1371/journal.pbio.1001332
R.A. Ulanowicz, The dual nature of ecosystem dynamics. Ecol. Model. **220**, 1886–1892 (2009)
Understanding Evolution, From prokaryotes to eukaryotes (2018), https://evolution.berkeley.edu/evolibrary/article/_0/endosymbiosis_03. Accessed 5 Jan 2021
A.M. Vandamme, *Basic Concepts of Molecular Evolution* (NYU Computer Science, 2019). https://cs.nyu.edu/mishra/COURSES/06F.COBIO/basic_concepts_mol_evo.pdf. Accessed 5 Jan 2021

M.P. Veldhuis, M.P. Berg, M. Loreau, H. Olff, Ecological autocatalysis: a central principle in ecosystem organization? Ecol. Monogr. **88**(3), 304–319 (2018)

M. Vellend, *The Theory of Ecological Communities. Monographs in Population Biology 57* (Princeton University Press, Princeton, 2016)

A. Vespignani, Twenty years of network science. Nature **558**(7711), 528–529 (2018)

T. Watson, The trickster microbes shaking up the tree of life. Nature **569**(7756), 322–324 (2019)

F.W. Went, N. Stark, Mycorrhiza. Bioscience **18**, 1035–1039 (1968)

J.S. Wicken, Thermodynamics, evolution, and emergence ingredients for a new synthesis, in *Entropy, Information, and Evolution*, ed. by B. H. Weber, D. J. Depew, J. D. Smith, (MIT Press, Cambridge, 1988), pp. 138–169

R.G. Wiegert, *Ecological Energetics. Benchmark Papers in Ecology 4* (Dowden Hutchison & Ross, Stroudsburg, 1976)

N. Wolchover, A new physics theory of life. Quanta (2014). https://www.quantamagazine.org/a--new-thermodynamics-theory-of-the-origin-of-life-20140122/#comments. Accessed 6 Jan 2021

Y. Wu, C. Wang, X. Dong, B. Zhao, L. Jiang, J. Chen, B. Li, Effects of saltmarsh invasion by Spartina alterniflora on arthropod community structure and diets. Biol. Invasions **11**, 635–649 (2009)

C.S. Wylie, C. Ghim, D. Kessler, H. Levine, The fixation probability of rare mutators in finite asexual populations. Genetics **181**(4), 1595–1612 (2009). https://doi.org/10.1534/genetics.108.094532

Chapter 13
Why Is Species Diversity Higher in the Tropics?

13.1 Tropical Explorations

In the spring of 1799, Alexander von Humboldt and his companion Aimée Bonpland set out for South America aboard the Spanish frigate Pizarro. The Spanish King financed the trip in the hope that Humboldt, trained as a mining engineer, might discover new gold and silver mines, but Humboldt's real purpose was to make scientific observations (von Humboldt 1810). "I shall endeavor to find out how nature's forces act upon each other, and in which way the geographic environment exerts its influence on animals and plants", said Humboldt in a letter to his professor at the mining academy at Freiberg (Jordan 1981).

In contrast to Humboldt, many of the nineteenth century explorers of South America were interested primarily in the potential of the tropics for providing species that were of economic interest (Jordan 1982). These explorer-scientists sent thousands of specimens of previously unknown kinds of animals and plants back to Europe for classification and evaluation (Wallace 1878). It became evident that the number of species in the tropics was much greater than the number in higher latitudes.

C. F. Jordan, *Evolution from a Thermodynamic Perspective*,
https://doi.org/10.1007/978-3-030-85186-6_13

13.2 A Few Theories

Box 13.1
Species richness is simply the number of species that occur in an area, while species diversity takes into account the relative abundance, or number of individuals of each species. The terms are often used interchangeably except in conservation studies where it is important to know if an endangered species contributes just one or many individuals to the diversity of an ecosystem.

It wasn't until the twentieth century that scientists began to ask why species diversity was higher in the tropics. When growing up, these scientists from northern latitudes had experienced only ecosystems where diversity is low, so of course they assumed that low diversity was the normal condition. This led them to frame the question as, "why are there so many species in the tropics?" Pianka (1966) reviewed some of the many hypotheses that included: The time theory (speciation not interrupted by glaciation as in higher latitudes); The theory of spatial heterogeneity (more topographic relief in the tropics); The competition hypothesis (niches are smaller in the tropics); The predation hypothesis (niches are larger in the tropics); The theory of climatic stability (tropical climates are more stable allowing for less extinction); The productivity hypothesis (the longer growing season of tropical regions allows the species to partition the environment temporally as well as spatially). Tomašových (2019) proposed that high levels of biodiversity in the tropics can emerge from temporally variable but spatially patchy precipitation regimes. Losos and Schluter (2000) suggested that area available for a species to evolve (such as an island) can be a factor. Brown (2013) argued that both phylogenetic niche conservatism and higher tropical productivity are responsible. Rohde (1992) suggested that the effect of temperature on physiological processes causes faster rates of evolution and more rapid responses to selection in the tropics. Mueller and Diamond (2001) speculated that an underlying factor in rapid animal evolution could be food availability because net primary productivity (NPP) is higher in the wet tropics.

Schluter and Pennell (2017) divided the theories on the origin of the latitudinal gradient into two categories.

- Speciation is faster in the tropics because higher temperatures lead to higher metabolic rates (in ectotherms), shorter generation times and more rapid genetic divergence between populations.
- Higher speciation rates in the tropics is due to the greater heterogeneity of resources and species interactions that are made possible by greater amounts of solar energy, higher annual productivity (on land, at least), reduced temperature seasonality and stronger biotic interactions.

Willig and Pressley (2018) surmised that it is unlikely that a single mechanism drives these patterns. Rather, a combination of mechanisms likely explains latitudinal gradients of biodiversity, with the relative importance of particular mechanisms being contingent on taxon, biogeographic region, or how biodiversity is measured.

Slobodkin after his 1969 presentation at the Brookhaven Symposium on diversity and stability in ecological systems (Slobodkin and Sanders 1969), wryly noted "there are more species in the tropics because its nicer there."

13.3 A Thermodynamic Explanation

13.3.1 The Latitudinal Energy Gradient

Figure 13.1 indicates that the hours of daylight per year varies very little along a latitudinal gradient. While summer days are longer at high latitudes, winter days are shorter, so on a yearly basis, it evens out. However at high latitudes, the sun is low in the sky for at least 6 months of the year, so the amount of solar energy in a beam of sunlight is spread out over the horizontal surface of the Earth (Box 13.2). As a result, along an equatorial-to-polar gradient the amount of energy delivered annually to a square meter of the surface of the Earth diminishes greatly (Fig. 13.2) and the heating ability of solar radiation during winter months at high latitudes is greatly reduced. Growing seasons are shorter because reduced solar energy entering the Earth causes a reduction in temperature below that needed for photosynthesis during the winter months. As a result, the net primary productivity in terrestrial ecosystems is highest in the tropics (Lieth 1975; Wright and Boorse 2016; Tomasik 2018), and lowest at high latitudes.

Box 13.2

The effect of angle of light interception on energy intercepted per unit area on a horizontal surface can be shown by the following demonstration: Hold a flashlight two feet above a table top, shine it directly downward and measure the intensity of reflection on the surface. Then hold it again a few degrees above horizontal, and again measure the intensity of light reflected per unit area. The intensity reflected will be much lower because the beam is spread out.

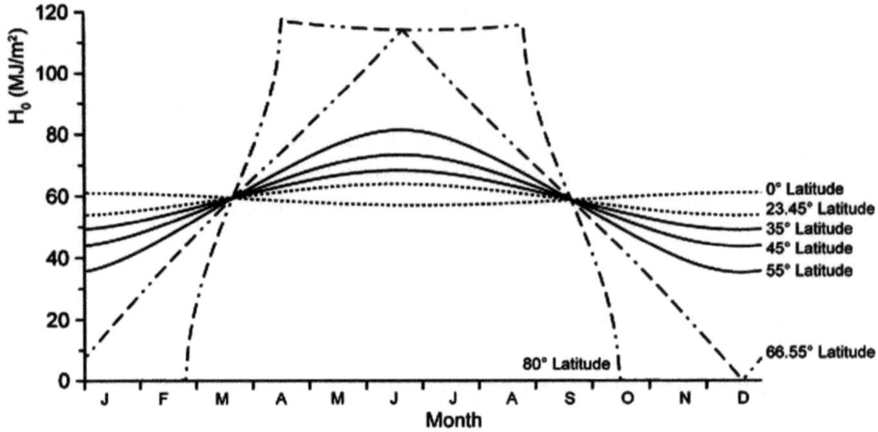

Fig. 13.1 The total daily amount of extraterrestrial irradiation on a plane perpendicular to the Earth's surface for different latitudes (H_o). Despite differences in day length, the areas under each respective curves are similar. (Figure courtesy of Itaca (2019))

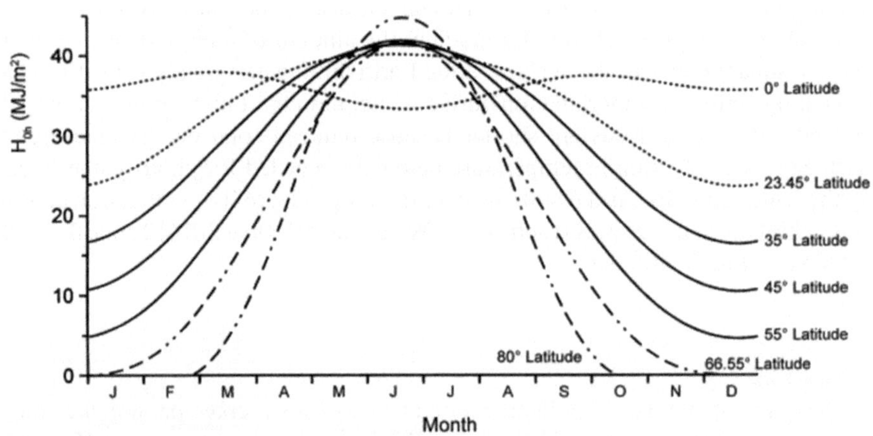

Fig. 13.2 The total daily amount of extraterrestrial irradiation on a plane horizontal to the Earth's surface for different North latitudes (H_{on}). Differences are due to the angle at which solar rays strike the plane. For South latitudes, there is a six-month offset in the horizontal axis. (Figure courtesy of Itaca (2019))

13.3.2 The Latitudinal Productivity Gradient

13.3.2.1 What Does Productivity Mean?

Productivity in agriculture means yield. For farmers, it means the total kg of corn grain produced per hectare per growing season. For agroecologists, it includes the dry weight of leaves, stem, and roots of the corn plant. For natural communities, productivity of secondary successional vegetation is the rate at which total biomass accumulates (Figs. 10.2 and 10.3). What does productivity for a mature ecosystem mean? A mature forest ecosystem does not increase in biomass because tree death offsets growth of living trees. Productivity is defined by biomass turnover. Biomass turnover is leaf production balanced by leaf fall and decomposition, plus death of old trees balanced by growth of new trees.

13.3.2.2 Forests as Sinks for Energy and Carbon Dioxide

Forests are said to be carbon sinks, and therefore should be preserved to alleviate the increasing carbon dioxide concentrations in the atmosphere that results in global warming. Forests are sinks only in the sense that if they are not cut, the carbon in their biomass remains stored. Once a forest is mature, it returns carbon dioxide to the atmosphere as fast as it removes the carbon. Only successional forests such as those following agricultural abandonment are sinks that decrease atmospheric carbon dioxide.

13.3.3 The Data

Figure 13.3 shows rate of biomass turnover due to leaf production and decomposition as a function of latitude and month of the year based on 8-day increases in leaf biomass. High biomass turnover does not mean yearly biomass accumulation. It means high leaf production followed by high rates of decomposition.

Box 13.3
Biomass turnover in mature forests also is growth of young trees that replace old trees that have died. That is not included in the Figure.

The figure shows that in January at high latitudes in the northern hemisphere, there is little to no leaf production (Fig. 13.3a) but in moist tropical regions, leaf production is ongoing. As the northern hemisphere starts to warm in April, a pulse of productivity is evident in the southeastern part of the United States and Southern Europe (Fig. 13.3b). By July (Fig. 13.3c), high leaf production has moved further northward. Although leaf production is high during the growing season, the growing season is shorter at higher latitudes, so total production on a yearly basis is lower. In the tropics with longer growing seasons, the amount of biomass produced is greater but the amount of biomass decomposed is also greater. Energy flow and productivity on a yearly basis is higher in the tropics. The July pattern in Southern latitudes should mirror the January pattern in Northern latitudes, but because land area is much less, the pattern is scarcely apparent.

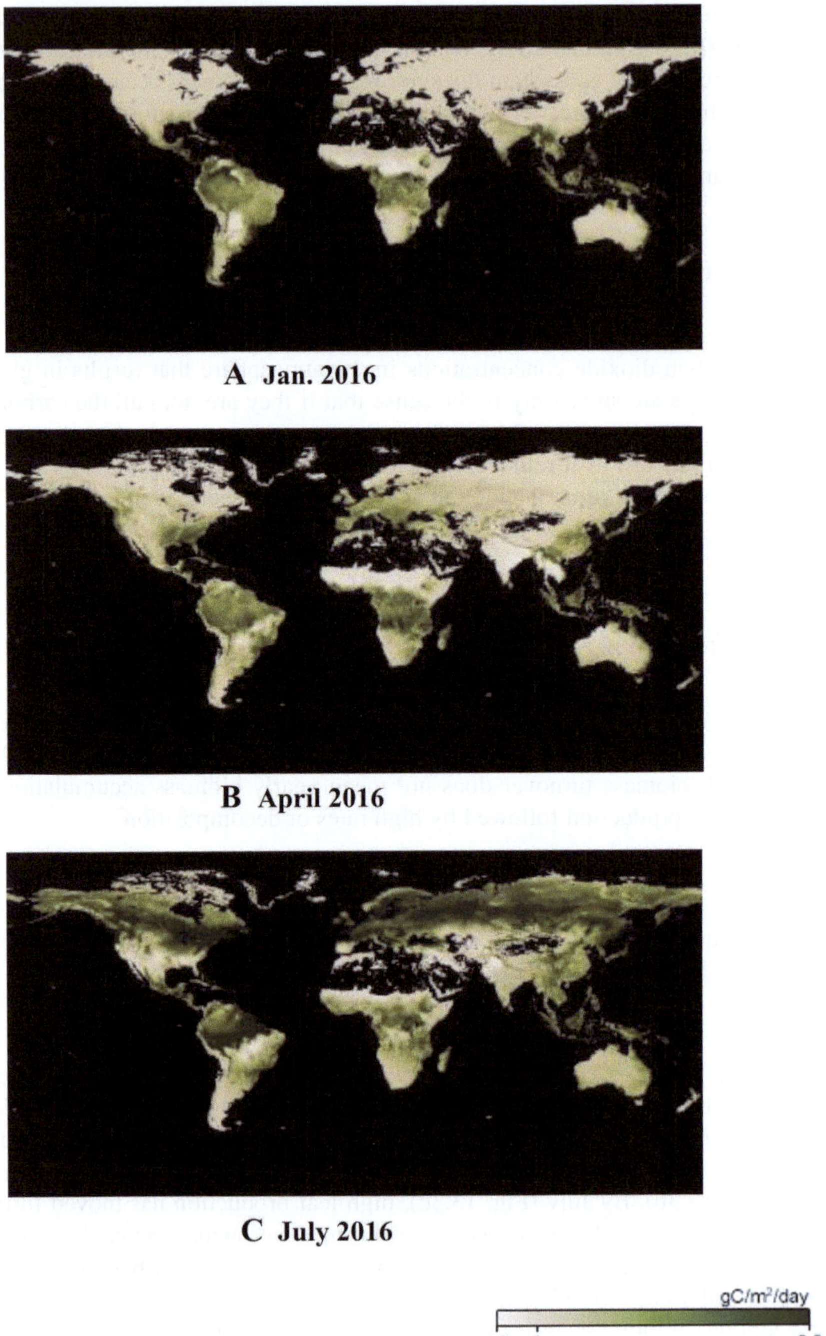

Fig. 13.3 Net Primary Productivity derived from 8-day changes in leaf biomass as a function of latitude in 2016. Productivity trends are more evident in the northern hemisphere because its greater land mass. (Figure reproduced from NOAA (2019))

13.3.4 *Other Factors Affecting Productivity*

13.3.4.1 Rapid Nutrient Recycling

In 1970 when I was compiling data on primary productivity on ecosystems of the world as a function of energy available during the growing season, I noticed that while biomass turnover of forest ecosystems was greater in the wet tropics, it was due to greater production of leaves, not of wood (Jordan 1971). Turnover of wood biomass as a function of light available during the growing season was not significantly different between latitudes. I speculated on the evolutionary reasons, but came to no final conclusion (Jordan and Murphy 1978). Now, after understanding how the rate of nutrient recycling affects primary productivity, the reason becomes clear. Leaves are decomposed faster than tree trunks, and so there is an evolutionary advantage to rapid leaf production and decomposition. It gives a greater power output to the ecosystems due to a greater rate of nutrient release to plants.

13.3.4.2 Photosynthesis: Respiration Balance

Early Spanish colonists in Central and South America often settled in mid-elevational valleys, because they discovered that agricultural productivity for most crops is highest there. In addition, they were the most pleasant places to live. Many of these cities have acquired the nickname “City of Eternal Spring”, including Caracas Venezuela, Cuernavaca Mexico, Quito Ecuador, Medellin Colombia, Cidra Puerto Rico, and Heredia, Costa Rica.

Productivity is influenced by the energy balance of an ecosystem, the difference between energy fixed by photosynthesis and lost through respiration (Gates 1969). If we start at the top of tropical volcanoes such as those described by von Humboldt (1810) and work downward, temperature increases as we descend. Photosynthesis and respiration also increase, but not at equal rates. Photosynthesis increases faster than respiration until we reach mid elevation where average temperature is about 22 °C (Fig. 13.4). At that point, there is the maximum difference between photosynthesis and respiration, and net primary productivity is highest. As we continue downward, respiration increases faster than photosynthesis and when we reach the elevation where temperature is about 30 °C, decreasing photosynthesis intersects with increasing respiration. Biomass can no longer increase. Respiration continues to increase until about 40 °C at which point the whole system begins to shut down. Harris et al. (2013) tested and confirmed this interaction by measuring photosynthesis and respiration from 0 to 1000 m along the slope of El Yunque in The Caribbean National Forest, Puerto Rico. Highest net primary productivity was at 500 m above sea level. That is a much lower elevation for maximum productivity than occurs in the Andes due to the Massenerhebung effect. Mountains surrounded by large ranges will tend to have maximum net primary productivity at higher elevations than more isolated mountains due to heat retention and wind shadowing.

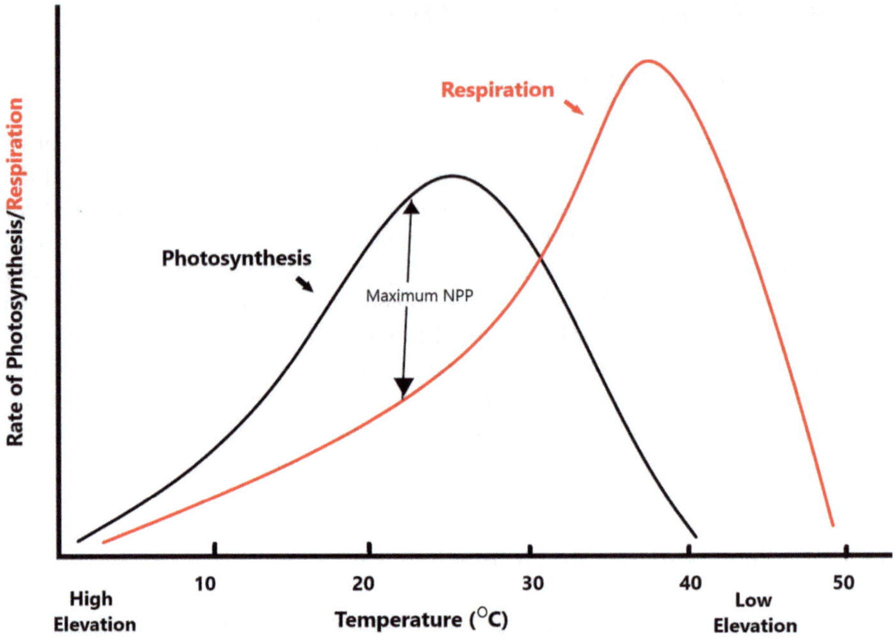

Fig. 13.4 The relative relationship between photosynthesis and respiration as a function of temperature on a tropical mountain. Highest net primary production occurs where the vertical difference between the two curves is greatest, in the vicinity of 22 °C. NPP = net primary productivity

In temperate regions we do not see a mid-elevational peak as we go down a mountain because average temperatures never reach 22°C. In the opposite direction, as we increase elevation and decrease temperature below 22 °C, photosynthesis decreases faster than respiration, till at the tree line they are equal.

13.3.4.3 Productivity and Stressful Environments

While productivity is correlated with latitude, productivity within each latitudinal belt is affected by locally stressful environments. A plant growing in a stressful environment will respire more heavily. Since net primary productivity is the difference between photosynthesis and respiration, NPP will be lower in stressful environments. Lack of moisture as in deserts or dry forests is a stress, so productivity there will be lower than in mesic environments. Geological variability results in factors such as low nutrient availability and high concentrations of elements such as sodium and selenium that lower productivity. Coastal salt marshes are an exception to stress reducing productivity. The major autotroph, *Spartina alterniflora*, is eaten by insects, then dies, decomposes, and as detritus, furnishes food for the other fauna of the marsh (Teal 1962). Despite high salt concentration, biomass turnover and nutrient cycling is rapid. Tides provide an energy subsidy to the ecosystem in the form of sulfates that fuel the mineralization of organic matter (Giblin and Weider 1992). Grasslands such as the Serengeti also are an exception, where fire rather than being a stress, is a subsidy that increases light energy available to resprouting leaves and stems.

13.4 Empirical Evidence for a High Productivity High Diversity Correlation

Adler et al. (2011) found no relationship between species richness and productivity in 48 herbaceous-dominated plant communities on five continents. However, from a global forest data set from 770,000 sample plots in 44 countries, Liang et al. (2016) found a positive and consistent relationship between tree diversity and ecosystem productivity. In an 8-year experiment, Huang et al. (2018) also found strong positive effects of tree diversity on forest productivity. In temperate-arctic marine ecosystems, Witman et al. (2008) found that diversity–productivity relationships were more prevalent at smaller spatial scales (landscape) than larger (regional, continental) spatial scales. Waide et al. (1999) reviewed 200 studies on the relationship between productivity and species richness. 30% were unimodal, 26% were positive linear, 12% were negative linear, and 32% were not significant. A unimodal relationship means that richness increased with productivity up to a point, but decreased with productivity beyond that point. They attributed results that did not conform to a positive linear relationship to differences in scale of the areas studied. There are other possible explanations. If the trend is due to an invasive species that is a better competitor for nutrients, diversity will decrease despite higher ecosystem productivity (Hooper and Vitousek 1997) (See Box 13.4). The effect is similar to that caused by artificial enrichment of an ecosystem. An experimental ecosystem that is artificially fertilized will lose diversity, as a fast-growing species adapted to resource rich conditions out-competes slower growing species that are better competitors when resources are limited. Over the long term however, the relationship between diversity and productivity is linear. Tilman (1987) pointed out that lower diversity in enriched plots occurs because the original plant communities do not have time to adapt to changed conditions. When plant communities have evolutionary time to adapt to soil conditions, those on richer soils will have higher diversity than those on poor soils. When looking at the relationship between diversity, productivity, and nutrient availability, researchers must take into consideration time for adaptation to new conditions (Kardol et al. 2018).

Box 13.4
On my farm in the Georgia Piedmont, Johnson grass (*Sorghum halepense*) an aggressive C-4 species has invaded my pastures beginning in about 2010. Its success is due to more efficient photosynthesis. While productivity per acre has increased, species diversity has decreased.

13.5 Humboldt's Enigma

For terrestrial ecosystems, species biodiversity is highest on tropical mountains (Rahbek et al. 2019a). An explanation had eluded ecologists for decades, so it was called "Humboldt's Enigma" because of Humboldt's view of the Andes as a vertical exhibit of the earth's diversity (Jordan 1981). The enigma was resolved when it was realized that on tropical mountains, the latitudinal range of species diversity is compressed vertically, so that the factors contributing to diversity are more easily distinguished.

13.5.1 Are Productivity and Species Richness Correlated on Tropical Mountains?

Because of the high productive potential in mid-elevational tropical valleys, settlers cleared the ecosystems for agriculture, so there is unfortunately no record of the native plant richness in these areas. On mountain slopes, the topographic heterogeneity of many tropical mountains causes a complex mosaic of mountain microclimates that confounds the correlation of diversity with elevation (Rahbek et al. 2019a, b). For the Amazon region, we do know that the number of plant species on highly weathered soils of the old, low elevation soils of the Rio Negro region in Venezuela (Clark and Liesner 1989) is lower than those found on in higher productivity ecosystems on younger richer soils at the base of the Andes (Gentry 1988a), and that at elevations around 100 m in Peru, diversity may be the highest in the world (Gentry 1988b). In páramos at the highest elevations, productivity again is low while diversity varies considerably because of human disturbance (Luteyn 2019).

Patterns of animal diversity on mountains generally follow the pattern of plant productivity. Rahbek (1995) in his review of diversity as a function of elevation pointed out mid-elevational peaks for birds and rodents on tropical mountains. On other tropical and temperate zone mountains, a mid-elevational peak in species richness has been shown for certain taxa: small mammals (McCain 2004); spiders (Chaladze et al. 2014); ants (Sanders 2002); and plants (Grytnes 2003; Grytnes and Vetaas 2002).

13.6 The Mechanism Linking Productivity and Diversity

Although productivity may be correlated with species diversity, that does not mean that high productivity is the cause of high species diversity. Correlation of high energy flow through ecosystems and species diversity in those ecosystems does not imply cause and effect. To understand why high diversity is *caused* by high annual energy flux and productivity, we need to take a viewpoint that stretches over the evolution of life to show that high diversity is caused by greater energy flows.

Two billion years ago, primary productivity consisted of the biomass produced by the first eukaryotic autotroph. There were no trophic food chains. Upon death of the autotroph, it went straight to decomposer bacteria (Fig. 12.1a). Because each new trophic link in an evolving food chain needs a minimum amount of energy to survive (Odum 1971), the first trophic chain evolved only after autotrophic eukaryotes began fixing enough solar energy to support not only themselves, but also an herbivore (Fig. 12.1b). As life spread throughout the Earth, trophic chains grew and became longest in regions where the growing season was longest because more energy was available to support them.

Longer trophic chains alone however cannot account for high species richness, in as much as each addition to the trophic chain adds only one species to the ecosystem diversity, and that addition occurs slowly. But once a new trophic level is established, diversity can increase much faster. Leigh et al. (2015) and Tian et al. (2019) have shown that life-history traits accrue mutational variance faster than morphological traits because morphological traits are more complex (i.e., influenced by a greater number of loci). The amount of plasticity in a trait also accrues mutational variance faster for life-history traits than for morphological traits. As a result, evolution of new species within trophic levels occur much faster than species that represent a new trophic level. For example, the evolution of a new species of herbivorous beetle a should be relatively fast compared to the evolution of a new morphology necessary for evolution of a predator on that beetle. Faster evolution within trophic levels as opposed to between levels could also be explained by epigenetic or phenotypic evolution (Nei 2007; Zhang 2018), where new species evolve through only minor changes in phenotype, that occur when sub-populations within a trophic level split off from the main population. Therefore, latitudes where food chains have more trophic levels will have greater diversity. Rapid evolution within trophic levels is the reason.

13.7 Answer to "Why Is Species Diversity Higher in the Tropics?"

Energy flow through ecosystems increased during evolution due to the feedback loop between populations and the energy flow through their ecosystems. With increasing energy flow there was greater opportunity for more species. Greater flow enabled the energy to be divided more ways. This enabled more complex food webs, longer food chains, and more development of new species. This deviation amplifying system increased until further increases were limited by the yearly amount of solar energy. Since that is highest in the tropics, that is where diversity is highest. Now we have more than a correlation. We have a cause and effect. High energy flow is the cause of high species diversity.

13.7.1 Differences Within the Tropics

Within the tropics there are a wide variety of ecosystems including deserts, grasslands, dry forests, rain forests, swamps, and montane forests. Productivity varies between them due to differences in rainfall, elevation, topography and soil type, and consequently diversity varies accordingly. But for each undisturbed ecosystem type, diversity is usually higher in the tropics. Diversity in tropical deserts is higher than diversity in high latitude deserts. An exception would be where tropical deserts have an additional stress such as saline soils.

Raven et al. (2020) looked at differences in the tropics at a regional scale. They compared diversity of vascular plants species in the Afrotropical Region, Tropical Latin America, and Tropical Asia. Latin America had the highest diversity, while Tropical Asia had the highest diversity per unit area. Floras of Tropical Africa and South America were similar before the breakup of Gondwanaland beginning more than 100 million years ago. Differences today can be attributed to differences in scale and time of mountain building that profoundly affected regional climates and presumably energy availability. Species diversity in the Indonesian archipelago is separated by the Wallace line that divides the region into two distinct parts: a western portion in which the animals are largely of Asian origin, and an eastern portion where the fauna reflects Australasia.

13.8 Why Is Species Diversity Low at High Latitudes?

Although ecosystems in temperate latitudes today have lower diversity than comparable ecosystems in tropical latitudes, it was not always so. Fossil evidence indicates that regions of the Earth that now have temperate climate had a more tropical climate and a higher species diversity during the Eocene period, than they have now (Harrington et al. 2012). During the Eocene beginning 55 million years ago, high diversity was the "normal" condition for mid-latitudinal ecosystems, so what must be explained is why diversity has decreased at high latitudes. Diminution of food chains occurred after the Eocene when global surface temperatures began to decrease and plunged to a low during the Quaternary period that extends through today. As growing seasons shortened, food chains at higher latitudes were diminished. There was insufficient energy to support more than a few links. Only of low latitudes could trophic chains of more than five or six links survive. Low species diversity today at high latitudes resulted from global cooling caused by reduced energy inputs that reduced food chains at higher latitudes. Low latitude ecosystems were spared, and retained their high diversity.

13.9 An Economic Perspective on Diversity

We know a lot more about economic systems than we do about ecological systems, so sometimes it is helpful to draw upon economic systems to shed light on questions about ecological systems. Economic systems resemble ecological systems in many ways. Both are controlled by information, and like ecosystems, economic systems are thermodynamic, cybernetic, hierarchical, and autocatalytic.

13.9.1 Energy Flow, Economic Growth and Professional Diversity

The gross domestic product of a nation increases as a result of technological innovation and population increase. The most important technological innovations for economic development of the world's nations were those that enabled the substitution of fossil fuel energy for draft horse energy in agriculture, and for human energy in factories (Smil 2017). As the economy of a country grew, the amount and rate of energy flowing through the system increased. As the amount of energy flow in one direction increased, the amount of money flow in the opposite direction also increased. As the amount of money flowing through an economic system increased, the diversity of professions in the system increased (Moretti 2013).

In the medical profession we have an example of how technological innovation spurred the growth of uninflated currency. In the early twentieth century, most doctors were generalists and many were country doctors. He or (rarely) she could treat some general ailments, but couldn't help people with eye cataracts, progressive deafness, and arthritis in the knees. But then several things happened. Technology provided cures, or at least relief from many ailments. At the same time, population of cities increased, and the amount of money flowing through cities increased. Medical students noticed. They realized that they could make more money by becoming specialists where there were many people with relatively rare ailments. As specialty centers developed in cities, more people moved in to take advantage of these specialists. As more people moved in, there were more opportunities for specialists, and so the feedback system continued with money flow providing the feedback. Metropolitan areas have more medical specialists than country towns because the greater flow of money can be divided in many more ways. Other professions from lawyers to carpenters followed a similar trajectory (Moretti 2013). Increase in professional diversity resulted from an increase in uninflated money.

Governments sometimes print more money to increase the economy. While this may result in a short-term stimulus, it has no long-term effect because the currency is inflated. Lasting effect comes only through technological innovation which means that human energy is used more efficiently, or fossil fuel energy to perform a particular task is used more efficiently. In ecosystems, power output increased during evolution as energy use required to circulate nutrients through food chains became more efficient. That was uninflated growth.

Box 13.5
For real growth of an economy, rate of energy flow has to increase faster than rate of inflation.

The analogy between economic systems and ecosystems has limitations. Individuals in economic systems do not always follow the money, while species always follow the energy. Since species always follow the energy, species diversity is higher in the tropics.

References

P.B. Adler, E.W. Seabloom, E.T. Borer, et al., Productivity is a poor predictor of plant species richness. Science **333**(6), 1750–1753 (2011)

J.H. Brown, Why are there so many species in the tropics? J. Biogeogr. **41**(1), 8–22 (2013). https://doi.org/10.1111/jbi.12228

G. Chaladze, S. Otto, S. Tramp, A spider diversity model for the Caucasus Ecoregion. J. Insect Conserv. **18**(3), 407–416 (2014). https://doi.org/10.1007/s10841-014-9649-1

H.L. Clark, R.L. Liesner, Angiosperms, in *An Amazonian Rain Forest:. The structure and function of a nutrient-stressed ecosystem and the impact of slash and-burn agriculture*, Man and the Biosphere Series, ed. by C. F. Jordan, vol. 2, (UNESCO, Parthenon Publishing Group, Paris/Carnforth, 1989), pp. 111–120

D.M. Gates, Climate and stability, in *Diversity and Stability in Ecological Systems. Report of a Symposium held May 26–28, 1969*, ed. by G. M. Woodwell, H. H. Smith, (Biology Department Brookhaven National Laboratory, Upton, 1969), pp. 115–127

A.H. Gentry, Changes in plant community diversity and floristic composition on environmental and geographical gradients. Ann. Mo. Bot. Gard. **75**, 1–3 (1988a)

A.H. Gentry, Tree species richness of upper Amazonian forests. PNAS **85**(1), 156–159 (1988b)

A.E. Giblin, R.K. Weider, Sulfur cycling in marine and freshwater wetlands, in *SCOPE 48 - Sulphur Cycling on the Continents*, ed. by R. W. Howarth, J. W. B. Stewart, M. V. Ivanov, (SCOPE, Wiley, 1992) Chapter 5

J.A. Grytnes, Species-richness patterns of vascular plants along seven altitudinal transects in Norway. Ecography **26**(3), 291–300 (2003) www.jstor.org/stable/3683370

J.A. Grytnes, R. Vetaas, Species richness and altitude: a comparison between null models and interpolated plant species richness along the Himalayan altitudinal gradient, Nepal. Am. Nat. **159**, 294–304 (2002)

G.J. Harrington, J. Eberle, B.A. Le-Page, M. Dawson, J.H. Hutchison, Arctic plant diversity in the early Eocene greenhouse. Proc. Biol. Sci. **279 B**, 1515–1521 (2012). https://doi.org/10.1098/rspb.2011.1704

N.L. Harris, C.A.S. Hall, A.E. Lugo, A test of the maximum power hypothesis along an elevational gradient in the Luquillo mountains of Puerto Rico. Ecol. Bull. **54**, 233–243 (2013)

D.U. Hooper, P.M. Vitousek, The effects of plant composition and diversity on ecosystem processes. Science **277**(5330), 1302–1305 (1997)

Y. Huang, Y. Chen, N. Castro-Izaguirre, et al., Impacts of species richness on productivity in a large-scale subtropical forest experiment. Science **362**(6410), 80–83 (2018). https://doi.org/10.1126/science.aat6405

Itaca, The sun as a source of energy. Part 2: Solar energy reaching the earth's surface (2019), https://www.itacanet.org/the-sun-as-a-source-of-energy/part-2-solar-energy-reaching-the-earths-surface/. Accessed 21 Dec 2020

C.F. Jordan, A world pattern in plant energetics. Am. Sci. **59**(4), 425–433 (1971)

E.L. Jordan, The birth of ecology: an account of Alexander von Humboldt's Vooyage to the equatorial regions of the new continent, in *Tropical Ecology. Benchmark Papers in Ecology/10*, ed. by C. F. Jordan, (Hutchinson Ross Publishing Co, Stroudsburg, 1981), pp. 4–15

C.F. Jordan, Amazon rain forests. Am. Sci. **70**, 394–401 (1982)

C.F. Jordan, P.G. Murphy, A latitudinal gradient of wood and litter production, and its implications regarding competition and species diversity in trees. Am. Midl. Nat. **99**, 415–434 (1978)

P. Kardol, N. Fanin, D.A. Wardle, Long-term effects of species loss on community properties across contrasting ecosystems. Nature **557**, 710–713 (2018). https://doi.org/10.1038/s41586-018-0138-7

C. Leigh, M. Peacock, D.J. Civitello, J.L. Dudycha, J.M. Meik, S. Schaak, The phenotypic effects of spontaneous mutations in different environments. Am. Nat. **185**(2), 243–252 (2015)

J. Liang, T.W. Crowther, N. Picard, et al., Positive biodiversity-productivity relationship predominant in global forests. Science **354**(6309), 196 (2016)

H. Lieth, Primary productivity in ecosystems: comparative analysis of global patterns, in *Unifying Concepts in Ecology*, ed. by W. H. van Dobben, R. H. Lowe-McConnell, (Springer, Dordrecht, 1975), pp. 67–88

J.B. Losos, D. Schluter, Analysis of an evolutionary species-area relationship. Nature **408**, 847–850 (2000)

J.L. Luteyn, Páramo ecosystems (2019), http://www.mobot.org/mobot/research/paramo_ecosystem/introduction.shtml. Accessed 26 Dec 2020

C.M. McCain, The mid-domain effect applied to elevational gradients: species richness of small mammals in Costa Rica. J. Biogeogr. **31**, 19–31 (2004)

E. Moretti, *The New Geography of Jobs* (Mariner Books, Houghton Mifflin Harcourt, Boston, 2013)

P. Mueller, J. Diamond, Metabolic rate and environmental productivity: well-provisioned animals evolved to run fast and idle fast. PNAS **98**(22), 12550–12554 (2001)

M. Nei, The new mutation theory of phenotypic evolution. PNAS **104**(30), 12235–12242 (2007). https://doi.org/10.1073/pnas.0703349104

NOAA, Net primary productivity (1 month – Terra/modis) (2019). https://neo.sci.gsfc.nasa.gov/view.php?datasetId=MOD17A2_M_PSN. Accessed 27 Dec 2020

H.T. Odum, *Environment, Power and Society* (Wiley, New York, 1971)

E.R. Pianka, Latitudinal gradients in species diversity: a review of concepts. Am. Nat. **100**(910), 33–46 (1966)

C. Rahbek, The elevational gradient of species richness: a uniform pattern? Ecography **18**(2), 200–205 (1995)

C. Rahbek, M.K. Borregaard, R.K. Colwell, B. Dalsgaard, B.G. Holt, N. Morueta-Holme, D. Nogues-Bravo, R.J. Whittaker, J. Fjeldså, Humboldt's enigma: what causes global patterns of mountain biodiversity? Science **365**(6458), 1108–1113 (2019a)

C. Rahbek, M.K. Borregaard, A. Antonelli, R.K. Colwell, B.G. Holt, D. Nogues-Bravo, C.M.Ø. Rasmussen, K. Richardson, M.T. Rosing, R.J. Whittaker, J. Fjeldså, Building mountain biodiversity: geological and evolutionary processes. Science **365**(6458), 1114–1119 (2019b)

P.H. Raven, R.E. Gereau, P.B. Phillipson, C. Chatelain, C.N. Jenkins, C.U. Ulloa, The distribution of biodiversity richness in the tropics. Sci. Adv. **6**(37), eabc6228 (2020). https://doi.org/10.1126/sciadv.abc6228

K. Rohde, Latitudinal gradients in species diversity: the search for the primary cause. Oikos **65**, 514–527 (1992)

N.J. Sanders, Elevational gradients in ant species richness: area, geometry, and Rapoport's rule. Ecography **25**(1), 25–32 (2002)

D. Schluter, M.W. Pennell, Speciation gradients and the distribution of biodiversity. Nature **546**(7656), 48–55 (2017)

L.B. Slobodkin, H.L. Sanders, On the contribution of environmental predictability to species diversity, in *Diversity and Stability in Ecological Systems. Report of a Symposium held May 26–28, 1969*, ed. by G. M. Woodwell, H. H. Smith, (Biology Dept., Brookhaven National Laboratory, Upton, 1969), pp. 82–95

V. Smil, *Energy and Civilization. A History* (MIT Press, Cambridge, 2017)

J.M. Teal, Energy flow in the salt marsh ecosystem of Georgia. Ecology **43**(4), 614–624 (1962)

X. Tian, H. Ohtsuki, J. Urabe, Evolution of asexual *Daphnia pulex* in Japan: variations and covariations of the digestive, morphological and life history traits. BMC Evol. Biol. **19** (2019) Article number: 122. https://bmcevolbiol.biomedcentral.com/articles/10.1186/s12862-019-1453-9

D. Tilman, Secondary succession and the pattern of plant dominance along experimental nitrogen gradients. Ecol. Monogr. **57**, 189–214 (1987)

B. Tomasik, Net primary productivity by land type (2018), https://reducing-suffering.org/net-primary-productivity-land-type/. Accessed 4 Jan 2021

A. Tomašových, Nat. Ecol. Evol. **3**, 1376–1377 (2019)

A. von Humboldt, *Personal Narrative of Travels to the Equatorial Regions of the New Continent (30 volumes)* (Longman, London, 1810)

R.B. Waide, M.R. Willig, C.F. Steiner, G. Mittelbach, L. Gough, S.I. Dodson, G.P. Juday, R. Parmenter, The relationship between productivity and species richness. Annu. Rev. Ecol. Syst. **30**, 257–300 (1999). https://doi.org/10.1146/annurev.ecolsys.30.1.257

A.R. Wallace, *Tropical Nature and Other Essays* (Macmillan. Reprinted AMS Press, London/New York, 1878)

M.R. Willig, S.J. Pressley, Latitudinal gradients of biodiversity: theory and empirical patterns. Encycl. Anthropocene **3**, 13–19 (2018)

J.D. Witman, M. Cusson, P. Archambault, A.J. Pershing, N. Mieszkowska, The relation between productivity and species diversity in temperate-arctic marine ecosystems. Ecology **89**(11), S66–S80 (2008)

R.T. Wright, D.F. Boorse, *Environmental Science: Toward a Sustainable Future* (Pearson, Boston, 2016)

J. Zhang, Neutral theory and phenotypic evolution. Mol. Biol. Evol. **35**(6), 1327–1331 (2018). https://doi.org/10.1093/molbev/msy065

Chapter 14
What Have We Learned by Viewing Evolution from a Thermodynamic Perspective?

14.1 What We Have Learned

What have we learned by viewing evolution from a thermodynamic perspective that we could not have learned from the Neo-Darwinian view?

14.1.1 Fitness Means Maximization of Power Output

We have learned that higher forms of life could not have evolved through evolutionary pressures acting on species alone. It is too unlikely. Higher forms could have evolved only if there was continuous autocatalytic feedback between species and the ecosystems in which they existed that increased power output of both. Competition for energy gave species with the greatest power output an evolutionary advantage. An increase in power output by species resulted in an increase in power output of their ecosystems. Power output increased complexity. Complexity increased power output. Complexity is a characteristic of higher forms of life and more powerful systems. *Fitness leading to evolution of higher forms of life means survival of those species and ecosystems that can best maximize power output.*

14.1.2 Feedback Is Essential for Control

Feedback is essential for control. Control is necessary for sustainability. Flows of energy through an ecosystem vary according to season, and following disruptions caused by pollution and climate change. Sustainability depends on the ability of the system to adapt to these changes. We already know that species adjust through

C. F. Jordan, *Evolution from a Thermodynamic Perspective*,
https://doi.org/10.1007/978-3-030-85186-6_14

changes in metabolic rate, and changes in birth and death rates. We also know that species will migrate, become extinct, or adapt through selection of characteristics better adapted to the new environment. What thermodynamic theory tells us is that the mechanism through which ecosystems adapt to a changed environment is feedback. Ecosystems adapt simultaneously and reciprocally with species. That is the essence of control and sustainability.

14.1.3 Control of Energy Flow Occurs Both Top-down and Bottom-up

Top-down energy flow through ecosystems determines dynamics of species populations that comprise the ecosystem. As species adjust to the incoming energy flow, they change the outgoing flow, thereby changing ecosystem level flow bottom-up.

14.1.4 Storage of Entropy Is a Powerful Characteristic of Living Systems

"Entropy storage" means that the entropy gradient between solar energy entering the earth and leaving as heat is reduced by entropy storage in the structure of living organisms. Gradient reduction by entropy storage is a powerful tendency of nature. It has built trophic chains and food webs, and preserved them albeit in a diminished form throughout at least five major global climate changes that drove most species in the corresponding epoch to extinction. Entropy storage has preserved ecosystem structure and enabled evolution to resume a path towards higher complexity without having to revert to the simple system that characterized the origin of life.

14.1.5 Evolution Is the Storage of Entropy

Despite the second law of thermodynamics that says that all systems tend to decompose, living systems seem to violate the entropy law, not only by maintaining structure and function, but also by increasing their structure and function. They do it by storing entropy. Building and maintaining organization is the opposite of entropy. That means they use solar energy to create and maintain organization by countering the effects of entropy, much as a teacher uses energy to maintain order among unruly students (or at least used to when I was a schoolboy). When students behave in the classroom, they are storing entropy. When they go outside for recess, their entropy is released. Civilization is the storage of entropy.

Chapter 15
Objections to the Ecosystem Concept

15.1 Criticisms of the Ecosystem Concept

15.1.1 Ecosystems Are Abstractions

A frequent criticism of the ecosystem concept is that an ecosystem is merely an abstraction (Schaffer 1981). For something to be real and not simply an abstraction such as beauty, truth, pornography and God, it has to have parameters that can be quantified (Stanford 2015). But ecosystems have parameters that can be quantified such as biomass, productivity, nutrient content, and species diversity.

15.1.2 Ecosystems Are Ephemeral

Levin (1999, p. 183) questioned whether ecosystems even exist, because "their identities are ephemeral in space and time", ephemeral meaning lasting only for a short time. "Species" also are ephemeral in space and time. Does this mean that species do not exist?

15.1.3 Ecosystems Are Oversimplifications

"Real world ecosystems (as opposed to their mathematical counterparts) are enormously complex associations of species which interact in diverse ways. As a matter of practical necessity, field ecologists can rarely specify, much less quantify all of the interactions in an ecosystem. Consequently, ecosystem models generally consider fewer than the total number of interacting species" (Schaffer 1981). Every

C. F. Jordan, *Evolution from a Thermodynamic Perspective*,
https://doi.org/10.1007/978-3-030-85186-6_15

parameter has a statistical distribution, and as more functions of species are added, errors accumulate (Saltelli et al. 2020). Complexity is inversely related to accuracy of ecosystem models.

"Everything should be made as simple as possible, but no simpler" is a quote attributed to Albert Einstein, but what he really said is more complicated: "It can scarcely be denied that the supreme goal of all theory is to make the irreducible basic elements as simple and as few as possible without having to surrender the adequate representation of a single datum of experience" (Calaprice 2013).

In ecosystem analysis, each hierarchical level should be as simple as possible. Studies at each hierarchical level have different objectives that cannot be answered by combining levels. Studies should have the minimum complexity needed to answer the question asked (see Appendix 3).

15.1.4 The Ecosystem Concept Is Merely a Paradigm

The ecosystem concept as a flow of energy and nutrients is a paradigm, a non-self-evident theorem that can be justified on the basis of postulates, axioms, and thermodynamic laws (Appendix 9). While the paradigm claim is true (O'Neil 2001), it should not diminish the usefulness of the concept. Economic systems too are a paradigm. Does that mean that the field of economics is not useful?

15.1.5 The Ecosystem Concept Is Not Based on Facts

A scientific fact is a statement about some fundamental quality or quantity of the universe that is true regardless of when or where the observation is made. The first law of thermodynamics – that energy cannot be created or destroyed – is a scientific fact. Gravity also is a fact. Other than these facts in physics, there are no facts in science, only probabilities (Appendix 10).

15.2 Ecosystems Are Not Cybernetic

The ecosystem concept depends on ecosystems being cybernetic, that is feedback controls the system. Engelberg and Boyarsky (1979) argued that ecosystems are not cybernetic since they lack internal control mechanisms that maintain the system at a discrete set point, like a thermostat does in an air conditioning system. In response, Patten and Odum (1981) answered that a fixed set point maintained by a discrete controlling unit is not necessarily a criterion for the existence of feedback control. The interplay of material cycles and energy flows under information control, generates self-organizing feedback with no discrete controller required. Thermodynamic theory says that ecosystems actually do have a set point, but it varies according to season of the year. Unlike air conditioning systems in which the temperature has narrow bounded limits around a set point, ecosystems have broad bounded limits but nevertheless exhibit stable equilibrium over a course of four seasons (refer to Fig. 4.1b).

15.3 Inappropriate Machine Analogies

Many of the early criticisms of the ecosystem concept resulted from comparing ecosystems to machines (Soule and Lease 1995). "The classical paradigm in ecology, with its emphasis on the stable state, its suggestion of natural systems as closed and self-regulating, and its resonance with the nonscientific idea of balance of nature, can no longer serve as an adequate foundation for conservation." (Pickett et al. 1992). While Pickett was correct about machine analogies, he was wrong about self-regulation. Self-regulation through feedback is the essence of ecosystem science. While pollution, weather, and other factors outside the feedback system influence ecosystem dynamics, feedback dampens the effect of perturbations as well as normal seasonal variation in energy availability, and maintains ecosystem stability.

15.4 Objections to Ecosystem Evolution

15.4.1 No Measure of Fitness

Levin (1999) as well as Lewontin (1970) argued that there is no sensible measure of fitness for evolutionary selection of ecosystems. "The properties of ecological communities emerge from the selfish evolution of potential antagonists rather than from a cooperative process among members of a monopolistic trust. What is good for the ecosystem is neither well defined nor relevant evolutionarily" Levin (1999 p. 186). Thermodynamic theory says that maximum power of biological systems emerges from selfish interactions between species (competition for energy and selection of the best competitor) that increase energy flow through the ecosystem. As species are selected on the basis of power output, power output of the ecosystem increases. Fitness is maximization of energy flow (Odum 1989). "Survival of the fittest" means persistence of those forms which can command the greatest useful energy per unit time (Odum and Pinkerton 1955). Even predator-prey interactions are cooperative interactions. While they are not selfish for the organism preyed upon, they are for the prey's population. They are selfish for the prey population because they increase stability of ecosystems and ecosystem stability is necessary for species existence.

15.4.2 Evolution Has No Goals

Schneider and Sagan (2005) laid a foundation for thermodynamic theory by saying life was generated because nature abhors a gradient. Corning (2015) criticized the theory as being teleological because it could be interpreted as saying that evolution occurred because of nature's need to decrease the energy gradient between the Earth's incoming solar radiation and outgoing heat. Nature does not have a need to reduce entropy, any more than nature has a need to reduce gravity by compelling apples to fall from trees. Entropy existed long before life on Earth. A supposed "goal" seemed to emerge only as a result of natural selection among species competing for energy in an entropic gradient.

15.4.2.1 Diversity as a Goal

Stephen Jay Gould (1996) was one of the strongest defenders of dogma that evolution has no inherent drive towards long-term "progress". He criticized ecosystem evolution as a ladder of progress, leading towards bigger, faster, and smarter organisms, the assumption being that evolution is somehow driving organisms in that direction. Academic orthodoxy insists that evolution has no fixed goal, but rather occurs through random selection of genes that facilitate survival. However a tendency for evolution along a pathway toward maximum power output does not require a fixed goal nor natural selection operating at levels beyond the individual (Phillips 2008). It results solely from natural selection resulting from competition for energy. Although Gould argued against goal-directedness in evolution, he theorized that evolution's drive is towards diversification. Diversification is as much a "goal" as is developing higher forms of life, and both result from the selection for increased power output.

15.4.3 The Theory Can't Be Tested

The thermodynamic theory of evolution could be experimentally tested, but because of the time scales needed, it would not be practical. It can be tested theoretically with models, using the hierarchical approach and data from the multitude of process level (Jørgensen and Fath 2006).

15.4.4 No Mechanisms

Lewontin (1970), Dawkins (1976), Levin (1999), Whittaker and Woodwell (1972) and Farmer (2005) claimed that for ecosystem evolution there has to be evidence for underlying mechanisms, and that there is no such evidence. Feedback loops discussed in appendix 9 are the mechanisms underlying ecosystem evolution.

15.4.5 Contradicts Neo-Darwinism

Ecosystem evolution does not conflict with the neo-Darwinian view of evolution. Ecosystem evolution results from genetic modification of species that constitute part of the ecosystems' energy and nutrient feedback loops (Jørgensen et al. 2016). As species play their role in these feedback loops, they adapt genetically to variations in these cycles. Their responses are fed back into ecosystem level cycles which change accordingly. Species populations then re-adjust to the new dynamic. Ecosystems change reciprocally and simultaneously with the evolution of species embedded in them. As Lotka (1922) explained, evolution in the thermodynamic sense means that the evolution of "armies of energy transformers – accumulators (plants), and engines (animals)" in what we now call ecosystems is the result of evolution of individual species connected by intra-system feedback loops of nutrients and energy. *"Thermodynamics, correctly interpreted, does not just allow Darwinian evolution; it favors it"* (Ludwig Boltzmann Quotes 2019).

15.4.6 Restricted Definition

Rejection of ecosystem evolution by biologists also stems from their very restricted definition of the term "evolution". In systems terminology, the term evolution means the gradual development of any kind of system from a simple to a more complex form. That is exactly what has transpired in ecosystems. Economists often talk about evolution of economic systems, and political scientists discuss evolution of political systems when they are simply talking about how their systems change, usually from simple to complex.

15.5 Setting Up a Straw Man

"If nothing in biology has meaning except in the light of evolution and if evolution is about individuals and their descendants – i.e. fitness – we should not expect to reach any depth of understanding from studies that are based at the level of the super-individual, whether it be the flora and fauna of the descriptive ecologist or the efficiency of production and resource cycling studied by the production ecologist, or the gross population phenomena of the population biologist. What we see as the organized behavior of systems is the result of the fate of individuals" (Harper 1977). Casting ecosystems as "super-individuals" sets up a straw man that ecosystems are alive. Ecosystems are not "alive". The straw man fallacy occurs when someone takes another person's argument or point, distorts it or exaggerates it in an extreme way, and then attacks the extreme distortion, as if that is really the claim the first person is making.

15.6 Harmony in Nature?

The idea of an ecosystem reflecting a harmony in nature may have originated with Humboldt's description of vegetation zonation on the slopes of volcanoes in Ecuador where he noticed that vegetation was highly synchronized with the local environment. The idea of harmony in nature was furthered by Clement's description of "climax" vegetation in an apparently balanced state with nature. The idea of harmony in nature is reflected in the view that indigenous peoples supposedly live in harmony with nature (Díaz et al. 2018). Botkin (1992) pointed out that this is a myth, and that environmental policy is too often based upon myth.

15.7 Conservatism

McIntosh (1981) observed that the ecosystem concept was slow to be adopted. It threatened academic specialization. A hundred years after ecosystem evolution was first proposed, it is still far from mainstream evolutionary thought. It threatens evolutionary dogma.

References

D. Botkin, *Discordant Harmonies: A New Ecology for the Twenty-first Century* (Oxford University Press, New York, 1992)

A. Calaprice, *The Ultimate Quotable Einstein* (Princeton University Press, Princeton, 2013)

P. Corning, (2015), Review of "Into the Cool". Institute for the Study of Complex Systems. Review withdrawn in 2019

R. Dawkins, *The Selfish Gene* (Oxford University Press, Oxford, 1976)

S. Díaz, U. Pascual, M. Stenseke, et al., Assessing nature's contributions to people. Science **359**(6373), 270–272 (2018)

J. Engelberg, L. Boyarsky, The noncybernetic nature of ecosystems. Am. Nat. **114**, 317–324 (1979)

J.D. Farmer, Cool is not enough. Nature **436**, 627–628 (2005)

S. Gould, *Full House: The Spread of Excellence from Plato to Darwin* (Harmony Books, New York, 1996)

J.L. Harper, The contributions of terrestrial plant studies to the development of the theory of ecology, in *The Changing Scenes in the Natural Sciences 1776–1976*, Special Publication 12, ed. by C. E. Goulden, (Academy of Natural Sciences, Philadelphia, 1977), pp. 139–157

S.E. Jørgensen, B.D. Fath, Examination of ecological networks. Ecol. Model. **196**(3,4), 283–288 (2006). https://doi.org/10.1016/jecolmodel.2006.02.029

S.E. Jørgensen, J.C. Marques, S.N. Nielsen, *Integrated Environmental Management: A Transdisciplinary Approach* (CRC Press, Taylor & Francis Group, Boca Raton/London/New York, 2016)

R. C. Lewontin, The Units of Selection. Ann. Rev. Ecol. Syst. **1**, 1–18 (1970)

S.A. Levin, *Fragile Domain: Complexity and the Commons* (Persus Publishing, Cambridge, 1999)

A.J. Lotka, Natural selection as a physical principle. PNAS **8**(6), 151–154 (1922)

Ludwig Boltzmann Quotes, (2019), https://www.azquotes.com/author/29309-Ludwig_Boltzmann. Accessed 26 Dec 2020
R.P. McIntosh, Succession and ecological theory, in *Forest Succession: Concepts and Application*, ed. by D. West, H. H. Shugart, D. B. Botkin, (Springer, New York, 1981), pp. 10–23
R.V. O'Neill, Is it time to bury the ecosystem concept? (With full military honors, of course!). Ecology **82**(12), 3275–3284 (2001)
E.P. Odum, *Ecology and our Endangered Life-Support Systems* (Sinauer, Sunderland, 1989)
H.T. Odum, R.C. Pinkerton, Time's speed regulator: the optimum efficiency for maximum power output in physical and biological systems. Am. Sci. **43**, 331–343 (1955)
B.C. Patten, E.P. Odum, The cybernetic nature of ecosystems. Am. Nat. **118**, 886–895 (1981)
J. Phillips, Goal functions in ecosystem and biosphere evolution. Prog. Phys. Geogr. **32**(1), 51–64 (2008)
S.T.A. Pickett, V.T. Parker, P.L. Fiedler, The new paradigm in ecology: implications for conservation above the species level, in *Conservation Biology: The Theory and Practice of Nature Conservation, Preservation and Management*, ed. by P. L. Fiedler, (Chapman & Hall, New York, 1992), pp. 65–88
A. Saltelli et al., Five ways to ensure that models serve society: a manifesto. Comment. Nature **582**(7813), 482–484 (2020)
W.M. Schaffer, Ecological abstraction: the consequences of reduced dimensionality in ecological models. Ecol. Monogr. **51**(4), 383–401 (1981)
E.D. Schneider, D. Sagan, *Into the Cool* (University of Chicago Press, Chicago, 2005)
M. Soule, G. Lease, *Reinventing Nature?* (Island Press, Washington, DC, 1995)
Stanford, *Measurement in Science* (The Metaphysics Research Lab, Stanford University, Stanford California, 2015). https://plato.stanford.edu/entries/measurement-science/ Accessed 4 Jan 2021
R.H. Whittaker, G.M. Woodwell, Evolution of natural communities, in *Ecosystem Structure and Function*, ed. by J. A. Wiens, (Oregon State University Press, Corvallis, 1972), pp. 137–159

Chapter 16
What Has Thermodynamics Taught Us About Conservation?

16.1 "Habitat" Is Not Synonymous with "Ecosystem"

The Endangered Species Act of 1973 is the primary law in the United States for protecting imperiled species. The act provides a framework to conserve and protect endangered and threatened species and their habitats. But protecting species' habitat is only half of the requirement for conserving species. "Habitat" is not synonymous with "ecosystem". Habitat is only the location where a plant or animal naturally or normally lives and grows. It is like saying that the house at 321 Elm Street, is John Smith's habitat. His ecosystem in contrast would include his family, their neighbors and the town where their house is located. It would include the flows of the food into and out of the house. Also included would be: all incoming information in the mail, television, and internet; all the repairmen for plumbing and other house repairs; appliances that make the house comfortable; cars, or other transportation; all security devices. Critical to Smith's ecosystem is feedback in the form of taxes which pays for government services in the form of police protection, schools and universities, pollution control, health services and maintenance of roadways. It is all the energy flows that allow Smith to survive.

"Ecosystem" for species includes the food chains that cycle the energy and nutrients necessary for sustainable ecosystem function. It includes predators that control the species, and prey that the species eats. It includes the function of all the other species that sustain the ecosystem – herbivores that break down recalcitrant plant material, the decomposers that recycle nutrients, and the vegetation that supports the entire food web. It includes fire that maintains the ecosystem structure to which indigenous species have become adapted. It includes safeguards to prevent invasive species from monopolizing resources or attacking endangered species. It is part of a landscape that includes several ecosystems required by migrating species. For primates, it includes their culture that teaches them to avoid danger. Most critically

C. F. Jordan, *Evolution from a Thermodynamic Perspective*,
https://doi.org/10.1007/978-3-030-85186-6_16

it includes feedback loops that continually adjust populations to changes in ecosystem energy flow, and ecosystem energy flow to changes in populations.

16.2 Conservation and Feedback

The most critical aspect of species conservation is protection of their life support systems – the flows of energy through their ecosystem. The most important factor is protecting the system's life support system is feedback. Feedback is necessary to keep the energy flow upon which species depend. Whenever feedback is interrupted or distorted, endangered species are particularly threatened because their low numbers make them particularly vulnerable to extinction. Factors that distort feedback through ecosystems include pollution, invasive species, predator control, and ill-advised management programs that destroy ecosystem feedback functions.

16.3 A Few Case Studies

The following are examples that illustrate the role of feedback in conservation of endangered species.

16.3.1 The Serengeti

In the Serengeti, wildebeests and other herbivores exist in a predator-prey dynamic where an increase in prey is followed by an increase in predator followed by a decrease in prey followed by a decrease in predator. If predators are eliminated, there is an unsustainable increase in prey populations. Protection of lions from Maasai cattle herders has been justified because lions are considered a "keystone species", a species that has a disproportionate effect on its food chain relative to its abundance. While maintaining a healthy population of lions certainly is important in keeping wildebeests from overgrazing their food source, the lion-wildebeest interaction is only part of the conservation story. The rest depends on several interacting hierarchical feedback cycles.

One feedback cycle is between the autotrophs, the herbivores, and the decomposers. The herbivores chew and digest the grass, rendering the biomass readily decomposable and nutrients more readily recyclable. As they are taken up by autotrophs, they interact with photosynthate and the autocatalytic interaction stimulates further production of the grass. Superimposed on this cycle is another cycle between lions and FIV (feline immunodeficiency virus), a virus similar to HIV (human immunodeficiency virus) in humans. FIV attacks a feline's immune system and may be transmitted from one cat to another, usually through bite wounds. FIV is very common in Serengeti lions (Dybas 2009). More fights occur when the population is large. FIV spreads and the lion population decreases. FIV becomes less rampant in the diminished lion population, their decline reverses and they begin to slow the swelling of herbivore populations. Population booms and busts are less dramatic than those of lynx-hare populations in northern coniferous forests. Nevertheless, the same elements may be at work in Canadian ecosystems. Canadian lynx are susceptible to at least 11 different genome viruses (Defenders of Wildlife 2020) and it seems possible that outbreaks when populations are high could be behind the lynx-hare population cycles.

Adding to the complexity of these feedback cycles is the occurrence of fires which are more frequent when there is an accumulation of dead grass resulting from lack of grazing when the herbivore population is diminished. Feedback is initiated by lightning strikes, and the resultant fire clears the undecomposed dead grass and allows sunlight to stimulate new growth. Fire solubilizes many of the nutrients when grass is turned to ash, and knocks back woody species that compete with grass. Fires result in increased grass productivity that enables the herbivore population to recover that in turn enables the lion population to recover.

16.3.2 Black Footed Ferret

Black footed ferrets are a weasel-like endangered species. Their habitat consists of burrows of prairie dogs that live in the short-grass prairies of western United States. Protecting their habitat entails forbidding of plowing or grazing near burrows where prairie dogs live, but that is not enough. Protecting the prairie dogs when populations decline and controlling them when they expand is also necessary. Prairie dogs are herbivores. Like all herbivores, prairie dog populations if not controlled, have the potential to degrade the plant resource upon which their trophic chain depends. Black footed ferrets are the carnivore that regulates ecosystem stability by preying upon prairie dogs. To protect the prairie dog-ferret interaction, the whole short-grass prairie ecosystem must be protected. Protection of the prairie includes allowing fire. Occasional fire is necessary to kill invasive shrubs and to allow sunlight to penetrate to the apical meristem of the regrowing grasses. The short-grass prairie once teemed with massive herds of free-ranging bison and pronghorn antelope, as well as huge prairie dog colonies. The ecosystem was maintained by migration of bison and periodic fires, much as the Serengeti ecosystem is maintained by wildebeest migration and fires that favor grasses over trees. Just as fire is necessary for preservation of the Serengeti ecosystem, fire is necessary for the preservation of the short-grass prairie ecosystem. Just as light grazing by wildebeests benefits Serengeti grasses, light grazing by bison on the shortgrass prairie helps maintain the habitat for the black footed ferret.

The American Prairie Reserve in the Missouri Breaks region of central Montana is a reserve whose objective is to restore the short grass prairie with the original complement of species. It consists of 420,000 acres that are next to 2.75 million acres of public land that will provide the ecosystem (not just habitat) for prairie dogs, sage, grouse, coyote, bighorn sheep, bison, other species of native plants, birds and mammals, and particularly black footed ferrets. Controlled burns are part of the conservation effort (The Economist 2020; American Prairie Reserve 2020). No mention is made however, for plans to control the bison population. Without the feedback initiated by wolves or other top predators, there will be no control that stabilizes the system.

16.3.3 Golden Lion Tamarin

Learning is a feedback system where a primate can learn from someone else's mistake or learn from his or her own mistake. The Atlantic forest of Brazil is home to the golden lion tamarin, a New World monkey about 14 inches long with a full reddish-gold mane encircling its head. Tamarins have been threatened by deforestation. Efforts to save them have taken two approaches. One is to breed them in captivity, train them to survive, radio tag individuals and then release them into remaining patches of forest on privately owned ranches (Ackerman 1991). Training involves many factors. When I visited a release site in the 1980s, we found a dead individual. The analysis of cause of death revealed that it had eaten a poisonous coral snake. The wild tamarins eat non-poisonous snakes and know how to identify them, but that phase had not been part of the training for that unfortunate fellow. That knowledge is part of the tamarin feedback learning that is part of the tamarin culture.

16.3.4 Whooping Cranes

Whooping cranes are an endangered species. One of the more spectacular efforts at species conservation was the project to teach captively raised whooping cranes the ability to find their way from their nesting habitat in central Wisconsin to their winter-feeding ground in Florida. Chicks were hatched in a nursery and when they became strong enough to fly, were led by a crane-costumed pilot flying an ultralight aircraft from Wisconsin to Florida. While the training was successful, the birds did not reproduce in the wild, because the birds were too accustomed to the artificial habitat where they were born. As a result, "Operation Migration" decided to allow crane parents to raise chicks in the wild and teach the chicks how to survive in deep-water marshes, their natural ecosystem (Dalley 2016). Recovery is now slow but steady. Here, the feedback learning was for humans, not for the endangered species.

16.3.5 Puerto Rican Parrot

The critically endangered Puerto Rican parrot is one of the rarest birds in the world. Efforts to save them include captive breeding and restoration of nest cavities in old trees after devastation of habitat by hurricanes (Lopez 2020). Predation by black rats has been a major impediment to success. Rats were unintentionally introduced to North America and Puerto Rico as "stowaways" in the bilges of ships arriving from Europe. The earliest mentioning of rats in Puerto Rico dates back to 1493 and the Black (Roof) Rats that arrived with Christopher Columbus on his second voyage of discovery. At the present time there is no effective predator for black rats. Initially it was hoped that the small Indian Mongoose could control them. The mongoose was first brought to Puerto Rico from the Malay Peninsula in 1877 in an attempt to control the black rat on sugar cane plantations (Mowbray 2020). The effort was not completely successful, because mongooses sleep at night and forage during the day, while the rats sleep during the day and forage at night (Ojeda 2014). Black rats are now controlled through the use of poison bait stations strategically located near active parrot nests (Núñez-García and Torres-Báez 2020). Humans have become the indispensable controlling agent. The point here is that for control to be effective, the top predator must be synchronized in the system, a key factor in the evolution of ecosystems.

16.4 Conserving Feedback Loops

Regardless of ecosystem, conservation of endangered species depends on conservation of energy flow and nutrient feedback loops. Conserving feedback loops implies conserving the food chains that make up the feedback system. The critical link in food chains is the decomposer link that returns nutrients back to the autotrophs. Despite the fact that this step was the very first to become established in the evolution of ecosystems, many scientists and resource managers still do not realize its importance. This lack of understanding impedes conservation efforts. An example is the northern spotted owl. Northern spotted owls indigenous to old growth forests of northwestern United States have been on the endangered species list for many years, and the intense forest fires in the summer of 2020 have pushed them closer to extinction (Urness 2020). Because there are many standing and fallen dead trees after fires, calls for salvage logging are inevitable. The problem with salvage logging is it removes dead trees that supply the organic matter which fuels the soil microbial community that recycles nutrients. Without these recyclers, nutrients from the system are lost via erosion or leaching. The ecosystems can become unstable and revert to chapparal type vegetation as happened with logged ponderosa pine stands in Arizona in the nineteenth century (Dieterich and Hibbert 1990). The heathlands of Pennine England once were forests, but deforestation followed by frequent fires turned them into heaths or moors (Rackham 1986) with much reduced feedback loops.

16.5 The Importance of Reservoirs for Recovery of Feedback Loops

Mt. St. Helens in the Gifford Pinchot National Forest of southwestern Washington State is a volcano that had been quiet for over a hundred years, and its slopes were covered with forest typical of the Cascade Range. On the morning of May 18, 1980, Mt. St. Helens erupted. The resulting landslide smothered 60 km^2 of river valley to an average depth of 45 m. A vigorous vertical eruption plume followed, raining shards of volcanic debris (tephra) downwind and producing pumice-rich pyroclastic flows that settled atop the landslide deposit. Erosion, burial, heat, blunt force, and abrasion killed the forests on the volcano's slopes and produced a mosaic of landscape disturbances that extended tens of kilometers from the volcano. Major et al. (2020) described the progress of ecosystem recovery through 40 years. They found that surviving plants and animals from pre-eruption ecosystems persisted and were widely distributed. Even some of the most heavily affected landscapes were not as sterile as initially assumed. Lupines and alders flourished in the nutrient-impoverished volcanic substrates because of their ability to interact with root-borne, nitrogen-fixing bacteria. The pioneer species facilitated colonization of many other plants and animals by ameliorating inhospitable nutrient conditions and initiating soil development. Other species, such as the northern pocket gopher facilitated recovery by mixing inert tephra with buried nutrient-rich soil, thus improving conditions for plant growth. The unconsolidated texture of eruption deposits allowed animals to excavate burrows. Surviving plants penetrated thin deposits by sprouting new shoots upward, and roots of seedlings penetrated downward into nutrient-rich pre-eruption soils. Forests and aquatic systems surrounding the disturbed landscape were largely intact and supported plants and animals that served as source populations for post-eruption recovery. These populations were particularly important for the recovery of species that experienced complete mortality such as many large mammals and birds.

16.6 Biodiversity Hotspots

Biodiversity hotspots are landscapes within which top predators, often the most endangered species, can move depending on the season. For example, during the dry season in the rain forest, jaguars will prey upon fish concentrated in small pools along flood plains, but during the rainy season, they pursue peccaries foraging on roots and grubs in upland forests. A biodiversity hotspot includes both types of ecosystems and both types of food chains.

Approximately 36 regions around the world qualify as biodiversity hotspots and they support about 60% of the world's birds, reptile, mammal, plant and amphibian species with the majority of these species as endemics. Myers et al. (2000) have listed the top 15 of the hotspots that contain 2500 endemic plants species and 10 of them at least 5000. Topping the list is the tropical Andes, with ecosystems ranging from high-altitude paramos to rain forests at the edge of the Amazon Basin.

16.7 A Conservationist's Dilemma

Should conservation concentrate on the species or the ecosystem? Panda bears and tigers have charisma while flows of energy do not. But by focusing entirely on the species, conservationists miss the role of the ecosystem in supporting the species. Conserving feedback loops is critical in biodiversity hotspots, where interruptions in energy flows and nutrient cycles would predominantly effect threatened and endangered species. To save a species, save its ecosystem and its incorporated services of nature.

16.8 Conservation and Feedback: A Final Word

I began this book because I felt that ignoring the ecosystem perspective in formulation of a law to conserve species was a mistake. It resulted in a law that did not emphasize the most important factor for saving species: preservation of the species' life support system, but I realize that the role of an ecosystem as a species' life support system was a concept not generally appreciated in 1973. In the years since, the application of systems analysis to ecology has resulted in the understanding that feedback between species and ecosystems is critical for stability of species' populations. The stability resulting from thermodynamic feedback has basic significance for conservation: It preserves the life support system of endangered species. Conservation of endangered species depends on maintenance of thermodynamically powered feedback loops in the endangered species' ecosystem.

References

D. Ackerman, Golden monkeys. The New Yorker (24 June 1991), p. 36

American Prairie Reserve, (2020), https://en.wikipedia.org/wiki/American_Prairie_Reserve. Accessed 7 Dec 2020

J. Dalley Whooping crane migration changes course (2016). https://www.sierraclub.org/sierra/2016-5-september-october/green-life/whooping-crane-migration-changes-course. Accessed 14 Dec 2020
Defenders of Wildlife, Viruses threaten wildlife too (2020). https://defenders.org/blog/2020/03/viruses-threaten-wildlife-too. Accessed 20 Feb 2021
J.H. Dieterich, A.R. Hibbert, Fire history in a small ponderosa pine stand surrounded by chaparral, in *Effects of Fire Management of Southwestern Natural Resources: Proceedings of the Symposium, Tucson, AZ, November 14–17, 1988. General Technical Report RM-GTR-191*, ed. by J. S. Kramer, (USDA Forest Service, Rocky Mountain Forest and Range Experiment Station, Fort Collins, 1990), pp. 168–173
C.L. Dybas, Infectious diseases subdue Serengeti lions. Bioscience **59**(1), 8–13 (2009)
Economist, Where the wild things are. De Economist (14 Nov 2020), pp. 26–27
M. Lopez, Puerto Rican parrot recovery (2020), https://www.fws.gov/southeast/pubs/facts/rioescon.pdf. Accessed 26 Dec 2020
J.J. Major, C.M. Crisafulli, F.J. Swanson, Lessons from a post-eruption landscape. Eos.org, a weekly publication of the American Geophysical Union (2020), https://eos.org/features/lessons-from-a-post-eruption-landscape. Accessed 17 Dec 2020
A. Mowbray, Small Indian mongoose (2020). https://www.fs.usda.gov/detail/elyunque/learning/nature-science/?cid=fsbdev3_042886. Accessed 26 Dec 2020
N. Myers, R.A. Mittermeier, C.G. Mittermeier, G.A.B. Fonseca, J. Kent, Biodiversity hotspots for conservation priorities. Nature **403**(24 Feb), 853–858 (2000)
F. Núñez-García, P. Torres-Báez, Recovery plan for the Puerto Rican Parrot (Amazona vittate) (2020), https://esadocs.defenders-cci.org/ESAdocs/recovery_plan/090617.pdf. Accessed on 13 Dec 2020
A. Ojeda, Never try to kill a rat with a mongoose (2014), http://www.blessedbutstressed.com/2014/04/07/never-try-to-kill-a-rat-with-a-mongoose/. Accessed 27 Dec 2020
O. Rackham, *The History of the Countryside. The Classic History of Britains's landscape, flora, and fauna* (J.M. Dent Publishers, London, 1986). Cited in Moorland, Wikipedia. https://en.wikipedia.org/wiki/Moorland
Z. Urness, Spotted owls pushed closer to extinction vortex by Oregon wildfires. Salem Statesman J. (2020). https://www.usatoday.com/story/news/nation/2020/12/01/northern-spotted-owl-habitat-threatened-oregon-fires/6462923002/ Accessed 1 November, 2021

Part II
Application

Thermodynamic Laws and Agriculture

Continuous economic expansion is necessary for survival of a nation-state that competes with other nations for resources and ultimately survival. Economic expansion causes pressure on all sectors of the economy to expand, but this is especially difficult for agriculture. For other sectors, the energy needed for expansion is still available – fossil fuel, nuclear, hydroelectric, and wind. For agriculture, the energy needed to expand is photosynthetic energy, but expansion is not possible because the rate of photosynthesis has already reached the maximum possible through the use of fertilizers and pesticides. More petroleum will not help because fossil fuel energy cannot be converted into food energy. Thermodynamic laws are creating problems:

The first law says that you can't win. You can't get more energy from a crop than what photosynthesis puts in. All that can be done is to rearrange how the crop uses that energy.

The second law says that you can't break even. Entropy will cause your farm to degenerate. Weeds will invade, insects will eat your crop, and your tractor will break down.

The third law says you can't even get out of the game. You can't quit farming because you owe the bank.

Chapter 17
A Farmer's Dilemma

17.1 How Ecosystems and Economic Systems Are the Same

Ecological systems and economic systems are similar in many ways, and insights from one system can increase understanding of the other.

- Systems grow by maximizing power output.
- Growth is faster when power is concentrated in a few products or species.
- Immature systems use energy inefficiently.
- Growth of systems is eventually limited by available energy.
- Feedback enables systems to adapt to energy limitations through stabilization.
- Stabilization is abetted by an increase in diversity that increases efficiency.
- A highly efficient system is stable but resistant to change.
- Resistance to change results in systems decline.

17.2 How Ecosystems and Economic Systems Are Different

In ecological systems the transition to stabilization results from feedback and change is gradual. When natural ecosystems reach the point of energy limitation, feedback causes a switch from a strategy of growth to a strategy of stability. In contrast, economic systems do not stabilize. Politicians keep trying to keep economic systems growing by printing more money and decreasing lending rates from the Federal Reserve. There are important reasons why economies are compelled to remain in the growth stage. Competition between nations for world dominance forces a national economic system to maintain growth. Rogue nations continually strive for expansion by incorporating smaller or weaker nations. To defend against

C. F. Jordan, *Evolution from a Thermodynamic Perspective*,
https://doi.org/10.1007/978-3-030-85186-6_17

this expansion, democratic societies must increase their economy to support a stronger economy in general and an increase in military capacity in particular. An equally important reason is that population growth necessitates economic growth to provide jobs for all. So there is an inherent conflict between human efforts to increase the economy and thermodynamic laws that limits perpetual growth.

17.3 Planet Earth Is a Feedback System

Global warming is occurring simultaneously with rising levels of carbon dioxide and methane in the atmosphere caused by economic activity in which fossil fuels are burned. That does not necessarily mean that economic activity is causing global warming. Global warming has occurred in the past. During the early Eocene Epoch about 50 million years ago, the world was basically ice free (Keating-Bitonti et al. 2011). Sea levels were much higher than they are today. The difference is that warming is occurring much faster today than it did 50 million years ago. That suggests increasing economic activity *is* the factor causing global change.

Sea-level rise is the feedback that tells us the world economy is unstable. But it hasn't risen much, so we aren't paying much attention. The world will increase its rate of energy use for a while yet. Fossil fuel reserves have not been depleted. Nuclear energy, and wind and solar energy are possibilities. Radioactive waste disposal will eventually pose a problem for increase in atomic energy. For wind and solar energy, it is not clear if wind turbines and solar panels can produce more energy than it takes to manufacture them. In any case, we are not yet nearing the end of energy growth.

For biological systems, the situation is different. Exosomatic energy (fossil fuels, nuclear, wind power) cannot be converted into food. To produce food, endosomatic energy is needed, the energy supplied by photosynthesis. Exosomatic energy merely facilitates endosomatic energy production by photosynthesis. Genetic engineers have increased agricultural yield by creating plants that divert endosomatic energy used for defense against insect pests and for competition against weeds to energy used for yield, but now there is little endosomatic energy left to divert. (Hamilton et al. 2013). Most is already used for yield. Try as they may, farmers cannot increase yield on existing land. Nevertheless, economic and political pressures are forcing them to try. Adding to the problem is depletion of aquifers and pollution of waterways caused by their efforts.

Farmers are in a difficult position. They are part of an economic system that demands continuous growth. But they are also part of a thermodynamic system that tells them continuous growth is impossible. The following chapters look more closely into this dilemma, and examine how some farmers are dealing with it.

References

A. Hamilton, S.B. Balogh, A. Maxwell, C.A.S. Hall, Efficiency of edible agriculture in Canada and the U.S. over the past three and four decades. Energies **6**(3), 1764–1793 (2013). https://doi.org/10.3390/en6031764

C. Keating-Bitonti, L. Ivany, H. Affek, P. Douglas, S. Samson, Warm, not super-hot temperatures in the early Eocene subtropics. Geology **39**(8), 771–774 (2011). https://doi.org/10.1130/G32054.1

Chapter 18
Agricultural Problems Are Systems Problems

18.1 The Morrill Land-Grant Acts

As cities in the United States grew in the nineteenth Century, there was a need to increased food production. The Morrill Land-Grant Acts created the agricultural colleges in 1862 to *"teach such branches of learning as are related to agriculture and the mechanic arts, in such manner as the legislatures of the States may respectively prescribe, in order to promote the liberal and practical education of the industrial classes in the several pursuits and professions in life.*" An anticipated result of the Land-Grant Acts was to increase agricultural production through better farming practices. An unexpected result of establishing agriculture as an academic discipline was the proliferation of isolated specialties. Whereas farmers take a holistic view of their farm and base their management strategy on that view, academic researchers take a reductionist view. Reductionism has been the essence of science and has contributed greatly to improving the human condition, so it is no surprise that researchers in agricultural colleges followed that path. The problem with reductionism in studies of systems is that it does not consider indirect interactions that produce unexpected "side effects".

Box 18.1
Farmers sometimes wonder, why do scientists do the things they do? Agricultural researchers wonder, why do farmers often ignore their advice?

C. F. Jordan, *Evolution from a Thermodynamic Perspective*,
https://doi.org/10.1007/978-3-030-85186-6_18

18.2 Agricultural Colleges

18.2.1 The Evolution of Agricultural Research

Agricultural research has experienced several stages.

18.2.2 Reductionism

A farm can be reduced into many component parts. To increase the effectiveness of each component, agricultural colleges created a department for each component: soils, crops, animal science, dairy science, poultry science, horticulture, insect control, weed control, plant pathology, agricultural economics and food science. It did not take long however, for specialties to arise within each department. Soil science fissioned into soil physics, soil chemistry, and soil morphology. Within these specialties there were sub-specialties. In soil chemistry there emerged inorganic chemistry that dealt with fertilizer application, and organic chemistry focused on the nature of soil organic matter. It was understood however that the goal of the college and of each department, specialty and sub-specialty was to maximize the effectiveness of respective components so that when all components are simultaneously maximized, overall agricultural productivity would be increased.

Science is an endless frontier. No matter the degree to which a phenomenon has been analyzed, it can always be broken down further. Accompanying reductionism is an increase in esoteric jargon necessary to facilitate communication between sub-specialists. Soil taxonomy provides an example. Soils were first classified based on parent material, climate, organisms, topography and time (Jenny 1941), a system similar to the old Russian system (Basinski 1959). Since then, soils have been classified according to soil chemical and physical properties: Entisols, Inceptisols, Vertisols, Mollisols, Alfisols, Aridisols, Spodosols, Ultisols, Oxisols, Histosols, Andisols, and Gelisols (Soil Survey Staff 1999; Munna 2019). If a soil is a Mollisol, we know that it has a high content of organic matter and is relatively fertile. They are the "breadbasket" soils of most continents. If a soil is a Spodosol, that means it is infertile and is unsuitable for agriculture. Nutrient elements have been leached away by weathering processes. This level of knowledge works fine for a farmer who wants to buy a piece of land in Wisconsin with good cropping potential. Many hardwood forests in humid areas like Wisconsin grow on soils within the Alfisol soil order. Alfisols have a clay-enriched subsoil and a relatively high fertility that is conducive for agriculture.

Box 18.2
Many immigrants from Scandinavia chose to settle in regions of Wisconsin and Minnesota that reminded them of their homeland, areas covered with conifers. Late-comers were relegated to hardwood sites. In the Northern United States, pines dominate on infertile Spodosols, so the latecomers actually fared better on the Alfisols. Aldo Leopold's (1949) book A Sand County Almanac has been a touchstone of modern conservation science, policy, and ethics. It was based on experience on a sandy Spodosol in Wisconsin.

It soon became apparent however that Jenny's classification could be further refined as studies revealed more details about physical and chemical properties of soils. The question that has never been asked is how far should reductionism be carried. Is there a limit where understanding and usefulness is optimum, or can effectiveness always be increased by further reductionistic studies? In pursuit of more detail, taxonomists refined the classification system. Soil orders were broken down into suborders, great groups, subgroups, family, and series. For example, the soil in the San Joaquin valley of California is classified as: Order: Alfisols; Suborder, Xeralfs; Great Group, Durixeralfs; Subgroup, Abruptic Durixeralfs; Family, Fine, Mixed, Active, Thermic Abruptic Durixeralfs; Series: San Joaquin. Does that help *promote the liberal and practical education of the industrial classes in the several pursuits and professions in life*? The esoteric jargon inherent and focus on ever-more detail in reductionistic science inhibits communication and understanding that could reduce agricultural problems. Don't scientists realize this? Many do, but reductionism is what gets them grants, publications, promotion, and tenure.

18.2.3 The Empirical Approach

Agricultural research began with the empirical approach – "Let's throw on another ton of fertilizer and see what happens." Today, production problems are typically analyzed by specialists who apply the traditional "cause and effect" approach. Too many insect pests? Increase pesticides! Low yields? Increase nitrogen fertilizers! (Jayne and Sanchez 2021). In recent years, sustainability has become a priority. The cause-and-effect approach cannot solve sustainability problems because solving a problem at one part of a system creates new problems in other parts of the system. Agricultural problems result from changes in interactions between all system components that are impossible to predict by cause-and-effect models.

The cause-and-effect approach to agricultural problems is epitomized by a statistical method known as analysis of variance (ANOVA). Agronomists lay out a series of plots randomly arranged on the same soil type and with the same crop, and to half of the plots they give a treatment, for example a new type of pesticide. The other half is kept as a control. With ANOVA, a conclusion as to whether a treatment has an effect is based on a 95% level of confidence, the level of confidence used by statisticians as being high enough to be convincing. It also means there is a 5% probability that there are no differences. Confidence can be improved through the use of the Bayesian approach that uses sequential analysis to include data from earlier trials. One of my projects in graduate school was to determine if a brush fire increased potassium concentrations in the soil compared to a non-burned site. The data did not quite meet the 95% level of confidence that it was higher, but it was clear that if I went back into the field to take a few more samples and incorporated them into my original data set, I could clear that hurdle. The statistics professor would not allow that. In the Bayesian model, it would be permissible.

A problem for either approach is the assumption is that all the plots are identical except for treatment, and that any difference in results is due to the treatment. The system works well when differences are large, but a problem is that in natural ecosystems, paired plots are never completely identical. One never knows if differences are due to treatment or to differences between plots. In agricultural experiments where an effort is made to homogenize the plots, confidence can be higher. Yet there are always differences between plots, if only because some are upwind and others are downwind of a potential chemical source.

Box 18.3
For a farmer, the effectiveness of a treatment must be obvious without resorting to statistics. As a cynical farmer once said, "If you have to use statistics to prove your point, it can't be very important."

18.2.4 The Analytical Approach

In the twentieth Century, many scientists began to believe that the key to increasing productivity was through the analytic approach that would uncover the mechanisms that controlled production. Consequently there followed decades of research on functional processes that underlie agricultural productivity. It was assumed that each component of a farming system should optimize its respective function. If a pest control model indicated that for optimum insect control a field should be sprayed every two weeks, that would be the recommendation for the farmer. For maximizing yield, determine how much fertilizer is needed and then suggest that. The presumption is that when performance of each of these sub-models is maximized, they can be added together into an all-encompassing model that indicate the best way to maximize agricultural yield. It is linear thinking based on the assumption that the effects of each treatment do not interact. This assumption has led to the failure of many development projects.

18.3 Agricultural Development Models

The assumption of the analytical approach has been that optimization of all components individually would lead to optimization of the whole. The fallacy of this assumption is demonstrated by The Green Revolution. The "First Green Revolution" refers to a series of research, development, and technology transfer activities occurring in the decades after WWII that resulted in dramatic increases of agricultural production. The system is often referred to as "industrial agriculture." Researchers developed pesticides, herbicides, inorganic fertilizers and intensive irrigation methods that increased crop production when used with high yielding varieties of seeds. This "Green Revolution" quickly benefitted farmers who could afford the price of these innovations. As the Green Revolution spread, it was hailed as the key to banishing world hunger. It required a number of years for the down-side of the revolution to appear, but a price has now been paid through erosion, depleted aquifers, ground water pollution, and pests and diseases that have increased resistance to control. The most serious unintended consequence in India was that the increase of food availability led to an increase in population to the point where hunger again is becoming a problem (Pingali 2012). To solve the problem, the same cause and effect approach is being used, except genetic engineering now is the basic tool. It is being called the "Second Green Revolution" (Piesse 2017). The belief that increased yield this time will solve world hunger is reflected in statements of geneticists and agricultural scientists. *"The current trajectory for crop yields is insufficient to nourish the world's population by 2050. Greater and more consistent crop production must be achieved against a backdrop of climatic stress that limits yields, owing to shifts in pests and pathogens, precipitation, heat-waves and other weather extremes. Accelerated crop improvement must leverage naturally evolved traits and transformative engineering driven by mechanistic understanding to yield the resilient production systems that are needed to ensure future harvests in a post-Green Revolution"* (Bailey-Serres et al. 2019).

Blind faith is repeating the same mistake and expecting a more favorable result. Another example is from sub-Saharan Africa. After a 30-year hiatus, large-scale irrigation projects have returned to the development agenda of the region. Higginbottom et al. (2021) found *"overwhelming evidence that investments have failed to deliver promised benefits."* Reasons included: an emphasis on national food security promotes low-value crops that reduce economic viability; proposals are unrealistically large, driven by optimism bias and political incentives; centralized bureaucracies lack the technical expertise, local knowledge and financial resources to ensure long-term maintenance. The authors conclude that there is a need for greater learning from past investments' outcomes if improvements in agricultural productivity and water security across sus-Saharan Africa are to be realized.

The situation is no better in India. Jain et al. (2021) report that due to groundwater depletion, cropping intensity may decrease by 68% in some regions.

Combatting world hunger requires the realization that food systems are feedback systems, and that any change in one part of the system can result in unanticipated changes in another part of the system. Myron Tribus who was an organizational theorist and director of the Center for Advanced Engineering at MIT, expressed it this way: *"If you try to improve the performance of a system of people, machines, and procedures by setting numerical goals for the improvement of individual parts of the system, the system will defeat your efforts and you will pay a price where you least expect it"* (Tribus 1992).

18.3.1 The Ratchet Effect

The ratchet effect is a situation or process that tightens steadily in a series of small but irreversible steps. The battle to control pest insects illustrates the effect. A pesticide has a limited useful life span before pests evolve immunity. New or more powerful pesticides are needed to out- maneuver nature which is continually fighting back. Science must increase weapons faster and faster just to maintain parity. This has been called the "Red Queen effect" (Dyer 2014) from Lewis Carroll's Through the Looking-Glass where the Red Queen tells Alice she has to run faster and faster just to stay in the same place. Once society becomes dependent on increased yield, there is no alternative but to continually tighten the ratchet. Tightening becomes exacerbated when genetically modified crops are bred to depend on these subsidies.

Further tightening of the ratchet comes about due to competition between farmers. If new agricultural technologies result in increased yield, an early adopter farmer will employ them. To keep competitive, other farmers will soon adopt these technologies, and energy input rates to cropland will increase. Increased yield will result in lower prices and more food availability, so world hunger will be temporarily lessened, but world population will grow. Because of the larger population that depends on increased yield, the technologies that permitted the yield increase cannot be abandoned. The ratchet has tightened. The outcome is that there will be an increased impact on the farmland supporting this increased yield, and as yield for all farms increases, there will be a decrease in individual farm profits due to increased supply and resultant competition in the marketplace. The introduction of hybrid corn in Midwestern United States exemplifies this sequence. The increase in yield by early adopters required an increase in fertilization rates, but that increased pollution from nitrogen leaching. When late adopters employed hybrids, the greater supply of corn lowered prices and profits decreased, while nitrogen pollution increased. Thus, problems of nutrient pollution in the Gulf of Mexico persist (Manning et al. 2020). Similar dynamics have occurred in the Baltic Sea, the Adriatic Sea, and the Black Sea (Mitcsh et al. 2001).

18.3.1.1 Jevons' Paradox

Much of agricultural research is dedicated to improving efficiency. Drip irrigation has improved the efficiency of water use. Precision agriculture has improved the efficiency of fertilizers. Governments and economists often assume that efficiency gains will lower resource depletion. This assumption is called Jevons' paradox. It occurs when technological progress or government policy increases the efficiency with which a resource is used, but that does not help conserve the resource. The demand for that resource rises due to cheaper cost. Improvements in air conditioners increased their energy use efficiency, and this was applauded as an energy saving measure. What happened though was that the decrease in cost of running air conditioners enabled many more people to buy them. The result was an increase in demand for energy.

18.4 The Systems Approach

Conventional approaches cannot solve agricultural development problems because solving a problem at one part of a system creates new problems in other parts of the system. A systems approach is now sometimes used to understand agricultural problems (Donatelli et al. 2017). It considers that agricultural systems consist of many sub-systems, all interacting at the same time. To understand agricultural pollution, it is necessary to understand the interaction of the sub-systems. An example is the problem of nitrogen pollution in waterways resulting from excess fertilization of agricultural fields. Nitrogen reacts in different ways with each part of the ecosystem. Soil scientists know that some is held on the surface of clay particles, some is dissolved in soil water, and some is volatilized. Hydrologists can determine how it is moved by ground water. Physiologists can model how it moves up through the crops in a gradually broadening pulse. Ecologists can measure the rate at which it is transferred from crops to the soil organic matter via root decomposition, and the rate that it becomes part of the soil organic matter. Microbial ecologists can determine the rate it is released from soil organic matter and leached into soil water. Each of these processes can be modeled. However, synthesis of these sub-models cannot predict the rate that it will move from fertilizer spreader to nearby streams because there is no top-down control that reduces deviation amplifying interactions of the subsystems. A systems model that incorporates subsystems has been used to compare nitrogen leaching from conventional versus organic farming systems (Hansen et al. 2000). Such models cannot predict actual outcomes, but with the use of sensitivity analysis, they lead to an understanding of what controls nitrogen transfer. They can lead to recommendations as to what should be done to reduce nitrogen pollution. To predict actual outcomes, a top-down model is needed A top-down model requires data from field studies of the whole system. That is done by fertilizing a field with a known amount of nitrogen, and then measuring its concentration in the stream. That will tell you how much pollution comes from a given amount of fertilizers. However it will not indicate the factors that control nitrogen transfer, and cannot tell what steps need to be taken to minimize pollution. Both bottom-up and top-down analyses are needed formulate better management strategies.

18.4.1 Adaptive and Deterministic Cycles

In agricultural systems as in ecological systems, adaptive cycles are nested within deterministic cycles. The slower and larger levels set the limits within which faster and smaller ones function. The top-down, thermodynamic deterministic dynamic sets the overall trend, while within, the bottom-up, stochastic dynamic searches for more efficient and effective ways to adapt to the thermodynamic imperative. Much of agricultural research consists of analyzing ecosystem processes that represent the smaller faster adaptive levels within the slower, larger ecosystem dynamics that control them. Process models tell how each component of the system works. Efforts to synthesize these sub-models is a bottom-up approach. The problem with synthesis is that errors in each of the sub-models are cumulative (Saltelli et al. 2020). Agricultural development models need top-down control to prevent bottom-up influences that send the system out of control. The Panarchy model in Chap. 11 exemplifies top-down control. It shows how growth is limited by biophysical constraints. When systems growth reaches a biophysical limit, management should change to consolidation, that is establishing a bureaucracy that puts restrictions on the amount of pollution allowed by agriculture. If instead the system continues to grow, the system will become unsustainable.

18.4.2 Business Cycles and New Paradigms

The interaction between smaller faster adaptive levels nested in larger slower levels occurs in business cycles. In a capitalist economy, businesses are initiated to make money by offering a service or product that serves the needs or desires of society. To remain profitable, they must adapt to changes in society that often results from technical advances. They cannot keep doing the same thing but in ever-more refined detail. We can see that in the automotive industry. Gasoline engines have become increasingly more efficient and less polluting. Advances however have become slower as the top-down control limits further efficiency. There has been resistance to abandon the gasoline powered engine, but early-adaptive companies have recognized that ever-more refinement of the gasoline engine is a dead-end and have embraced the electric approach that mandates the direction the engine designers must take.

When output of businesses, of government agencies or of agricultural colleges no longer adapt to the needs of society, it's time for a new paradigm.

References

J. Bailey-Serres, J.E. Parker, E.A. Ainsworth, G.E.D. Oldroyd, J.I. Schoreder, Genetic strategies for improving crop yields. Nature **575**(7781), 109–118 (2019)

J.J. Basinski, The Russian approach to soil classification and its recent development. J. Soil Sci. **10**(1), 14–26 (1959)

M. Donatelli, R.D. Magarey, S. Bregaglio, L. Willocquet, J.P.M. Whish, S. Savary, Modelling the impacts of pests and diseases on agricultural systems. Agric. Syst. **155**, 213–224 (2017)

A. Dyer, *Chasing the Red Queen: The Evolutionary Race Between Agricultural Pests and Poisons* (Island Press, Washington, DC, 2014)

B. Hansen, E.S. Kristensen, R. Grant, H. Hogn-Jensen, S.E. Simmelsgaard, J.E. Olesen, Nitrogen leaching from conventional versus organic farming systems – a systems modelling approach. Eur. J. Agron. **13**(1), 65–82 (2000)

T.P. Higginbottom, R. Adhikari, R. Dimova, et al., Performance of large-scale irrigation projects in sub-Saharan Africa. Nat. Sustain. (2021). https://doi.org/10.1038/s41893-020-00670-7

M. Jain, R. Fishman, P. Mondal, G.L. Galford, N. Bhattarai, Groundwater depletion will reduce cropping intensity in India. Sci. Adv. **7**(9) (2021). https://doi.org/10.1126/sciadv.abd2849

T.S. Jayne, P.A. Sanchez, Agricultural productivity must improve in sub-Saharan Africa. Science **372**(6546), 1045–1047 (2021)

H. Jenny, *Factors of Soil Formation* (McGraw Hill, New York, 1941)

A. Leopold, *A Sand County Almanac* (Oxford University Press, Oxford, 1949)

D.W.P. Manning, A.D. Rosemond, J.P. Benstead, P.M. Bumpers, J.S. Kominoski, Transport of N and P in U.S. streams and rivers differs with land use and between dissolved and particulate forms. Ecol. Appl. Published online (2020) https://doi.org/10.1002/eap.2130. Accessed 26 Dec 2020

W.J. Mitcsh, J. Day, J.W. Gilliam, P.M. Groffman, D.L. Hey, G.W. Randall, N. Wang, Reducing nitrogen loading to the Gulf of Mexico from the Mississippi River Basin: strategies to counter a persistent ecological problem. Bioscience **51**, 373–388 (2001)

N.H. Munna, 12 soil orders in soil taxonomy with their major characteristics. Earth Review, Environmental Research and Training (2019), https://www.earthreview.org/12-soil-orders-in-soil-taxonomy/. Accessed 22 Feb 2021

M. Piesse, Genetically modified food and the second green revolution. Future Directions International (2017). http://www.futuredirections.org.au/publication/genetically-modified-food-second-green-revolution/. Accessed 30 Dec 2020

P.L. Pingali, Green revolution: impacts, limits, and the path ahead. PNAS **109**(31), 12302–12308 (2012)

A. Saltelli et al., Five ways to ensure that models serve society: a manifesto. Comment. Nature **582**(7813), 482–484 (2020)

Soil Survey Staff, *Soil Taxonomy: A Basic System of Soil Classification for Making and Interpreting Soil Surveys*, 2nd edn. (Natural Resources Conservation Service. U.S. Department of Agriculture Handbook 436, Washington, DC, 1999)

M. Tribus, *Quality First* (National Society of Professional Engineers Myron Tribus Quotes, Washington, DC, 1992). https://www.azquotes.com/author/31772-Myron_Tribus. Accessed 18 Apr 2021

Chapter 19
Instability in Economic Food Systems

19.1 Pressures for Economic Expansion

Economic growth is an irresistible force. To provide full employment for a growing population, the economy must continually expand. In addition, national economic expansion is a necessity for survival in a world where all nations are competing to increase power through increased economic output. Continual economic expansion assumes that resources are unlimited or are interchangeable and that pollution can be eliminated or reduced to tolerable levels. Although some economists have questioned these assumptions and argued that continual economic expansion is unsustainable and must be stopped (Daly 2014), the argument has fallen on deaf ears. No politician has ever been elected on a platform of no economic growth. People want both economic growth and a clean environment, but regulations that limit pollution put a brake on the economy. The conflict between those who insist that the agricultural economy must expand and those who argue that continual expansion is unsustainable is particularly hard to resolve because of political pressures, humanitarian challenges, and invested academic interests.

19.1.1 Political Pressures

Economic expansion is necessary for a capitalist society. Unemployment is always a potential problem in a non-agrarian society, so politicians are elected when they promise to increase jobs through economic growth. Expanding the economy is also politically desirable at the international level. Globalization of trade helps to increase the economy of participating countries, and also reduces risk of warfare between stake-holder countries, since nations would be reluctant to risk a conflict that endangers economic growth.

C. F. Jordan, *Evolution from a Thermodynamic Perspective*,
https://doi.org/10.1007/978-3-030-85186-6_19

19.1.2 Humanitarian Challenges

The FAO (2017) report "The Future of Food and Agriculture: Trends and Challenges" emphasizes humanitarian challenges as a driver of expansion for the agricultural economy.

> *"Persistent and widespread hunger and malnutrition remain a huge challenge in many parts of the world. The current rate of progress will not be enough to eradicate hunger by 2030, and not even by 2050. Looking ahead, the core question is whether today's agriculture and food systems are capable of meeting the needs of a global population that is projected to reach more than 9 billion by mid-century and may peak at more than 11 billion by the end of the century."*

19.1.3 Invested Academic Interests

The mission of agricultural colleges has been to increase crop production. Scientists are doing the job that gets them grants and promotion.

19.2 Instability of Economic Food Systems: External Factors

19.2.1 Booms and Busts

Booms and busts in the global economy are an external cause of instability for economic food systems (Carter et al. 2011). They result from unpredictable social and political factors caused by spikes and drops in food prices that bankrupt farmers at the base of economic food systems (Mishra and Sandretto 2002; Global Harvest 2018).

19.2.2 Vulnerability of Farmers

Farmers are more vulnerable to economic busts than are many other professions. Many workers belong to unions that protect their members. Others are in jobs that require professional certification. Farmers in contrast are in direct competition with each other to keep prices low so as to maintain market share.

19.3 Instability of Economic Food Systems: Internal Factors

19.3.1 Source of Energy for Yield

While exosomatic energy increases cop yield, crop yield itself comes only from endosomatic energy. Endosomatic (photosynthetically derived) energy is used for both crop yield, and for maintenance such as providing crop resistance to insects and disease and enabling plants to compete against weeds. While global agricultural output continues to grow (FAO 2020), much of the growth has come about through diversion of endosomatic plant energy from maintenance to yield. The decline in ratios of energy output (yield) /fossil fuel energy input of farming systems (Hamilton et al. 2013) indicates that the endosomatic pool of maintenance energy may soon be depleted. This means that increasing exosomatic energy inputs (fertilizers, pesticides) will change agricultural yield very little, but will increase strongly the negative impacts of fossil fuel inputs.

19.4 Lack of Feedback: Case Study

Part 1 of this book showed that feedback is necessary for system stability. A USDA Economic Research Report given by Canning et al. (2010) shows how lack of adequate feedback between a cropping system and the economic system in which it is embedded causes instability of both. The research report broke down energy flow in the U.S. food supply chain into seven levels (Table 19.1).

Table 19.1 was used to construct Fig. 19.1. In that diagram, the food system of the U.S. (unshaded inverse pyramid) is embedded in a higher hierarchical level, the National Economy (shaded rectangle). At each food system level, there is a feedback loop consisting of money flowing out of the level into the national economy (black arrows - demand), and exosomatic energy flowing into the level (brown arrows - supply) from the national economy. The exchange is controlled by supply

Table 19.1 Annual Energy consumption by stage of production in 2002 for U.S. Food Systems (Canning et al. 2010)

Level	$J \times 10^{18}$
1. Farm production and agribusiness (agriculture)	2.1
2. Food processing	2.7
3. Food and ingredient packaging (packaging)	0.7
4. Freight services (transportation)	0.5
5. Wholesale and retail trade and marketing services (wholesale/retail)	2.2
6. Away-from-home food and marketing services (food service)	1.5
7. Household food services (households)	3.9

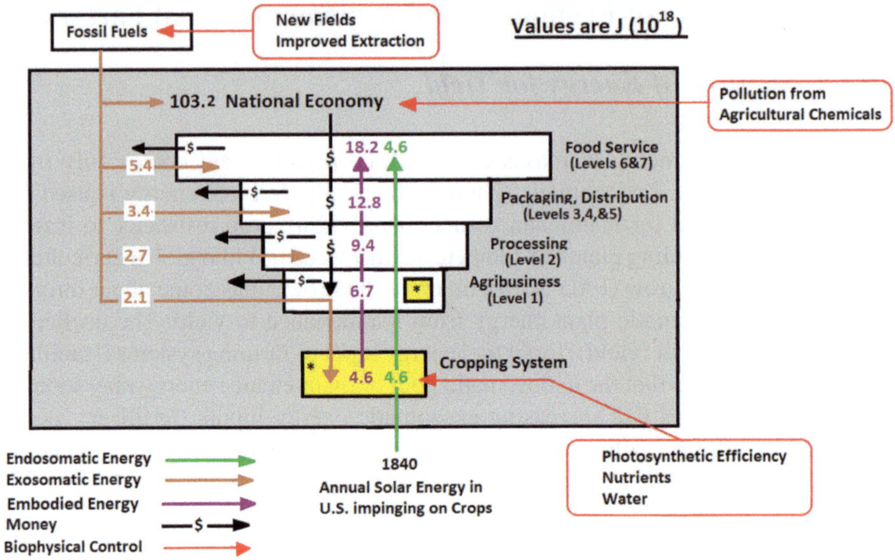

Fig. 19.1 Energy flow in 2002 through the U.S. food system embedded in the national economy. (Reprinted from Jordan (2019) with permission from Elsevier)

and demand. Beginning at the agribusiness level, food energy (green arrow) is propelled upward by demand from the processing level (dollars coming downward). That money buys exosomatic input for the agribusiness level. Feedback in this part of the system is working as it should. Food processing systems expand as capital becomes available. As a result of feedback between supply and demand at higher levels, food energy moves upward through the levels, and the amount of embodied energy (purple arrow) increases at each step due to the energy input at each level.

At the agribusiness level however, chemicals are purchased and then fed into the cropping system, but there is no feedback from the cropping system to the food and economic systems to warn of instability. Black arrows that indicate initiation of a feedback loop do not exist. The yellow-shaded cropping system is embedded in the agribusiness level of the economic food system but is shown as vertically connected to show that feedback at this interaction is lacking. The system keeps pouring fertilizer and other exosomatic energy subsidies in the anticipation they will increase production, but the economy does not know that the cropping system cannot expand. The message of instability at the crop level does not reach the economic food system.

Energy flow and nutrient recycling no longer remain within bounded limits, and the pollution in rivers and bays causes the economic system to become unstable. Feedback here should be information from scientists monitoring waterways, but feedback is not working because the information that an increase in fossil fuel inputs will no longer produce an increase in crop yield is a message that the economic system does not want to hear. It conflicts with the economic goal of permanent growth.

19.4.1 Data

The data from Table 19.1 were used for exosomatic energy inputs to each level. Food service and households (Levels 6 & 7 in Table 19.1) were combined, as they both represent consumption by humans. Packaging, transportation and wholesale/retail (Levels 3, 4, & 5) were also combined into one level, as could occur with vertical integration of the food industry. Processing and agribusiness are separate levels as in Table 19.1.

19.4.1.1 Energy Inputs

Canning et al. (2010) estimated energy consumption of the National Economy in 2002 to be 103.2 exajoules (EJ) or 10^{18} joules.

The quantity of edible food energy output for the U.S. agricultural sector in 2000 was 4.6 EJ (Hamilton et al. 2013). This is the energy in edible food supplied by photosynthesis of plants in the cropping system.

Solar energy input to food energy generated by the cropping system was estimated by assuming that in industrial agriculture, approximately 0.25% of solar radiation is converted into grain (Wikipedia 2018). Solar radiation impinging on crops therefore is 4.6/0.0025 = 1840 EJ.

19.4.1.2 Energy Returned on Energy Invested

Energy returned on energy invested (EROI) (Hall 2017) is a useful guide for comparing management systems in agriculture and natural resources. Many of the studies of "efficiency" in agricultural systems reviewed by Pimentel and Pimentel (2008) and Jordan (2016) are actually measures of endosomatic food energy returned to the farmer on exosomatic energy invested. The measure of energy returned on energy invested at the cropping level is the energy content of the grain harvested (not the entire plant) divided by the exosomatic energy invested by the farmer in growing the crop. In Fig. 19.1, the cropping-level ratio (EROI) is endosomatic energy (4.6 J EJ) divided by the exosomatic subsidies at the crop level (2.1 EJ) equals 2.19.

The green arrow in Fig. 19.1 represents endosomatic energy, the energy derived from solar radiation through photosynthesis that becomes food energy (4.6 EJ). It is embedded energy, the amount of energy that is released when the food is oxidized. This quantity does not change during passage upward through the food system. The energy returned to the consumer at the top of the food system is still 4.6 EJ. The energy embodied in the food product (purple arrow) increases from 4.6 EJ at the crop level, to 18.2 EJ when purchased by the consumer. Embodied energy in food product is the sum of all the energy required to produce the final food product bought by the consumer, considered as if that energy was incorporated or "embodied" in the food itself. It is the sum of the endosomatic energy from photosynthesis plus all the exosomatic energy inputs from the economy. The length of the bars at each level represents the embodied energy at that point. Only by counting all the embodied energy in a product can we appreciate its true value.

To calculate EROI, only the sum of exosomatic energies invested is counted as input. The energy invested part of EROI means investment by economic entities from farmers to corporations. The energy returned as food energy (4.6 EJ) divided by the sum of the exosomatic energies invested (13.6 EJ) equals an EROI efficiency of 0.34. Although adding exosomatic energy to food makes EROI at the consumer level low compared to EROI at the crop level (2.19), it results in greater benefits for humans by freeing them from growing and processing their own food. It allows humans to pursue activities that enrich themselves as well as society, and leave farming to those who like to farm.

A major problem for the embodied energy concept (Odum 1988) has been disagreements about assumptions necessary for calculating the exosomatic energy embodied in a final product. (Cleveland 2013; Murphy et al. 2016). The values in Fig. 19.1 can be disputed, and values change from year to year (Poore and Nemecek 2018), but different values will not change feedback interactions and how they affect stability.

19.5 Control in Ecological vs. Economic Systems

Rappaport (1971) in his study of the Tsembaga tribe in the mountains of New Guinea, showed how feedback maintained stability of the energy flow in their system of shifting cultivation. *"In such systems, abuses seldom need to be repaid by declining yields because they are quickly signaled to those responsible by subtler signs of environmental degradation. Information feedback from the environment is sensitive and rapid in such small autonomous ecological systems, and such systems are likely to be rapidly self-correcting"*. Subtle signs of environmental degradation would include leaf discoloration, reduced harvest, crop loss to insects, weed competition, and erosion.

In contrast to control carried out by indigenous farmers, control in economic systems is by supply and demand. In modern economic food systems, there is feedback control at each level of the food system by money flowing out and exosomatic energy flowing in. Endosomatic energy flowing up through the food system is demanded by money flowing down from the economic system that regulates exosomatic inputs into the agribusiness level and above. Expansion at each level is controlled by the money flowing into and out of each level. In contrast, there is no money flowing into and out of the cropping system to regulate chemical input. Chemicals that flow into the cropping system have no endogenous system control. Chemical inputs are not bought by the cropping system, but are thrust upon it by the economic system. Demand and price are not influenced by what the cropping system can produce but are dictated by the rate the national economy is expanding. The problem is lack of negative feedback between the cropping system and the national economy to warn the economy that pollution is becoming a problem. When the irresistible force of economic demand confronts immovable biophysical limits, something's got to give.

19.6 The Emergence of Feedback and Control

It is unrealistic to ask, as some economists have done (Daly 2014), that the economy be managed for no-growth. What is reasonable to ask is for more and better feedback to the economic and political systems from the cropping system. States have made some progress in this direction from feedback information generated by air and water quality monitors. This has resulted in nutrient management plans such as Maryland's "Nutrient Management Program" (Maryland 2017). The program protects water quality in the Chesapeake Bay and its tributaries by regulating the rate at which farmers and urban land managers apply fertilizers, animal manure and other nutrient sources to their fields. Other resource management programs such as the Mississippi River Basin Healthy Watersheds Initiative (USDA 2018a) have had more limited success. While several tributaries have been scheduled for delisting from the impaired waters lists, in 2017 the hypoxic zone in the Gulf of Mexico was the largest ever measured (NOAA 2018). Excess irrigation also remains a problem. While many states in the Great Plains region have projects to decrease agricultural water use (USDA 2018b), farmers are still draining the Ogallala Aquifer faster than it replenishes (Miller 2018).

Pollution control programs increase the cost of food production. When the information that pollution control is needed is transmitted to the economic system, it is often ignored or rejected because costs of pollution control increase the cost of food. That is a development that politicians and their economic advisors do not want to see because it is thought to reduce economic growth. That is a false idea. People pay more for food, but in return they get a clean environment. That's the way the system is supposed to work. In addition, if the growth in the pollution control industry is counted as economic growth, it would be welcomed.

19.7 Stability of Economic Food Systems

Box 19.1
Theory assumes that buyers always act to maximize their own economic utility, but this is not always the case, for example when someone places a conservation easement on a forest they own to preserve nature's services for the benefit of the population at large.

One of the first lessons in an economics course is that the unobservable market force that helps the demand and supply of goods in a free market to automatically reach equilibrium is what Adam Smith called the invisible hand of the marketplace. When the interaction is between a seller and buyer, it is obvious how supply and demand determine the price of an object or service (but see Box 19.1).

What was less obvious is how the sum of millions of these interactions produce the unique conditions under which society's total utility is maximized by efficiently allocating resources. The field of systems analysis had not yet developed in the time of Adam Smith. But modern systems analysis tells us that feedback in the form of information is the invisible hand that enables an economic system to function smoothly, just as feedback in the form of energy flow and nutrient recycling is the invisible hand that stabilizes an ecosystem. *Sustainability of agricultural systems depends on an information feedback loop between the farm and the economy in which the farm is embedded.*

References

P. Canning, C. Ainsley, S. Huang, K.R. Polenske, A. Waters, Energy use in the U.S. Food System, ERR-94. U.S. Dept. Agric. Econ. Res. Serv. (2010). https://www.ers.usda.gov/publications/pub-details/?pubid=46377

C.A. Carter, G.C. Rausser, A. Smith, Commodity booms and busts. Annu. Rev. Resour. Econom. **3**, 87–118 (2011). https://doi.org/10.1146/annurev.resource.012809.104220

C.J. Cleveland, Net energy analysis. Encyclopedia of earth (2013). Available at: http://www.eoearth.org/view/article/154821/. Accessed 7 June 2018

H.E. Daly, *From Uneconomic Growth to a Steady-State Economy* (Edward Elgar, Northampton, 2014)

FAO, *The future of food and agriculture – Trends and challenges* (Food and Agriculture Organization of the United Nations, Rome, 2017). http://www.fao.org/3/a-i6583e.pdf. Accessed 17 Dec 2020

FAO, Food and agriculture data (2020), http://www.fao.org/faostat/en/#home. Accessed 17 Dec 2020

Global Harvest, T*he* agricultural business cycle: managing through booms and busts (2018), https://globalagriculturalproductivity.org/case-study-post/the-agricultural-business-cycle-managing-through-the-booms-and-busts/. Accessed 19 Dec 2020

C.A.S. Hall, *Energy Return on Investment*, Lecture Notes in Energy (Springer, Cham, 2017). https://doi.org/10.1007/978-3-319-47821-0. Accessed 19 Dec 2020

A. Hamilton, S.B. Balogh, A. Maxwell, C.A.S. Hall, Efficiency of edible agriculture in Canada and the U.S. over the past three and four decades. Energies **6**(3), 1764–1793 (2013). https://doi.org/10.3390/en6031764

C.F. Jordan, The farm as a thermodynamic system: Implications of the Maximum Power Principle. Biophys. Econ. Res. Qual. **1**, 9 (2016). https://doi.org/10.1007/s41247-016-0010-z

C.F. Jordan, Energy flow and feedback control in ecological and economic food systems. Ecol. Econ. **156**, 91–97 (2019). https://doi.org/10.1016/j.ecolecon.2018.09.015

Maryland, Maryland's nutrient management program (2017), http://mda.maryland.gov/resource_conservation/Pages/nutrient_management.aspx. Accessed 26 Dec 2020

C. Miller, The conversation: environment and energy (2018), https://theconversation.com/farmers-are-drawing-groundwater-from-the-giant-ogallala-aquifer-faster-than-nature-replaces-it-100735. Accessed 26 Dec 2020

A.K. Mishra, C.L. Sandretto, Stability of farm income and the role of nonfarm income in U.S. agriculture. Rev. Agric. Econ. **24**(1), 208–221 (2002). https://doi.org/10.1111/1058-7195.00014

D.J. Murphy, M. Carbajales-Dale, D. Moeller, Comparing apples to apples: why the net energy analysis community needs to adopt the life-cycle analysis framework. Energies **9**(11), 917 (2016). https://doi.org/10.3390/en9110917

NOAA, Gulf of Mexico "dead zone" is the largest ever measured (2018), https://www.noaa.gov/media-release/gulf-of-mexico-dead-zone-is-largest-ever-measured

H.T. Odum, Self organization, transformity and information. Science **242**(4882), 1132–1139 (1988)

D. Pimentel, M.H. Pimentel, *Food, Energy and Society*, 3rd edn. (CRC Press, Boca Raton, 2008)

J. Poore, T. Nemecek, Reducing food's environmental impacts through producers and consumers. Science **360**(6392), 987–992 (2018)

R.A. Rappaport, The flow of energy in an agricultural society. Sci. Am. **225**, 1117–1132 (1971)

USDA, Mississippi River Basin Healthy Watersheds Initiative (2018a), https://www.epa.gov/sites/production/files/2015-10/documents/htf_report_to_congress_final_-_10.1.15.pdf. Accessed 5 Jan 2021

USDA, Ogallala aquifer initiative (2018b), https://www.nrcs.usda.gov/wps/portal/nrcs/detailfull/national/programs/initiatives/?cid=stelprdb1048809. Accessed 5 Jan 2021

Wikipedia, Photosynthetic efficiency (2018), https://en.wikipedia.org/wiki/Photosynthetic_efficiency. Accessed date: 5 June 2018

Chapter 20
Energy Efficiency in Agricultural Systems

20.1 Two Kinds of Energy

There are two types of energy involved in agricultural systems. One is endogenous energy, the energy internal to the system captured by photosynthesis. It is the energy embedded in complex organic compounds that becomes food. Food energy is released when complex organic compounds are broken down in a digestive system.

Exogenous energy is the energy supplied to the system in forms derived from fossil fuel such as fertilizers and pesticides. It is used to increase the output of endogenous energy. The efficiency of exogenous energy has been called Energy Returned on Energy Invested (EROI, Hall 2017). EROI is measured by the ratio of endosomatic energy output of the system divided by the exogenous energy supplied to the system.

20.2 Early Comparisons of Energy Use Efficiency

In the 1970s, scientists became aware of a looming problem of energy scarcity and the importance of society becoming more energy efficient. Ecological and agricultural scientists began calculating the efficiency with which fossil fuel energy is used to produce food energy: (Black 1971; Steinhart and Steinhart 1974; Cox and Atkins 1979; Fluck and Baird 1980; Pimentel and Pimentel 2008; Gelfand et al. 2010). Fossil fuel energy used for input in agriculture includes the energy content used to synthesize fertilizers used on the crop, and energy content of the fuel used to spread the fertilizers. Similar calculations are used for pesticides and herbicides. Also included are the energy costs of plowing the fields, and harvesting the crops. Pimentel et al. (1973) calculated that in the U.S. between 1945 and 1970, an

C. F. Jordan, *Evolution from a Thermodynamic Perspective*,
https://doi.org/10.1007/978-3-030-85186-6_20

investment of 1.0 calories of fossil fuel energy returned an average of 3.05 calories of corn energy for an EROI of 3.05.

An interesting study was by Steinhart and Steinhart (1974). They determined that the ratio for paddy rice was 20:1. This extraordinarily high EROI is a result of a very unique service of nature not included in the calculation. Rice paddies are flooded to control weeds, and also to allow the growth of the water fern *Azolla sp.*, which has a symbiotic relationship with a blue-green alga that fixes nitrogen from the atmosphere. It is the weed control and nitrogen availability that make paddy rice so energy efficient from the viewpoint of energy invested.

20.3 Energy Returned on Energy Invested

Hall (2017) defined the concept of energy returned on energy invested as:

$$\text{EROI} = \frac{\text{Energy returned from an energy gathering activity}}{\text{Energy used to get that energy}}$$

In a natural system that means the endogenous energy content of biomass produced divided by the photosynthetic energy used to produce that biomass (Harris et al. 2013). In modern agricultural systems, we are interested in the efficiency of exogenous energy in producing endogenous energy. The ratio is food energy output divided by fossil fuel energy input.

EROI varies greatly. In systems that depend on services of nature, ratios can be 10:1 or greater. For systems dependent of fossil fuel subsidies, the ratio is less. For peanuts, it is 1.4:1, for apples, 1.6:1. Energy yield efficiency for meat and poultry is low because energy subsidies are used to grow feed for animals, and then there is another thermodynamic energy loss when feed is converted to animal biomass. For broiler chickens, the yield is 0.25:1, and for turkey, 0.1:1. Beef is 0.03:1 and lamb is 0.02:1 (Pimentel and Pimentel 2008). Although energy efficiency for meat is less than 1.0, meat is not eaten for energy content. The ratios are still useful because they tell us what kinds of meat production are most energy efficient. Hydroponic agriculture for vegetables can be very energy inefficient, especially when grown in towers with artificial light, but like meat, one does not eat vegetables for their energy content.

Comparisons of food energy-produced/fossil-fuel energy invested are only meaningful for energy efficiency when energy output of crops is the main objective of management (Table 20.1).

The data suggests that indigenous agriculture is highly efficient, with energy returns on investment of better than 10:1. That is because only the labor of cutting down the trees was counted as input energy. The energy costs of the services of nature were not included. In swidden agriculture, trees are cut down and burned. The ash and decomposing trunks fertilize the crops. When the energy costs of nature's services (energy content of tree biomass) are included for slash and burn agriculture, the ratio of energy returned to energy invested varies between 0.02 and 0.34 (Chap. 23).

Table 20.1 Energy ratios for agricultural crops used to produce food energy in different management systems

Type of agriculture	Energy- out/Energy -in (EROI)	Yield: KJ/ ha(10^6)
Swidden (Shifting cultivation)		
Swidden Agriculture in New Guinea – sweet potato, taro, cassava, yam	15.4:1	47.6
Swidden Agriculture in Mexico – corn	12.6:1	28.9
Corn		
Corn production in Mexico using only human power	10.7:1	28.9
Corn production in Nigeria, human power only	6.41:1	14.9
Corn production in Philippines using carabao	5.06:1	14.0
Corn production in Guatemala, human power	4.84:1	15.8
Corn production in Mexico using oxen	4.34:1	14.0
Corn production in Guatemala using oxen	3.11:1	15.8
Conventional corn production in the U.S.	3.84:1	130.3
Rice		
Rice production in Borneo using only human power	7.08:1	30.6
Rice production in the Philippines using carabao	3.29:1	25.1
Conventional rice production in Japan	2.80:1	96.1
Conventional rice production in the U.S.	2.24:1	111.0
Sorghum		
Sorghum production in the Sudan using human power	14.43:1	12.4
Conventional sorghum production in the U.S.	1.96:1	44.1

Data from Pimentel and Pimentel (2008)

Studies of traditional and family farms include energy inputs from human and animal labor as well as from fossil fuel subsidies. Table 20.1 shows that traditional and family farms have a higher energy-out/energy-in ratio than "conventional" agriculture. Much of the energy input is in the form of human labor. Because traditional agriculture and small family farms lack economies of scale, they might be expected to have an output/input ratio that is lower than that for conventional, but it is actually higher. The lower ratio in industrial systems is due to the intensive use of fossil fuel energy that results in higher yield per hectare but at a much lower efficiency. Efficiency is sacrificed for power output. In recent years, agriculture has become more efficient through the use of precision farming, where computers in tractors are guided by satellites so that each square meter of crop land receives only the exact amount of fertilizer and pesticide that it needs. That decreases the operational energy costs but does not include infrastructure costs.

Box 20.1

Human energy is usually ignored in calculating EROI for industrial farming. Energy spent in driving a tractor is infinitesimally small compared to the energy used by the tractor.

20.4 EROI for Industrial Corn

20.4.1 Production Functions

The energy-out/energy-in ratios for conventional farming in Table 20.1 are valid for only a specified rate of input. However, a farmer has a choice about rate of energy inputs. If their cost is low, he may use a high rate, but if cost is high, he may restrict input. To decide, he has to know how his crop responds to a range of inputs. Agricultural production functions relate quantity of yield in terms of mass or volume to a range of input subsidies. The curve in Fig. 20.1 represents output in terms of kilograms of corn produced per hectare per year as a function of input of nitrogen fertilizer (data from Debertin 2012). The dots on the curve represent the efficacy with which nitrogen fertilizer produces corn at each input rate (kg corn per ha output/kg N input per ha). As input increases from zero, the efficacy of nitrogen to produce corn increases until the inflection point of the curve where it changes from concave upward to concave downward. That is the point where nitrogen produces corn most effectively. It is the efficacy of the nitrogen at this point and is determined by a line drawn from the base to where it is tangent to the curve. The efficacy at this point is 5900/112 = 52.7. That means that 112 kg/ha of nitrogen on this farm yields 5900 kg corn/ha. With greater nitrogen inputs, efficacy at the margin decreases but yield still increases. Because farmers are competitive, they will continue to add nitrogen until yield is maximum, using 201.6 kg/N/ha to produce 8593 kg corn/ha, even though the efficacy at that point is only 42.6. The reasoning is the same as the

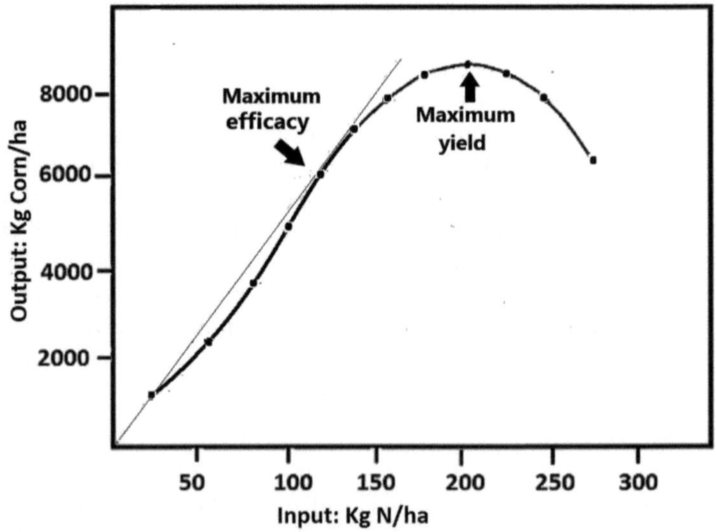

Fig. 20.1 Output of corn in kg/ha per year as a function of nitrogen input in kg/ha

truck driver in Chap. 9 who drove at the speed that gave maximum power output rather than maximum efficiency because maximum power gave him more income.

20.4.2 An Energy Production Function

Energy embedded in corn as a function of exogenous energy invested is plotted in Fig. 20.2. Also included are income to farmer and environmental costs.

The black line in Fig. 20.2 is a theoretical energy production function for corn monocultures in the United States using input energy equivalents of fertilizers, pesticides, and tractor fuel and output values for corn (Jordan 2016). The energy production function indicates both biomass of corn yield and the energy embodied in corn yield as a function of the sum of energy subsidies into the system. As in all biological production systems, the function is sigmoid (Smathers et al. 1983), and energy efficiency for maximum yield is always lower that that for maximum possible efficiency. That is in accordance with the maximum power principle (Odum and Pinkerton 1955) that says energy efficiency for maximum yield is the optimum efficiency for maximum power output.

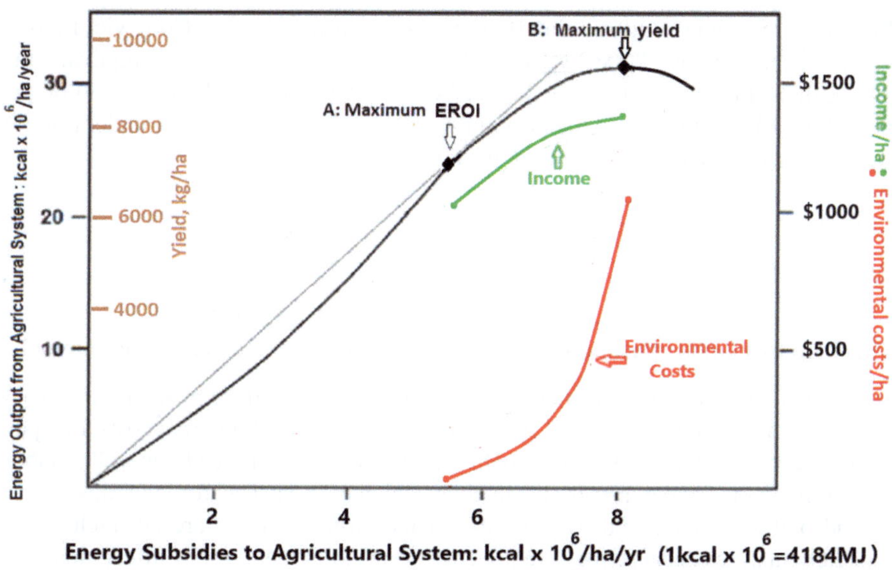

Fig. 20.2 Energy output, yield, income for farmer, and costs of pollution per hectare as a function of exogenous energy input (energy subsidy) into an energy-intensive corn production system. A indicates the point where energy use efficiency (energy output/ energy input) is highest. B indicates the point of maximum yield. The absolute values for farmer income and environmental costs will vary on a case-by-case basis, but as input increases beyond point A, environmental costs will always be concave upward, while farmer income will always be concave downward. ((Figure, calculations and data are from Jordan 2016). Reprinted from Jordan (2016) with Permission from Springer)

The green line (scale on the right) is the income from sale of the corn as a function of energy input. The red line (scale on the right) is the environmental cost of energy pollution as a function of energy input. Energy input at point B is 8.115 (10^6) kcal/ha, and output in the form of corn is 31.158 (10^6) kcal/ha or 8655 kg/ha (data from Pimentel and Pimentel 2008). These data were representative of average corn monocultures in the USA around the year 2000, and we assume that the average industrial corn farmer in the USA managed for maximum yield. The energy use efficiency ratio (the energy-out/energy-in ratio or EROI) for subsidies at maximum yield is 3.84 (energy output/energy input at point B). The maximum energy use efficiency (EROI) is highest at point A, where a line drawn from the base of the graph is tangent to the production function. Data for point A are from de Ponti et al.'s (2012) review of low-input agriculture. The energy use efficiency at point A is 4.45. Complete data and calculations on which the curve is based are given by Jordan (2016).

20.5 Economic Considerations

20.5.1 Income

Income (green line) under point B was calculated as follows: Corn yield (from Pimentel and Pimentel 2008) was 8655 kg/ha, equivalent to 19,041 lb, equivalent to 340 bushels. Income per ha from 340 bushels (8655 kg) at the US price of $4.00 per bushel is $1360. Since yield under point A is 80% of point B, income under point A was assumed to be 80% of that at point B or $1088.

20.5.2 Costs

There are two kinds of economic costs in a farming system: external and internal.

- Internal costs are investments in energy inputs and are paid for by the farmer. They are the energy subsidies of the horizontal axis in Fig. 20.2. Internal costs are operating costs and include supplies such as chemicals, fuel, and seeds, as well as labor. Internal costs for the farmer also include capital costs such as for land, machinery, and buildings amortized over the life of the farm, but these are not included in EROI analyses because they are not energy subsidies that directly affect yield.
- External costs are costs for the environment and society (environmental costs, red line in Fig. 20.2,) caused by the farming system and paid for by the public. An example would be the decrease in revenue for fisherman due to hypoxia in the Gulf of Mexico caused by pollution from nitrogen runoff in midwestern corn fields. The costs of nitrogen pollution from agriculture that causes hypoxia in the Gulf of Mexico are difficult to calculate because the damage is diffuse. We used the cost of removing nitrate from drinking water as a proxy. Environmental costs from nitrogen pollution for energy-intensive farms operating at maximum yield (point B)

were calculated by taking the point where yield is greatest (201.6 kg/ha from Table 2 in Jordan (2016) and cost per hectare at that point from Tables 3 and 4 in Jordan (2016) ($1013.42/ha).

20.5.3 Profit

Profit is income minus costs, but accounting for profit is problematic because certain costs such as for farm machinery depend on an arbitrary depreciation rate. If we assume that profits are directly proportional to income, a profit curve would parallel the income curve.

Under economic rules where the farmer is not charged for pollution, he would gain the greatest profits by selecting inputs at point B. If the farmer were charged for the pollution costs from his runoff, he would select inputs at point A where the vertical distance between the red and green lines is maximum.

20.6 A Farmer's Dilemma

The economic values for income and environmental cost in Fig. 20.2 represent only one set of many possible values. Income to a farmer depends on the price of a bushel of corn which can vary year to year, month to month, or day to day. Losses to society from costs of environmental cleanup could vary depending on the type of pollution cleanup used. However, biophysical costs and yields are independent of economic fluctuations, and the energy production function resulting from these costs and yields will always follow the Maximum Power Principle as applied to agriculture: Efficiency at maximum power output is always less than maximum possible efficiency. Farmers can maximize yields or maximize energy use efficiency, but they cannot do both at the same time. Maximum yield will always provide the greatest income to the farmer. Maximum efficiency will always result in the least pollution per unit output. Point A would be the most beneficial long-term economic choice for society and the environment. It would be a farmer's choice if he envisioned himself as part of the environment in which the farm was embedded. However, conventional economics prevent the industrial farmer from taking this holistic viewpoint. Most industrial farmers and corporate farms will maximize yields because to stay in business, they need to compete for market share.

20.7 The Maximum Power Principle and Economic Theory

From the sustainability point of view, and from the point of view of a conservative investor, the system should operate where there is maximum return on investment, that is at point **A**. The maximum power principle insists the system must operate at point **B**, where EROI is lower than at **A**. Systems always operate to maximize power output to win the competition for energy and for survival. If a system does not win the competition for short-term profits, it will not survive to become sustainable in the long term. Because of environmental beliefs, some farmers operate at point **A.** They are violating the maximum power principle, and so they are punished with reduced profits. The maximum power principle says the same thing as economic theory. Both say that most individuals will operate to maximize their own self-interest (power output), even though that results in an efficiency that is less than that for maximum environmental quality. It is a variation on the tragedy of the commons.

References

J.N. Black, Energy relations in crop production—a preliminary survey. Ann. Appl. Biol. **67**, 272–278 (1971)

G.W. Cox, M.D. Atkins, *Agricultural Ecology* (Macmillan, London, 1979)

T. de Ponti, B. Rijk, M.K. van Ittersum, The crop yield gap between organic and conventional agriculture. Agric. Syst. **108**, 1–9 (2012)

D.L. Debertin, *Agricultural Production Economics* (University of Kentucky, Department of Agricultural Economics, 2012). http://www.uky.edu/~deberti/prod/agprod5.pdf. Accessed 17 Dec 2020

R.C. Fluck, C.D. Baird, *Agricultural Energetics* (AVI Publishing, Westport, 1980)

I. Gelfand, S.S. Snapp, G.P. Robertson, Energy efficiency of conventional, organic, and alternative cropping systems for food and fuel at a site in the US Midwest. Environ. Sci. Technol. **44**, 4004–4011 (2010)

C.A.S. Hall, *Energy Return on Investment*, Lecture Notes in Energy (Springer, Cham, 2017). https://doi.org/10.1007/978-3-319-47821-0. Accessed 19 Dec 2020

N.L. Harris, C.A.S. Hall, A.E. Lugo, A test of the maximum power hypothesis along an elevational gradient in the Luquillo mountains of Puerto Rico. Ecol. Bull. **54**, 233–243 (2013)

C.F. Jordan, The farm as a thermodynamic system: implications of the maximum power principle. Biophys. Econ. Res. Qual. **1**, 9 (2016). https://doi.org/10.1007/s41247-016-0010-z

H.T. Odum, R.C. Pinkerton, Time's speed regulator: the optimum efficiency for maximum power output in physical and biological systems. Am. Sci. **43**, 331–343 (1955)

D. Pimentel, M.H. Pimentel, *Food, Energy and Society*, 3rd edn. (CRC Press, Boca Raton, 2008)

D. Pimentel, L.E. Hurd, A.C. Bellotti, M.J. Jorster, I.N. Oka, O.D. Sholes, R.J. Whitman, Food production and the energy crisis. Science **182**, 443–449 (1973)

W.M. Smathers, C.F. Jordan, E.G. Farnworth, T.H. Tidrick, An economic production function approach to ecosystem management. Bioscience **33**, 642–646 (1983)

C. Steinhart, J. Steinhart, *Energy: Sources and Role in Human Affairs* (Duxbury Press, North Scituate, 1974)

Chapter 21
The First Law of Thermodynamics and Genetic Engineering (There Is No Free Lunch)

21.1 Source of Energy for Increased Crop Yield

The dramatic increase in agricultural production after WWII has been due a combination of increased availability of synthetic fertilizers and pesticides, and plant breeding that enabled plants to take advantage of these subsidies. Fertilizers and pesticides have increased the food yield of plants, but these subsidies in and of themselves are not the energy pools from which emerged the increase in grain yield. Grain yield is derived only from endosomatic energy, the energy derived from photosynthesis. Endosomatic energy consists of organic carbon that contains free energy (exergy) in a form available to be transformed into carbohydrates. Fertilizers, pesticides and machinery represent exosomatic energy, energy that can increase yield but these are not part of the flow of endosomatic energy. Precision farming, integrated pest management and automated irrigation also represent exosomatic energy. Exosomatic energy cannot be converted to food. It is not the source of energy used for increased crop yield, better tasting fruit, fruit resistant to mechanical harvesting, herbicide resistance, and frost resistance. Exosomatic energy merely enhances the production of endosomatic energy in agricultural systems (Hamilton et al. 2013), but that ability is limited by the capacity of leaves to capture solar energy, as was illustrated in Fig. 20.2.

There are two possible ways that endosomatic energy for food production can be increased: greater photosynthetic efficiency; or diversion of photosynthetic energy away from disease resistance, defense against insects, and nutrient acquisition. The last increase in photosynthetic efficiency occurred about 24–35 million years ago (Sage 2004), so that could not be the explanation. That leaves diversion of endosomatic energy in plants, and hybridization.

C. F. Jordan, *Evolution from a Thermodynamic Perspective*,
https://doi.org/10.1007/978-3-030-85186-6_21

21.1.1 Endosomatic Energy

During the domestication of crops from wild progenitors, there has been a sacrifice of energy used to grow fine roots in a symbiotic relation with mycorrhizae to take up nutrients from the soil. These sacrifices increased the availability of energy for yield (Chaudhary 2013). Loss of fine root growth (a loss of competitive ability to acquire nutrients) was first compensated for by fertilizing with nutrient supplements such as manure and guano. During the Green Revolution following WWII, an increased availability of chemical fertilizers, pesticides, pivot irrigation, and mechanization occurred simultaneously with the development of crop varieties that responded vigorously to these subsidies. Since the exosomatic subsidies do not contain potential energy (organic carbon), the increase in edible yield, input must have been due to changes in the way that endosomatic energy is used, a change that resulted from plant breeding. While trade-offs in endosomatic energy use have been mentioned in the scientific literature (Fineblum and Rausher 1995; Purrington and Bergelson 1999; Conway and Sechler 2000; Jordan 2002), they have never been quantified.

21.1.2 Hybridization

Higher yield in hybrid corn results from hybrid vigor that has been attributed to greater nitrogen use efficiency (Haegele et al. 2013) and increased stress resistance (Mastrodemenico et al. 2018). However the genetic reason for increased nitrogen use efficiency and stress resistance is unclear (Birchler et al. 2006). Perhaps the reason is hybrids divert energy used for reproduction to increased yield. While hybrids often have higher yield, they have lost the ability to reproduce, or to produce offspring with the same yield as the parents. If genetic engineers could produce enhanced vigor beyond the F1 stage, then perhaps we could say that genetic manipulation could actually increase endosomatic energy output, but alas they have been unable to do so.

21.1.3 High Yield Rice

Plant breeders have improved yields of rice by developing "dwarf" varieties that do not "lodge". Lodging means bending over at the base following heavy wind or rain. Was there a trade-off here? Paddy rice has been cultivated in Asia for centuries where part of the plant is underwater for part of the year. Paddy rice has long stems to allow the grain to remain above water. When dry-land rice was developed, there was no need for long stems, so dwarf varieties were developed. What was the trade-off? When rice paddies are flooded in the spring, they can be inoculated with *Azolla*.

Azolla is a floating pteridophyte, which contains as an endosymbiont the nitrogen-fixing cyanobacterium *Anabaena azollae* which then multiplies to cover the water, suppressing weeds. The rotting plant material releases nitrogen to the rice plants. The trade-off for dwarf varieties is that dry land or upland rice must be fertilized and weeded.

21.2 Domestication of *Balsas teosinte*

Maize (*Zea mays L. ssp. Mays*) was domesticated from its wild species ancestor, *Balsas teosinte* (*Zea mays ssp. Parviglumis*) in the central plateau of Mexico about 9000 years ago (Warburton et al. 2011). Figure 21.1 illustrates endosomatic energy redistribution during the transition from *Balsas teosinte* to *Zea mays* through selective breeding.

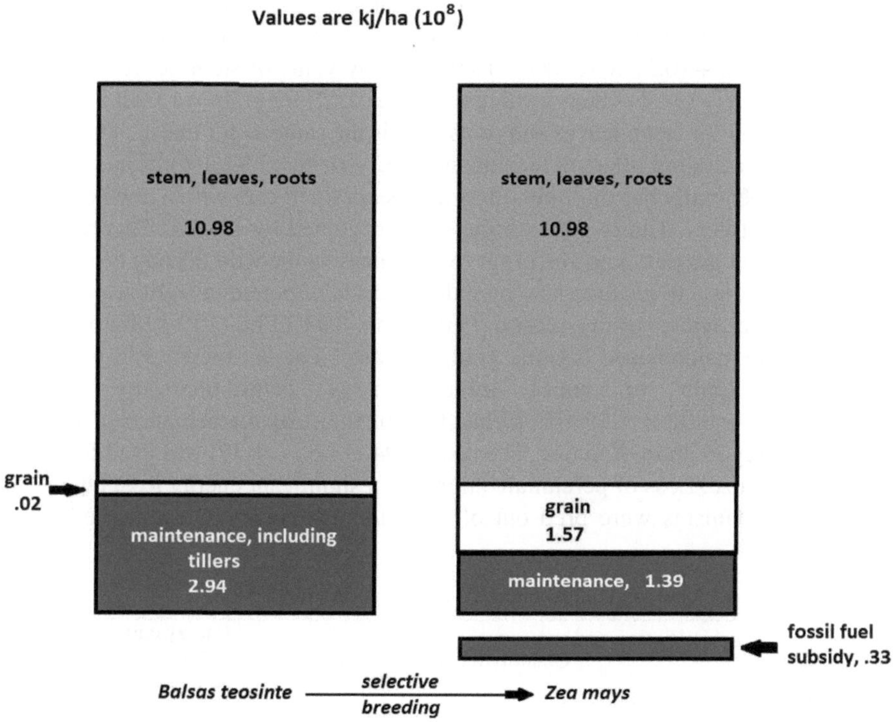

Fig. 21.1 Annual change in distribution of energy in maize resulting from selective breeding of teosinte

21.2.1 Calculations

The amount of solar energy striking the earth's surface at mid latitudes during the growing season is ~627 (10^8) kJ/ha (Jordan 1971).

Sunlight to biomass energy conversion is 1–2% for crop plants (Wikipedia 2017a). Since modern maize has a C4 photosynthetic pathway which has a relatively high efficiency, we assume the upper limit of the range (2%). Yield of total biomass energy of modern corn plants should be: 0.02 (627×10^8) = 12.55 (10^8) kJ of biomass energy per hectare per growing season.

In industrial agriculture, approximately 0.25% of solar radiation is converted into corn kernels (Wikipedia 2017a). Therefore, in sunlight to grain conversion = 0.0025(627×10^8) = 1.57 (10^8) kJ/ha of corn energy per ha is produced. Considering the variability in corn yields, this is reasonably close to the 1.30 (10^8) estimate of Pimentel and Patzek (2005) (Table 21.1). Biomass increment per year of stem, leaves and roots alone equals the total biomass increment (12.55×10^8) minus grain (1.57×10^8) = 10.98 (10^8) kJ/ha (stem, leaves, roots and grain in Fig. 21.1).

The grain yield from teosinte (no fertilizers) is only about 1% of domesticated corn grain (Doebley 2004) or 0.02(10^8) kJ/ha. Biomass of teosinte is 27% greater than maize under optimum conditions for growth because teosinte produces tillers while maize does not (Niazi et al. 2015). Because of tillers, teosinte is larger and captures more photosynthetic energy than maize. Biomass of teosinte therefore should be 10.98 + (0.27 × 10.98) = 13.94. Energy value of stem, leaves and roots plus tillers then is 13.94 – energy value for grain (0.02) for grain = 13.92(10^8) kJ/ha. Assume energy for stem, leaves and roots only is the same as for maize. Then maintenance cost including tillers in teosinte is 13.92–10.98 = 2.94(10^8) kJ/ha.

Teosinte generally has multiple tillers and bears small ears with a few kernels on the nodes of tillers. Tillers increase the space occupied by a plant. This represses growth of other grasses, a powerful survival strategy in the wild (Zhang et al. 2019). Maintenance costs in teosinte also include the costs of perennialization, the ability to survive over winter or dry season. Part of this 2.94 kJ/ha (10^8) kJ/ha of energy from teosinte maintenance became grain in maize. The net increase in grain was energy in grain of maize minus energy contributed by teosinte grain = 1.57 − 0.02 = 1.55(10^8) kJ/ha. 1.55 of the 2.94 maintenance of teosinte became energy in grain of maize. The rest, (2.94 − 1.55 = 1.39) was used for maintenance of maize. Loss of perennialization was a significant energy loss, but not all survival mechanisms were bred out of teosinte since 1.39 (10^8) remained. Total

Table 21.1 Energy inputs and outputs for U.S. industrial corn production

	Quantity/ha, dry wt. biomass	Energy equivalents kJ/ha (10^6)
Total inputs		33.953
Output: Corn yield	8655 kg	130.365
kJ output/kJ input (EROI)		3.84:1

Data from Pimentel and Patzek (2005)

maintenance for maize then is 1.39 plus 0.33 (10^8) kJ/ha of fossil fuel subsidy (from total inputs, Table 21.1) = 1.72 (10^8) kJ/ha.

21.2.2 Mechanisms

The mechanisms by which mycorrhizae increase absorption of nutrients include some that are physical and some that are chemical. Physically, most mycorrhizal mycelia are much smaller in diameter than the smallest root or root hair, and thus can explore soil material that roots and root hairs cannot reach, and provide a larger surface area for absorption. Teosinte has a more "bushy-looking" root system compared to a modern maize hybrid. The much greater root surface area in teosinte compared to modern maize shown in Fig. 2 of Gaudin et al. (2014) may reflect greater mycorrhizal infection, a maintenance cost for teosinte. Mycorrhizae secrete organic acids that dissolve or chelate many ions, or release them from minerals by ion exchange (Wikipedia 2017b). Cost of this synthesis is passed along to the plant host that benefits from greater nutrient availability.

Mechanisms for nutrient acquisition in *Balsas teosinte* are replaced by fertilizers in *Zea mays*. Total and average root hair lengths are significantly higher in *Balsas teosinte* than in all of the modern maize B73 genotypes (Gaudin et al. 2014). The energy devoted to root hairs and to support mycorrhizae that infest root hairs in teosinte is hypothesized to be the source of energy that became corn kernels in modern corn as breeding changed the allocation of energy.

21.3 Other Tradeoffs

One of the goals of plant breeders at The Land Institute (2020) in Salina Kansas is to convert annual crops to perennials. Perennial crops are more environmentally sustainable than annuals because perennials do not require the annual plowing that destroys nutrient conserving soil organic matter. We do not say perennialization should not be done. We are just saying that perennialization is a cost that must be paid through decreased yield. Plant roots must be "frost-proofed" leaving less energy available for production. It is bringing plants back to a wild state.

Genetic engineers have increased the growth rate and biomass of species used in aquaculture. They have increased the survival rate of trout exposed to bacterial pathogens from 30% to 80%, and of Atlantic salmon plagued by sea lice. The success has come with some collateral damage, including pollution from fish pond waste, heavy catches of wild fish used to feed to penned salmon, and the destruction of coastal wetlands to build shrimp ponds.

Diverting energy used for reproduction to energy used for yield is another tradeoff. Creating infertile oysters has increased their growth rate, because they don't devote much energy to reproduction, so they reach harvest size sooner. Hybrid catfish have increased yield due to diversion of energy used for reproduction (Stokstad 2020).

21.4 The Free Lunch Has Already Been Eaten

Genetic engineers have suggested that by redirecting efforts in genetic engineering, the amount of chemical subsidies in agriculture can be reduced. They have predicted that "in the long run there will be a major role for genetic modification in organic farming, because resistance will mean that herbicides and pesticides can be avoided and plants will become better able to take up organic fertilizer" (Fresco 2016). It is assumed that they mean without a reduction in yield. The assumption may be unwarranted. Hamilton et al.'s (2013) study found that energy efficiency of US agriculture has more than doubled from 0.8:1 in 1970 to 2.2:1 by 2000, then increased more slowly to 2.3:1 by 2009. The energy efficiency of the agricultural sector in Canada has not changed appreciably since 1980, and has varied about a mean of 2:1 from 1981 to 2009. The increase in grain production resulted from a drawdown of energy used for maintenance (Fig. 21.1). The decrease in the rate that efficiency is increasing indicates that endosomatic energy stocks that can be tapped by genetic engineering have been nearly depleted. Increases in maintenance capabilities will be limited by a shortage of endosomatic energy within the crop itself.

Malthusian predictions that agricultural yields cannot keep up with population growth have so-far been proven wrong because of greater exosomatic energy input and more efficient technology. It is not limitations of exosomatic energy availability that limit further yield increases on a per unit area of farmland basis. Rather it is the limited availability of endosomatic energy within the crop itself that limits further yield. The first law of thermodynamics says that you cannot re-establish maintenance functions while at the same time maintaining yield. Energy cannot be created, only redistributed.

References

J.A. Birchler, H. Yao, S. Chudalayandi, Unraveling the genetic basis of hybrid vigor. PNAS **103**(35), 12957–12958 (2006). https://doi.org/10.1073/pnas.0605627103

B. Chaudhary, Plant domestication and resistance to herbivory. Int. J. Plant Genom. **2013**, 572784 (2013). https://doi.org/10.1155/2013/572784

G. Conway, S. Sechler, Helping Africa feed itself. Science **289**, 1685 (2000)

J. Doebley, The genetics of maize evolution. Ann. Rev. Genet. **38**, 37–59 (2004)

W.L. Fineblum, M.D. Rausher, Tradeoff between resistance and tolerance to herbivore damage in a morning glory. Nature **377**, 517–520 (1995)

L.O. Fresco, *Hamburgers in Paradise* (Princeton University Press, Princeton, 2016)

A.C.M. Gaudin, S.A. McClymont, S.S.M. Soliman, M.N. Raizada, The effect of altered dosage of a mutant allele of Teosinte branched 1(tb1-ref) on the root system of modern maize. BMC Genet. **15**, 23 (2014). https://doi.org/10.1186/1471-2156-15-23

J.W. Haegele, K.A. Cook, D.M. Nichols, F.E. Below, Changes in nitrogen use traits associated with genetic improvement for grain yield of maize hybrids released in different decades. Crop Sci. **53**, 1256–1268 (2013). https://doi.org/10.2135/cropsci2012.07.0429

A. Hamilton, S.B. Balogh, A. Maxwell, C.A.S. Hall, Efficiency of edible agriculture in Canada and the U.S. over the past three and four decades. Energies **6**(3), 1764–1793 (2013). https://doi.org/10.3390/en6031764

C.F. Jordan, A world pattern in plant energetics. Am. Sci. **59**(4), 425–433 (1971)

C.F. Jordan, Genetic engineering, the farm crisis, and world hunger. Bioscience **52**(6), 523–529 (2002)

Land Institute, *Transforming Agriculture Perennially* (2020). https://landinstitute.org/about-us/. Accessed 26 Dec 2020

A.T. Mastrodomenico, J.W. Haegele, J.R. Seebauer, F.E. Below, Yield stability differs in commercial maize hybrids in response to changes in plant density, nitrogen fertility, and environment. Crop Sci. **58**(1), 230–241 (2018)

I.A.K. Niazi, S. Rauf, J.A. Teixeira da Silva, H. Munir, Comparison of teosinte (*Zea mexicana L.*) and inter-subspecific hubrids (*Zea mays L. x Zea mexicana*) for high forage yield under two sowing regimes. Crop Pasture Sci. **66**, 49–61 (2015). https://doi.org/10.1071/CP14155

D. Pimentel, T.W. Patzek, Ethanol production using corn, switchgrass, and wood: Biodiesel production using soybean and sunflower. Nat. Resour. Res. **14**(1), 65–76 (2005). https://doi.org/10.1186/1471-2156-15-23

C.B. Purrington, J. Bergelson, Exploring the physiological basis of costs of herbicide resistance in *Arabidopsis thaliana*. Am. Nat. **154**(Suppl), 582–591 (1999)

R.F. Sage, The evolution of C4 photosynthesis. New Phytol. **161**, 341–370 (2004)

E. Stokstad, Tomorrow's catch. Genomic technologies promise dramatic gains for aquaculture by accelerating the breeding of better strains. Science **370**(6519), 902 (2020)

M.L. Warburton, G. Wilkes, S. Taba, A. Charcosset, C. Mir, F. Dumas, D. Madur, S. Dreisigacker, C. Bedoya, B.M. Prasanna, C.X. Xie, S. Hearne, J. Franco, Gene flow among different teosinte taxa and into the domesticated maize gene pool. Genet. Resour. Crop Evol. **58**(8), 1243–1261 (2011). https://doi.org/10.1007/s10722-010-9658-1

Wikipedia, *Photosynthetic Efficiency* (2017a). https://en.wikipedia.org/wiki/Photosynthetic_efficiency. Accessed 6 Jan 2021

Wikipedia, *Mycorrhiza* (2017b). https://en.wikipedia.org/wiki/Mycorrhiza. Accessed 6 Jan 2021

X. Zhang, Z. Lin, J. Wang, et al., The tin1 gene retains the function of promoting tillering in maize. Nat. Commun. **10**, 5608 (2019). https://doi.org/10.1038/s41467-019-13425-6

Chapter 22
Top-Down Vs. Bottom-Up Control in Resource Management Systems

22.1 Background

Until the late 1940s, coffee in Latin America was grown by the "traditional" method known as "shade grown coffee" where coffee bushes are planted beneath tall, forest trees. The trees produce organic matter that facilitates nutrient recycling. Yield is low, but cost to the farmer also is low, and the method causes no environmental pollution. In the early 1950s, there began a movement to modernize coffee production by planting varieties adapted to growing in full sun. This "sun-grown coffee" has high yield due to fertilizers, and to pesticides that control disease and insect pests (Perfecto et al. 1996). Local villages received little economic benefit. Organic production arose when pulp from the coffee beans was composted in community owned facilities and used to fertilize coffee plantations.

22.2 Experimental Site and Methods

A 10-year study of the three coffee production systems was carried out beginning in 2000 at Centro Agronómico Tropical de Investigación y Ensañanza (CATIE) Turrialba, Costa Rica to compare the costs and benefits of each system (Rossi et al. 2011; Virginho Filho et al. 2015).

22.2.1 Methods

Each of the systems analyzed represents a hierarchical level embedded in a higher economic level.

C. F. Jordan, *Evolution from a Thermodynamic Perspective*,
https://doi.org/10.1007/978-3-030-85186-6_22

- Traditional agroforestry coffee production system (shade grown coffee) is embedded in a local economic system.
- A community based organic coffee production system fertilized is embedded in the national economy.
- A sun coffee plantation that is part of a corporate economy embedded in the international economy.

Energy enters each system bottom-up through the services of nature, and top-down through exogenous energy in the form of capital. Calculations followed Pimentel and Patzek (2005). A range of costs for inputs and prices for outputs were used to calculate the profit to cost ratio for each system under various economic conditions.

22.2.1.1 Currency for Models

A model gives a simplified picture of reality and will never be able to contain all the nuances of the real system. It is, however important that a model contains the characteristics of the system which are essential for management of a particular problem (Jørgensen et al. 2016). In capitalist economic systems, money-out/money-in has been the primary characteristic used to judge an investment. In systems of food production, energy-out/energy-in is used to judge efficiency of production systems when the output is valued for its energy production Where the output is valued for protein, energy input comparisons are still useful because they show the efficiency with which different types of meat and fish are produced. To evaluate production systems in energy terms, Odum (1971) proposed Net Energy Analysis (NEA) that quantifies all energy flows associated with such systems. It calculates all the energy embodied in a product and energy embodied in inputs. It takes into account the long-term impact on biophysical systems resulting from extraction of resources, and the impact of pollution created by the production system (Giampietro 2004).

For an agricultural system, the energy embodied in a kilo of sugar produced is not just the calories released when the sugar is burned. Embodied energy contains the energy in the fuel of tractors used to cultivate the fields. Energy equivalent of fertilizers and pesticides is also part of embodied input energy. Should the energy expended by humans while driving a tractor also be included? Should the energy expended used in manufacturing the tractor be included? Attempts to standardize assumptions has been done using Life Cycle Analysis (Murphy et al. 2016). While NEA generated initial enthusiasm (Gilliland 1975), it has not been widely used because the assumptions required in calculations has limited its usefulness (Leach 1975; Huettner 1976; Brown and Ulgiati 2004). Nevertheless, NEA is still useful for comparisons if the same assumptions are made for of all systems compared.

22.2.1.2 Standardization of Currency

When the agricultural output is valued strictly for its energy content as in the case of corn, rice, wheat, sorghum, potatoes and manioc, energy-out – energy-in comparisons are straight forward. In coffee production systems, caffeine, not energy is the desired product. This study used dry weight of coffee beans in lieu of caffeine as output. It still results in an output-input ratio that is useful if we assume a constant proportion between weight and caffeine content. The ratios tell indicate the efficiency with which energy is used to produce caffeine.

22.2.2 *Traditional Agroforestry (Hierarchical Level – The Ecosystem)*

A traditional agroforestry system consists of overstory trees, coffee bushes, and a community of soil organisms that recycle nutrients (Fig. 22.1). An internal loop of energy flow runs from the overstory trees through the soil community by way of leaf litter and root sloughing. Nutrients are carried by the litter and sloughed roots to the soil community that releases them for uptake by roots of the coffee bushes as well as the overstory trees.

- Endosomatic energy in the traditional plots refers to energy conversions linked to physiological processes within the soil-crop system. Sunlight is intercepted by overstory trees and coffee bushes, and is transformed through photosynthesis to biomass of the trees and bushes, and to coffee cherries (the pulp surrounding the

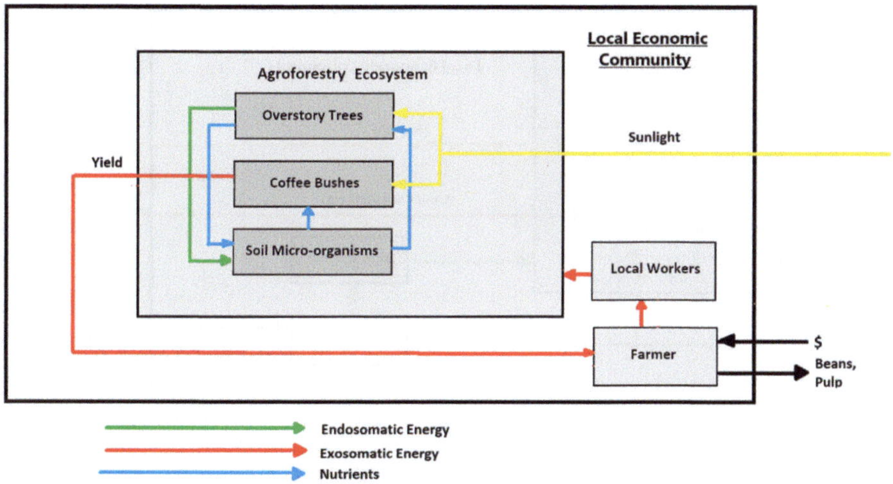

Fig. 22.1 Energy flow and nutrient feedback through a traditional agroforestry coffee production system embedded in a local economic system. (Modified from Jordan (2017))

beans). Energy is transferred from the overstory trees to the soil community by way of litter fall and root sloughing (Carrillo et al. 2011). Litter and dead roots are the energy source for the community of soil microorganisms that recycle nutrients to the coffee bushes and overstory trees. This nutrient feedback maintains the stability of the agroforestry plots. The blue arrows in Fig. 22.1 show the nutrient feedback system within the agroforestry plots.

- The system is embedded in the local economic community that supplies exosomatic energy. Exosomatic energy consists of labor for weeding and pruning by local workers to maintain structure of the system. Yield from the coffee bushes becomes exosomatic input to the traditional plot when part of the income from the sale of beans becomes wages for workers who maintain the plot.

22.2.3 Organic Production (Hierarchical Level – The Local Economic Community)

The organic system (Fig. 22.2) is an organic coffee production system embedded in the local economic community where community workers use pulp that surrounds the coffee bean to make compost that recycles nutrients back into the soil. It is not strictly an organic system because exosomatic inputs of potassium and phosphorus supplements are used to replace these nutrients lost when the coffee beans are removed from the community. It is an agroforestry system because prunings from the leguminous trees interplanted with the coffee supply nitrogen to the coffee bushes.

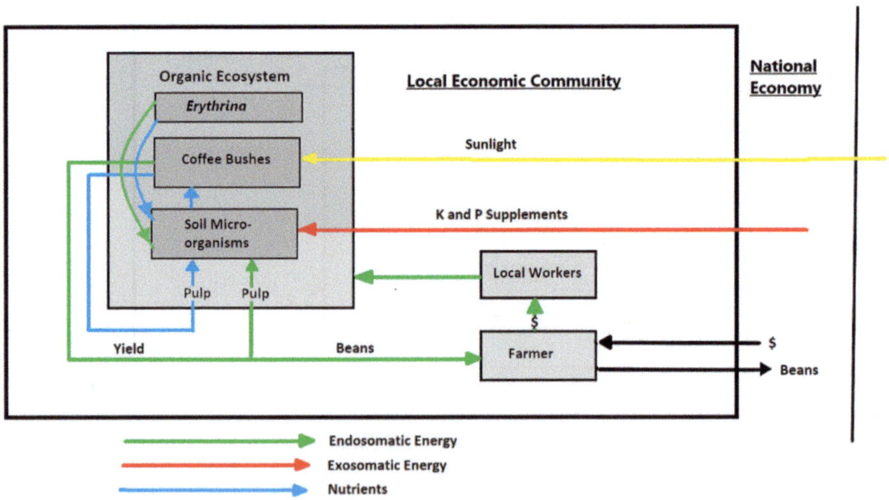

Fig. 22.2 Energy flow and nutrient feedback through a local economic community producing organic coffee. (Modified from Jordan (2017))

- The interplanted tree *Erythrina poeppigiana* has nitrogen-fixing symbionts.
- Part of the yield from the coffee bushes is pulp surrounding the bean. Pulp is endosomatic energy for the community when used as nutrient feedback into the organic plots (blue arrow). Local chicken manure is often added to the pulp.
- The feedback loop that included workers that was exosomatic for the traditional system becomes endosomatic when the boundary of the system is expanded to include the local community. Yield also is endosomatic energy for the community when it goes to the farmer to pay workers for separating pulp from beans and for weeding and pruning the agroforestry plots (green arrows).

Figure 22.3 shows a typical shade grown coffee plantation in Costa Rica with an overstory tree *Chloroleucon eurycyclum*, a species symbiotic with nitrogen fixing bacteria.

22.2.4 Sun Coffee Plantation (Hierarchical Level – The Corporate Economy)

Sun coffee plantations represent large scale, industrial type agriculture managed through a corporation or large-scale producer (Fig. 22.4). They are characterized by high energy inputs. All trees and shrubs other than the coffee bushes have been

Fig. 22.3 Shade grown coffee bushes in Costa Rica with *Chloroleucon eurycyclum* as the overstory tree. (Photograph by Florencia Montagnini, reproduced with her permission)

removed. There are no internal feedback loops to produce or sustain nutrient recycling. The corporate economy is closely linked to the national and international economy. The local economic community has little involvement.

- When the beans are sold internationally, there is monetary feedback to management and investors.
- Exosomatic inputs include fertilizers, pesticides, fungicides, and petroleum.
- When pulp is dumped into the environment, nutrients become pollutants.

22.3 System Comparisons

Input and production data from these studies were used in an economic analysis (Jordan 2017) to determine the trade-offs between short-term economic profitability and long-term sustainability.

Energy analysis cannot predict the economic costs and benefits from a particular coffee production system. But since (Energy-out) / (energy-input) ratios are independent of economic variables, they can always predict which type of system will give the greatest profit (or least loss) from investment, as prices of coffee and cost of inputs vary. They can be good indicators of which management system will give the highest net return on investment for a particular environment. Since the desired economic output of these systems is coffee beans, not energy, kg of yield is substituted for energy-out in calculations of economic trade-offs.

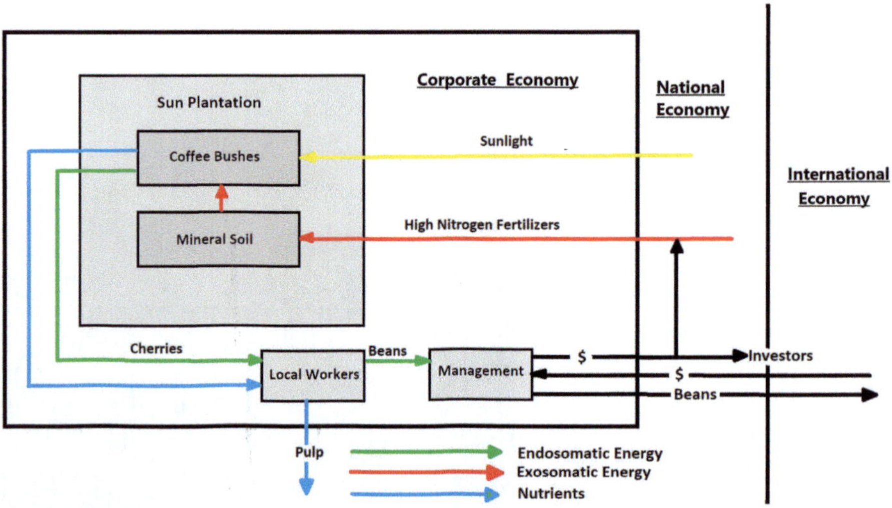

Fig. 22.4 Energy and nutrient flow through a corporate economy based on a sun coffee plantation. (Modified from Jordan (2017))

Table 22.1 Energy values for nutrient inputs

Type of input	Energy value of input
Coffee pulp	12.501 kJ/kg (Cubero et al. 2014)
Poultry manure	12.052 kJ/kg (Quiroga et al. 2010)
Nitrogen fertilizer	42 MJ/kg (Hulsbergen et al. 2001)
Phosphorus fertilizer	15.8 MJ/kg (Hulsbergen et al. 2001)
Potassium fertilizer	9.3 MJ/kg (Hulsbergen et al. 2001)
Total NPK	67 MJ/kg

Table 22.2 Energy inputs into the production systems

Management type	Intensity of input	Gross input/ha/year	Conversion factor	Net input/ha/year	Energy content	Energy input/ha/year
Chemical (sun monoculture)	Medium	400 kg fertilizer	10%[a]	40 kg	67MJ/kg	2.680 GJ
Organic	Medium	20 MT, coffee pulp	10%[b]	2000 kg	12.501 kJ/kg	25.0 GJ
		7.5 MT, manure	25%[c]	1880 kg	12.052 kJ/kg	22.6 GJ
		200 kg PO_4	10%[a]	20 kg	15.8 MJ/kg	0.316 GJ
		200 kg K	10%[a]	20 kg	9.3 MJ/kg	0.186 GJ
					Sum exogenous chemical	0.502 GJ
					Sum organic	48.102 GJ

[a]Assume 10-10-10 fertilizer
[b]Pulp is 90% water
[c]Chicken manure is 75% water

22.3.1 Energy Input

The energy content of the biomass that contains nutrient inputs, and the energy embedded in chemical fertilizers (Table 22.1) are multiplied times the input rate per hectare per year (Table 22.2) to give the exosomatic energy input costs for each management system.

22.3.2 Output

Price per kg yield of coffee beans times yield in kg per hectare equals income per hectare. Income minus costs equals profit. Profit/cost indicates a farmer's net return on investment.

22.3.3 Results

Table 22.3 and Fig. 22.5 use a range of prices to calculate profit from the yields of different treatments, and a range of costs to calculate exosomatic costs for each of these yields. Indicated values for price of coffee and costs of input are arbitrary. They do not necessarily reflect prices and costs at any point in time. They are used only to illustrate how changes in price of coffee and cost of chemical inputs affects income and return on investment (profit/cost) for sun, organic, and shade coffee systems. Substituting other values will change income and profit, but will not change trends.

Table 22.3 Income per hectare and return on investment (profit/cost) of sun, organic, and shade coffee as a function of coffee price and input costs

Economic trend	Management system	Price/kg yield $	Yield kg/ha x 10^3	Income $ from units produced	Energy cost $/Gj	Energy Input GJ/ha x 10^3	Cost /ha $	Profit/ha $	Profit/ cost
Coffee prices going up	Sun	2	1.921[a]	3842	3	2.680[a]	8040	-4198	–
	Organic	2	1.840[b]	3680	3	.502[b]	1506	2174	1.44
	Sun	5	1.921	9610	3	2.680	8040	1570	0.20
	Organic	5	1.840	9200	3	.502	1506	7694	5.11
Turrialba	Sun	10	1.921	19210	3	2.680	8040	11170	1.39
	Organic	10	1.840	18400	3	.502	1506	16894	11.22
Coffee prices going up[c]	Sun	2	1.600	3200	3	2.680	8040	-4840	–
	Shade	2	550	1100	3	.502	1506	-406	–
	Sun	5	1.600	8000	3	2.680	8040	-40	–
	Shade	5	550	2750	3	.502	1506	1244	0.83
World average	Sun	10	1.600	16000	3	2.680	8040	7960	0.96
	Shade	10	550	5500	3	.502	1506	3994	2.65
Input costs going up	Sun	5	1.921	9610	1	2.680	2680	6930	2.59
	Organic	5	1.840	9200	1	.502	502	8698	17.33
	Sun	5	1.921	9610	3	2.680	8040	1570	0.20
	Organic	5	1.840	9200	3	.502	1506	7694	5.11
Turrialba	Sun	5	1.921	9610	6	2.680	16080	-6470	–
	Organic	5	1.840	9200	6	.502	3012	6188	2.05
Input costs going up[c]	Sun	5	1.600	8000	1	2.680	2680	5320	1.99
	Shade	5	550	2750	1	.502	502	2248	4.48
	Sun	5	1.600	8000	3	2.680	8040	-40	–
	Shade	5	550	2750	3	.502	1506	1244	0.83
World average	Sun	5	1.600	8000	6	2.680	16080	-8080	–
	Shade	5	550	2750	6	.502	3012	-262	–

[a]Data from full sun monoculture, medium chemical intensity plots (10 year average)

[b]Exogenous chemical input only. Data from *Erythrina poeppigiana*, organic medium intensity plots (10 year average)

[c]Sun and shade data is from world data of the Audubon Society Shade-Grown Coffee Project (2004)

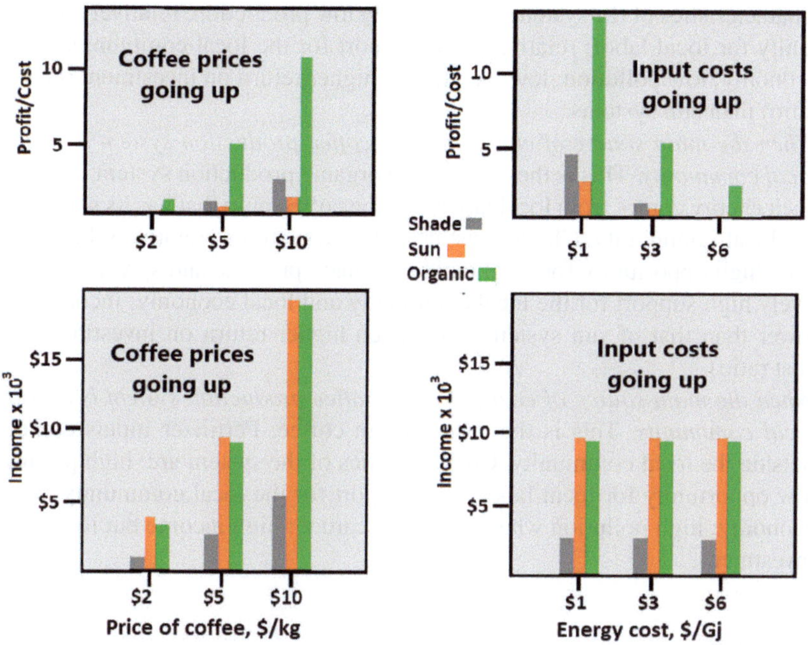

Fig. 22.5 Changes in income and return on investment (profit/cost) for sun, organic, and shade coffee systems as coffee prices increase and as costs of inputs to the production system increase. (Modified from Jordan (2017))

22.4 Discussion

The results show how the role of coffee prices and the price/cost ratio affect profit and investment opportunity. In both the Turrialba and world average data sets, sun grown coffee always produced the greatest gross income regardless of prices or costs because sun coffee produced the greatest yield, but when prices of coffee are low and when input costs are high, sun coffee is not profitable. Yields and income from organic coffee are always slightly less than from sun coffee, but their profit to cost ratio is always much higher than for sun coffee due to greater contribution of endosomatic energy and lower exosomatic energy costs. Shade coffee has lower yields and lower income than sun coffee but its profit to cost ratio is always higher than for sun coffee.

22.4.1 Effect of Energy Sources on System Outputs

- *When the main source of energy for the coffee production system is within the production plot itself.* This is the case for traditional agroforestry systems, where feedback energy comes from leaf and root litter of the overstory trees.

Characteristics of the system are: relatively low production; relatively low opportunity for local labor; relatively low support for the local community and local economy; low pollution; low income but higher return on investment (profit/cost ratio) than sun systems.

- *When the main source of energy for the coffee production system is within the local community.* This is the case for the organic production system, where feedback energy comes from local labor, and the coffee pulp that has its source within the local community. Characteristics of the system are: relatively high production; high opportunity for local labor to collect, process, and spread inputs; relatively high support for the local community and local economy; income slightly lower than that of sun systems, but much higher return on investment (profit/cost ratio).
- *When the main source of energy for the coffee production system is outside the local community.* This is the case for sun coffee. Fertilizer inputs come from outside the local community. Characteristics of the system are: high production; low opportunity for local labor; low support for the local community and local economy; high pollution when pulp is discarded; high income but low return on investment.

22.5 Conclusions

22.5.1 Feedback and Environmental Sustainability

The sustainability of a system depends on the proportion of energy driving the system that is endosomatically derived and therefore keeps cycles intact and does not cause pollution. Natural ecosystems and traditional coffee systems are sustainable systems because the energy that drives the nutrient supply is 100% endosomatic. The "organic" system is moderately sustainable because a large proportion of the energy that drives the nutrient supply is endosomatic. Sun coffee systems are environmentally unsustainable because the energy that drives nutrient supply is 100% exosomatic energy. The inefficiency with which exosomatically supplied nutrients (fertilizers) are used by plants causes unsustainability.

The traditional agroforestry system would be the choice for an individual farmer having little capital since the only exosomatic input required is labor. Organic coffee optimizes local economic sustainability through the energy feedback loop provided by local workers, and environmental sustainability through the nutrient feedback provided by pulp. It is ideal for a cooperative. Sun coffee would be the choice of a government wanting to increase exports to improve international balance of payments. However there is no hierarchical feedback from the field level to the economic system that signals when the cropping system is becoming stressed.

22.5.2 Economic Sustainability

Economic sustainability is measured by the ratio of profit to cost for each management system. A profit to cost ratio of 1.0 in Table 22.3 marks the difference between a sustainable and a non-sustainable system. Economic sustainability of sun systems occurs only with low costs of exosomatic inputs and high international coffee prices. Coffee production systems in which nutrient recycling depends on endosomatic energy are better investments when costs increase and/or prices decline.

References

Audubon Society Shade-Grown Coffee Project, *Cited in Wisconsin Natural Resources Magazine* (2004). http://dnr.wnrmag/html/stories/2004/feb04/shadechart.htm. Accessed 7 Dec 2020

M.T. Brown, S. Ulgiati, Energy quality, emergy and transformity: H.T. Odum's contributions to quantifying and understanding systems. Ecol. Model. **178**, 201–213 (2004)

Y. Carrillo, C.F. Jordan, K.L. Jacobsen, K.G. Mitchell, P. Raber, Shoot pruning of a hedgerow perennial legume alters the availability and temporal dynamics of root-derived nitrogen in a subtropical setting. Plant Soil **345**(1–2), 59–69 (2011)

R. Cubero-Abarca, R. Moya, J. Valaret, M. Tomazello Filho, Use of coffee (*Coffea arabaica*) pulp for the production of briquettes and pellets for heat generation. Cienc. Agrotec. **38**(5), 1–15 (2014)

M. Giampietro, *Multi-scale Integrated Analysis of Agroecosystems* (CRC Press, Boca Raton, 2004)

M.V. Gilliland, Energy analysis and public policy. Science **189**(4208), 1051–1056 (1975)

D.C. Huettner, Net energy analysis: An economic assessment. Science **192**(4235), 102–104 (1976)

K.J. Hulsbergen, B. Feil, S. Biermann, G.W. Rathke, W.D. Kalk, W. Diepenbrock, A method of energy balancing in crop production and its application in a long-term fertilizer trial. Agric. Ecosyst. Environ. **86**, 303–321 (2001)

C.F. Jordan, Energy analysis of coffee production systems: Implications for environmental and economic sustainability, in *Integrating Landscapes: Agroforestry for Biodiversity Conservation and Food Sovereignty*, ed. by F. Montagnini, (Springer, New York, 2017), pp. 151–164

S.E. Jørgensen, J.C. Marques, S.N. Nielsen, *Integrated Environmental Management: A Transdisciplinary Approach* (CRC Press, Taylor and Francis Group, Boca Raton, London, New York, 2016)

G.L. Leach, Net energy analysis – Is it any use? Energ Policy **3**(4), 332–344 (1975)

D.J. Murphy, M. Carbajales-Dale, D. Moeller, Comparing apples to apples: Why the net energy analysis community needs to adopt the life-cycle analysis framework. Energies **2016**(9), 917 (2016). https://doi.org/10.3390/en9110917

H.T. Odum, *Environment, Power and Society* (Wiley, New York, 1971)

I. Perfecto, R.A. Rice, R. Greenberg, M.D. Van der Voort, Shade coffee: A disappearing refuge for biodiversity. Bioscience **46**(8), 598–608 (1996)

D. Pimentel, T.W. Patzek, Ethanol production using corn, switchgrass, and wood: Biodiesel production using soybean and sunflower. Nat. Resour. Res. **14**(1), 65–76 (2005). https://doi.org/10.1186/1471-2156-15-23

G. Quiroga, L. Castrillon, Y. Fernandez-Nava, E. Maranon, Physico-chemical analysis and calorific values of poultry manure. Waste Manag. **30**(5), 880–884 (2010)

E. Rossi, F. Montagnini, E.M.V. Virginio Filho, Effects of management practices on coffee productivity and herbaceous species diversity in agroforestry systems in Costa Rica, in *Agroforestry as a Tool for Landscape Restoration*, ed. by F. Montagnini, W. Francesconi, E. Rossi, (Nova Science Publishers, New York, 2011), pp. 115–132

E.M.V. Virginho Filho, F. Casanoves, J. Haggar, G.S. Staver, J. Avelino, A. Tapia, M. Merlo, J. Salgado, M. Noponen, Y. Perdomo, A. Vásquez, La productividad util, la materia orgánica y el suelo en los primeros 10 años de edad en sistemas de producción de café a pleno sol y bajo varios tipos de sombra y niveles de insumos organicos y convencionales en de Costa Rica, in *Sistemas Agroforestales: Funciones Productivas, Socioeconomicas y Ambientales. Serie Tecnica Informe Tecnico No. 402.CATIE, Turrialba, Costa Rica*, ed. by F. Montagnini, E. Somarriba, E. Murgueitio, H. Fassola, B. Eibl, (Editorial CIPAV, Cali, Colombia, 2015), pp. 131–151

Chapter 23
Services of Nature in Agricultural Systems

23.1 Services of Nature

In natural ecosystems, services of nature are interactions between species that facilitate energy flow and nutrient recycling essential for life (Daily et al. 1997). Services of nature are fueled by endosomatic energy which means that they are linked to physiological processes within the ecosystem. These processes are fueled solely by photosynthetic energy in the form of reduced carbon (carbon stored as biomass). In agricultural ecosystems, many of these services are eliminated and replaced with fertilizers and pesticides (exosomatic energy sources, Giampietro 2004) to make the system easier to manage and more economically profitable. Odum et al. (1979) called them energy subsidies because they subsize solar energy to increase farm yield. Energy subsidies include the potential energy of tractor fuel, the kinetic energy of labor, and the energy embodied in fertilizers, pesticides, and herbicides. Substitutions of natural functions by exosomatic energy inputs in industrial agriculture cause system malfunctions (NOAA 2018) described in Appendix 4. To counter these side effects, organic farmers have reintroduced some of the interactions that characterize services of nature (Table 23.1).

23.2 The Nutrient Recycling Service of Nature

Nutrient recycling was the first service of nature to evolve when ecosystems first appeared more than two billion years ago. It was then, and still is now the most important service for maintaining ecosystem stability. The first step in recycling is done by decomposers as they consume labile carbon compounds in undecomposed organic matter. These compounds are readily utilizable by decomposers. As

C. F. Jordan, *Evolution from a Thermodynamic Perspective*,
https://doi.org/10.1007/978-3-030-85186-6_23

Table 23.1 Interactions and the services of nature they represent (Jordan 2013)

Interaction	Service of nature
Community of microorganisms × soil organic matter	Recycling of nutrients
Organic acids secreted by certain plants × P bound in soil	Phosphorus mobilization
Mycorrhizal fungi × plant roots	Facilitate nutrient uptake
Bacteria × leguminous plants	Nitrogen fixation
Beneficial insects × pest insets	Control of pest insects
Beneficial nematodes × harmful nematodes	Disease mitigation
Insects, some species of bats × plants	Pollination of crops
Birds, insects × plants	Dispersal of seeds
Leguminous plants × nitrogen-demanding crops	Over-yielding[a]
Plant roots × soil organic matter	Filtering of nutrient runoff
Certain soil bacteria × pollution	Detoxification of wastes
Microbial activity × soil water holding capacity	Mitigation of droughts

[a]Over-yielding means that a combination of crops has greater productivity per hectare than a one-hectare monoculture

compounds are broken down, nutrients are released and made available for recycling to autotrophs.

23.2.1 *Erosion Prevention*

Recalcitrant carbon is another broad category of carbon stored in biomass. It is composed of structural compounds such as lignin with a high content of humic acids and low concentration of nitrogen. These are difficult for microbes to metabolize (Kleber 2010). Recalcitrant carbon is beneficial to the system though, because it improves the structure of the soil and prevents erosion by increasing aggregate stability of the soil thereby maintaining its permeability to water and root growth (Ghabbour et al. 2017; Johnston et al. 2009).

The relative rate of release of both types of carbon from fresh organic matter is shown in Fig. 23.1.

23.3 Energy in Agricultural Systems

In this chapter we consider only the nutrient recycling service of nature in the context of managing for sustainability. Table 23.1 includes other interactions that are endosomatic energy transformations.

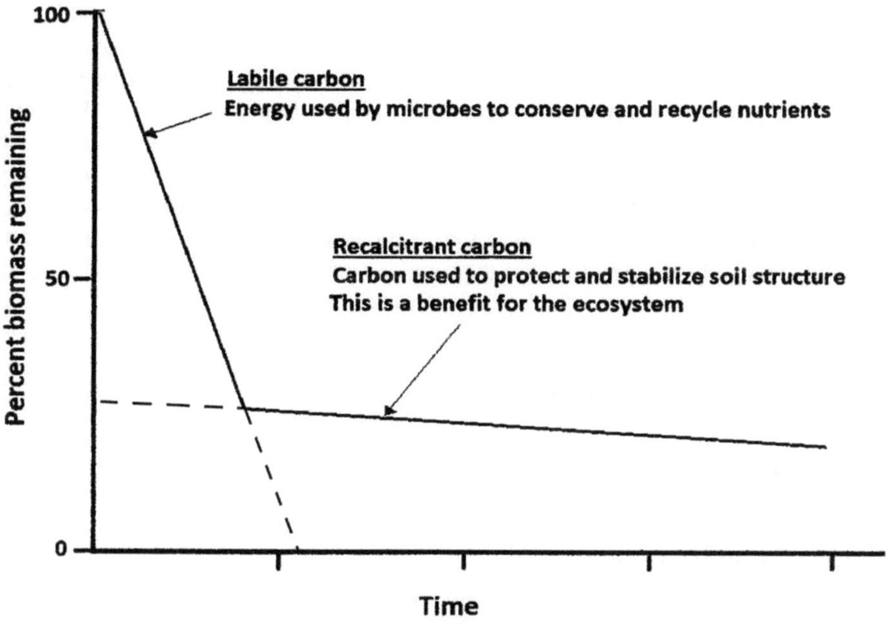

Fig. 23.1 Percent of initial biomass added to a soil that remains as a function of time during decomposition. The percentage of each type of carbon varies between systems

23.3.1 Embedded Energy

Embedded energy is the energy in biomass released by oxidation. When wood is burned, embedded energy is released. When we eat, the energy embedded in food is released. Embedded energy in exosomatic energy inputs is released very slowly as when steel machinery rusts, but very quickly when tractor fuel is burned.

23.3.2 Embodied Energy

The energetic cost of manufacturing or creating a product or work of art or of extracting a resource from the Earth is the product's embodied energy, a measure of the economic usefulness of transformed energy (Odum 1988). In agricultural systems, the embodied energy represents the sum of all the thermodynamic transformations that were necessary to achieve the final product. In industrial agricultural systems the endosomatic nutrient recycling service of nature has been replaced by exosomatic fertilizers. The most important exosomatic energy inputs in conventional agriculture by far are fertilizers, particularly nitrogen (Pimentel et al. 1973). The energy embodied in nitrogen fertilizer is equal to the amount of energy required by the Haber–Bosch process to synthesize ammonia from atmospheric nitrogen.

The energy embodied in phosphate fertilizer is the energy required to mine phosphate-bearing rock and extract the phosphate using industrial processes. The energy embodied in pesticides and herbicides is the energy required to synthesize them. The question is, how far back in the chain should one look? Definitely included is the energy content of the fuel that powers the tractor that cultivates the field. Should the energy used to build the tractor be included? How about the energy used to build the factory where the tractor was manufactured? Despite such ambiguities, embodied energy can be used for meaningful input comparisons of agricultural systems if the assumptions are the same for all systems compared.

23.4 The Systems Analyzed

1. Shifting cultivation on low fertility Oxisols in the Amazon Basin (Uhl and Murphy 1981)
2. Shifting cultivation on moderate fertility Entisols near a river valley in New Guinea (Rappaport 1971)
3. No-till cultivation of corn on medium high fertility Alfisols in Michigan (Snapp et al. 2015).
4. Conventional cultivation of corn on medium high fertility Alfisols in Michigan (Snapp et al. 2015)
5. Organic cultivation of corn (with aged manure) on medium high fertility soils at Rodale farms in Pennsylvania (Pimentel et al. 2005)
6. Organic cultivation of corn (with legumes) on medium high fertility soils at Rodale farms in Pennsylvania (Pimentel et al. 2005)
7. Conventional cultivation of corn on medium high fertility soils at Rodale farms in Pennsylvania (Pimentel et al. 2005)
8. Conventional cultivation of Iowa corn on very high fertility Mollisols having deep layers of organic residue from decomposed prairie grasses (Cox et al. 2011)
9. Conventional cultivation of corn on low fertility, low organic matter Ultisols in Georgia (Lee 2017)

Detailed site descriptions and the calculations used for Table 23.2 are in Appendix 11.

23.5 Summary of Results

Table 23.2 summarizes the comparisons. Rate of return means endosomatic energy returned on energy invested. Energy invested includes exosomatic investment by the farmer, and endosomatic investment represented by services of nature, in this case the recycling of nutrients.

Table 23.2 Summary of yields and rates of return on input energy for nine agricultural systems

Site	Food energy Yield, Gj/ha	% of input from services of nature	Rate of Return on endosomatic energy	Rate of Return on exosomatic energy[a]	Endosomatic/ exosomatic ratio
Shifting cultivation Amazon	35.1	99.9	0.02	13.8[b]	309
Shifting cultivation, New Guinea	95.3	98	0.34	16.4[b]	47
Conventional, Mich.	72.7	–	–	10.2	–
Conventional, Pa.	95.7	–	–	4.4[c]	–
Conventional, Ga	176.3	–	–	12.9	–
Conventional, Iowa	197.9	15	11.9	14.5	0.18
No-till, Michigan	78.5	44	9.1	16.0	0.78
Organic, legume Pa.	93.0	48	3.3	6.3[d]	0.91
Organic, manure Pa.	93.9	22	4.2	5.6[d]	0.29

[a]Rate of return is food energy yield divided by energy inputs
[b]Exosomatic subsidy in shifting cultivation is cutting down the trees which requires little energy compared to the food energy produced
[c]Low return due to high exosomatic costs
[d]Low rate of return from inputs in organic sites is due to high exosomatic costs, not low yield. Organic agriculture can have very high exosomatic costs (Schramski et al. 2013)

23.6 Discussion

These analyses focused entirely on nutrient cycling services of nature. Including other energetic services of nature such as that provided by beneficial insects would result in higher ratios for organic systems.

23.6.1 Shifting Cultivation

Despite the high rate of return on exosomatic subsidies (labor) in shifting cultivation, the energy return on endosomatic energy invested is very low because of the very low rate of return on services of nature. A lot of tree biomass accumulated over many years is destroyed to produce a small amount of crop yield for 2 or 3 years.

23.6.2 High Rates of Return on Exosomatic Inputs

The high rates of return (above 10.0) on exosomatic inputs (EROI) for conventional systems (Mich, Iowa, Ga.) is because the costs to society for pollution clean-up were not deducted from yield. Exosomatic energy inputs, particularly nitrogen and phosphorus have been indicted for polluting both fresh and salt water ecosystems and causing "dead zones" (NOAA 2019). If the costs of pollution were deducted from yield energy, the rates of return would be lower for all conventional systems.

23.6.2.1 The Iowa Case

The high rate of return on exosomatic inputs for conventional in Iowa is due in part to the contribution of the rich layer of organic matter derived from tall grass prairies that existed prior to agriculture. It has not been counted either as endosomatic or exosomatic energy input despite its contribution to productivity. The impact of those decomposed roots is being diminished yearly as plowing oxidizes another centimeter of organic soil.

23.6.3 Low Rates of Return on Exosomatic Inputs

For organic-manure in Pennsylvania, the low rate of return on exosomatic subsidies is because of the high energetic costs of colleting the manure and then redistributing it on the beds. Low rates of return on exosomatic inputs in the organic legume system is due to high tillage costs.

23.6.4 Rates of Return on Endosomatic Inputs

No-till and Iowa conventional have the highest rate of return on endosomatic energy inputs (9.1 & 12.7). No-till relies only on currently produced organic matter. The high production for the Iowa system probably results from inputs from the stored energy in the deep layers of organic soil that was not counted as part of the current year's input.

23.6.5 Sustainability

We have proposed that the ratio of endosomatic to exosomatic energy inputs is an index of sustainability. Endosomatic energy is that of the services of nature. It is non-polluting and it renews itself. The highest ratio is that of the organic legume system followed closely by the no-till system. Other practices besides no-till that use services of nature to reduce nutrient runoff include cover crops, green manure, conservation tillage, grass (buffer) strips between fields, and preservation of riparian vegetation (Minnesota Dept. of Agriculture 2017). Many other sustainable practices are promoted by Sustainable Agriculture Research & Education (SARE.org) and the U.S. Natural Resources Conservation Service (NRCS.USDA.gov). The ratio does not apply when current yield is dependent of energy stored over long periods of time, as in shifting cultivation and Mollisols (humus-rich soils) of Iowa.

23.6.6 Benefits and Costs of Herbicides

Organic agriculture is desirable because it causes little pollution. Organic agriculture is troublesome however, because plowing and plastic mulch are allowed for weed control. Plowing and rototilling break the nutrient cycling feedback system. Plastic mulch keeps oxygen and rain water out of the soil, and is difficult to dispose of.

No-till or minimum tillage systems are desirable because they keep the nutrient cycles intact, but they require the use of herbicides. Herbicides are more energy efficient than organic systems for controlling weeds. Pulling a sprayer behind a tractor requires less energy than pulling a plow. However, weeds quickly become herbicide resistant. A strategy to overcome resistance is to continually change the type of herbicide applied but it does not work for all weeds. Palmer amaranth (*Amaranthus palmeri)* commonly known as pigweed has developed resistance to multiple classes of herbicides and their different modes of action make them a challenge to use. Health issues are another problem of herbicides. There have been reports that the herbicide Atrazine is a potent endocrine disruptor that causes immunosuppression, hermaphroditism and even complete sex reversal in male frogs (Hayes et al. 2010) but the studies have been controversial (Jablonowski et al. 2011; Aviv 2014).

Once the world population becomes dependent on herbicides, we cannot forbid their use. Human populations have already swollen beyond that point. We have become addicted to their use and it is impossible to detoxify without causing a calamity.

References

R. Aviv, *A Valuable Reputation* (The New Yorker, 2014), Feb. 10, 2014. https://www.newyorker.com/magazine/2014/02/10/a-valuable-reputation. Accessed 23 Apr 2021

C. Cox, A. Hug, N. Bruzelius, *Environmental Working Group* (Losing Ground, 2011). https://www.ewg.org/losingground/report/executive-summary.html. Accessed 14 Dec 2020

G.C. Daily, S. Alexander, P.R. Ehrlich, L. Goulder, J. Lubchenco, P.A. Matson, H.A. Mooney, S. Postel, S.H. Schneider, D. Tilman, G.M. Woodwell, *Ecosystem Services: Benefits Supplied to Human Societies by Natural Ecosystems. Issues in Ecology 2* (Ecological Society of America, 1997)

E.A. Ghabbour, G. Davies, T. Misiewicz, R.A. Alami, E.M. Askounis, N.P. Cuozzo, A.J. Filice, J.M. Haskell, A.K. Moy, A.C. Roach, J. Shade, National comparison of the total and sequestered organic matter contents of conventional and organic farm soils. Adv. Agron. **146**, 1–35 (2017)

M. Giampietro, *Multi-Scale Integrated Analysis of Agroecosystems* (CRC Press, Boca Raton, 2004)
T. Hayes, V. Khoury, A. Narayan, et al., Atrazine induces complete feminization and chemical castration in male African clawed frogs (*Xenopus laevis*). PNAS **107**(10), 4612–4617 (2010)
N.D. Jablonowski, A. Schäffer, P. Burauel, Still present after all these years: Persistence plus potential toxicity raises questions about the use of atrazine. Environ. Sci. Pollut. Res. **18**, 328–331 (2011)
E. Johnston, P. Poulton, K. Coleman, Soil organic matter: Its importance in sustainable agriculture and carbon dioxide fluxes. Adv. Agron. **101**, 1–57 (2009)
C.F. Jordan, *An Ecosystem Approach to Sustainable Agriculture: Energy Use Efficiency in the American South* (Springer Verlag, Heidelberg, 2013)
M. Kleber, What is recalcitrant soil organic matter? Environ. Chem. **7**, 320–332 (2010). https://doi.org/10.1071/EN10006
Lee D, *A Guide to Corn Production in Georgia* (The University of Georgia College of Agriculture and Environmental Sciences, 2017), p. 101. http://caes2.caes.uga.edu/commodities/fieldcrops/gagrains/documents/2017_CornProduction_Final.pdf. Accessed 25 Feb 2021
Minnesota Dept. of Agriculture, *Conservation Practices* (2017). https://www.keyc.com/2019/12/24/minnesota-department-agriculture-program-promotes-conservation-practices-producers/. Accessed 26 Dec 2020
NOAA, *Gulf of Mexico "Dead Zone" is the Largest Ever Measured* (2018). https://www.noaa.gov/media-release/gulf-of-mexico-dead-zone-is-largest-ever-measured. Accessed 26 Feb 2021
NOAA, *Large "Dead Zone" Measured in Gulf of Mexico* (2019). https://www.noaa.gov/media-release/large-dead-zone-measured-in-gulf-of-mexico. Accessed 27 Dec 2020
H.T. Odum, Self organization, transformity and information. Science **242**(4882), 1132–1139 (1988)
E.P. Odum, J.T. Finn, E.H. Franz, Perturbation theory and the subsidy-stress gradient. Bioscience **29**(3), 349–352 (1979)
D. Pimentel, L.E. Hurd, A.C. Bellotti, M.J. Jorster, I.N. Oka, O.D. Sholes, R.J. Whitman, Food production and the energy crisis. Science **182**, 443–449 (1973)
D. Pimentel, P. Hepperly, J. Hanson, D. Douds, T. Seidel, Environmental, energetic, and economic comparisons of organic and conventional farming systems. Bioscience **55**(7), 573–582 (2005)
R.A. Rappaport, The flow of energy in an agricultural society. Sci. Am. **1971**, 1117–1132 (1971)
J.R. Schramski, K.L. Jacobsen, T.W. Smith, M. Williams, T.M. Thompson, Energy as a potential systems-level indicator of sustainability in organic agriculture: Case study model of a diversified, organic vegetable production system. Ecol. Model. **267**, 102–114 (2013)
S.S. Snapp, R.G. Smith, G.P. Robertson, Designing cropping systems for ecosystem services, in *The Ecology of Agricultural Landscapes: Long Term Research on the Path to Sustainability*, ed. by S. K. Hamilton, J. E. Doll, G. P. Robertson, (Oxford Univ. Press, New York, 2015), pp. 378–408
C. Uhl, P. Murphy, A comparison of productivities and energy values between slash and burn agriculture and secondary succession in the upper Rio Negro region of the Amazon Basin. Agro-Ecosystems **7**, 63–83 (1981)

Chapter 24
Optimizing Sustainability

24.1 Two Views of Sustainability

"Sustainability" has various, sometimes conflicting meanings. Economic sustainability requires permanent expansion at a maximum rate of power output, while ecological sustainability requires output stabilization at maximum energy use efficiency. Natural ecosystems are ecologically sustainable. They run entirely on services of nature. While hunter-gatherers can survive in a natural ecosystem, exosomatic energy must be invested in modern agricultural ecosystems to produce economically meaningful output.

24.2 A Compromise for Agriculture

It is not possible to maximize both ecological sustainability and maximum yield at the same time. (Jørgensen et al. 2016). A compromise is necessary, but to determine an objective compromise, endo and exosomatic inputs must be quantified. The greater the proportion of endosomatic energy inputs, the greater the sustainability. Deep Ecology (Ness 1973) is a philosophy that contends there is an inherent worth of nature's services, but it is incalculable. Nevertheless, there have been several attempts at quantification of nature's services.

C. F. Jordan, *Evolution from a Thermodynamic Perspective*,
https://doi.org/10.1007/978-3-030-85186-6_24

24.2.1 Value of Services of Nature (Endosomatic Inputs)

Ecological services have been quantified in terms of dollars and of energy. Costanza et al. (1997) put a dollar value of U.S. $33 trillion per year in 1997 dollars for 17 ecosystem services and functions that contribute to global ecological sustainability. The calculation was based on the replacement cost of the services. Jørgensen (2010) used the energy approach. He calculated the energy value of ecosystem services by determining the work capacity of various ecosystem types. The work capacity of trees for example is the amount of energy used to transpire the water from a hectare of forest. This transpiration cools the Earth and recycles water to the atmosphere. Work capacity values ranged from 2070 Gj/ha/year for deserts to 3.0 million Gj/ha/year for tropical rain forests. Using work capacity to valuate nature's services gives rates of energy use that are considerably higher than the energy embedded in the biomass of the species performing the services.

24.2.2 Value of Exosomatic Inputs

Exosomatic inputs also can be quantified in terms of dollars or energy. The dollar cost of a ton of fertilizer is the monetary value of an exosomatic input. The rate that the gross domestic product expands is a financial measure of economic expansion. The energy embodied in fertilizer is a energetic measure of exosomatic input (Macrotrends 2021). The rate that energy use expands is a thermodynamic measure of economic expansion (Kahan 2019).

24.2.3 Energy Vs. Dollars as a Measure of Sustainability

To evaluate sustainability of a system, energy is preferable over dollars. Dollar values for products, and for the inputs necessary to produce them vary according to supply and demand, but prices do not reflect the impact of demand on production ecosystems. For that we must use energetic values. For example, the economic value of soil organic matter was estimated by using the energy content of soil organic matter, determining the energy content of gasoline, and assuming a dollar cost of gasoline (Jordan 2013). If gasoline costs $3.50 per gallon, the value of soil organic matter per hectare is $104,000. If the price drops to $1.75 per gallon, the value would be $52,000 per hectare, even though the quantity of organic matter remains the same. The energetic value of the organic matter and its ability to facilitate nutrient cycling and agricultural production do not change as energy prices change. Regardless of supply or demand, the energetic value of soil organic matter to a depth of .3 m of an old-growth forest on the Georgia Piedmont is 940 kcal or 3933 kilojoules per hectare.

24.3 An Economic Model for Compromise

The optimal combination of endo- and exosomatic energy inputs for agricultural systems can be determined by an economic model based on production functions (Smathers et al. 1983). It is used in manufacturing systems to determine the optimal combination of variable inputs (Fig. 24.1). It can also be used to determine the optimum combination of endogenous and exogenous inputs in agricultural systems. Optimal compromise for agricultural systems means neither the ecological nor the economic benefits will be the highest possible, but the compromise will yield the highest possible combination of benefits.

In Fig. 24.1, X_1 is the exosomatic input and X_2 is the input derived from natures services. The relations denoted by Y_1, Y_2, and Y_3 are called isoquants (literally "equal products").

Isoquant Y_1 represents a farm with rich soil. It requires relatively low inputs of variables. Isoquant Y_3 is a farm on nutrient-poor soils. It requires high inputs. Farm Y_2 is on soil of intermediate fertility. Each isoquant represents a specific quantity of yield such as kilos of corn yield for each combination of inputs. The objective is to

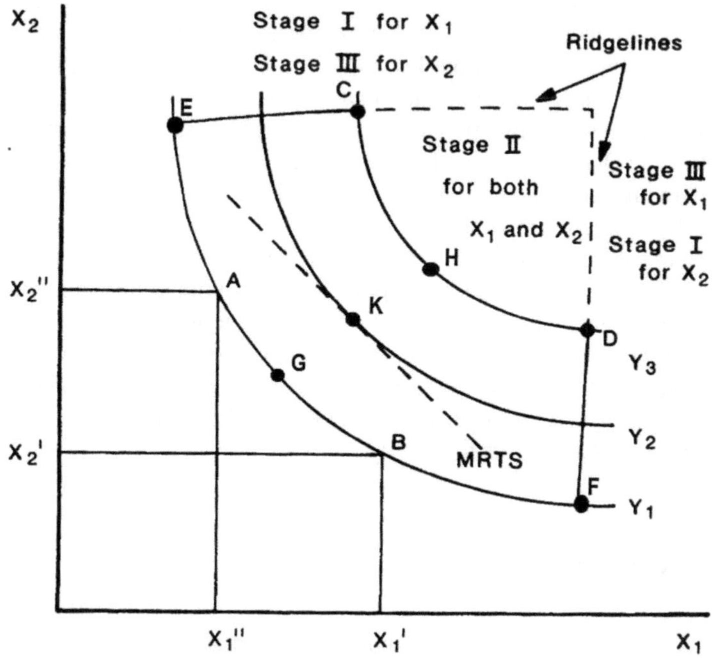

Fig. 24.1 Isoquant representation of input substitutability in production of a single output. X_1^{11} and X_2^{11} designate input rates at point A. X_1^{1} and $X2_1^{1}$ are input rates for point B. (Figure reproduced from Smathers et al. (1983) in BioScience with permission from American Institute of Biological Sciences)

produce a specific amount of product with the smallest outlay for inputs X_1 and X_2. The degree to which one input can substitute for another in economics is called the marginal rate of technical substitution (MRTS). MRTS is a concept of physical efficiency.

The point on a line with a 45° slope where the line is tangent to the relevant isoquant shows the respective amounts of X_1 and X_2 that reduces the cost of producing the output to a minimum. Those points are G, K, and H for the respective isoquants. These points would produce the optimum combinations, that is the combinations with the maximum efficiency. This would be the closest we could come to an optimum compromise in energy use between economic profits and ecological sustainability. Any combination between points A and B on line Y1 would produce the same yield but with different combinations of inputs. Beyond A and B, yield is declining, and beyond E and F, one of the inputs is returning a negative marginal product. Point E could be an organic system with maximum endosomatic inputs and minimum exosomatic inputs and F could be an industrial system with maximum exosomatic inputs and minimum endosomatic inputs like the sun-coffee system in Chap. 22.

"Ridgelines" C-E and D-F represent the limit beyond which only one input variable is operative. Above line C-E, further increases in X_2 would not increase output. Beyond line D-F, further increases in X1 would not increase output. When combinations of inputs are above line C-E, it represents stage 1 for X_1 where an increase in X_1 results in an increase in output but for X_2, it is stage 3 where in increase in X_2 does not increase output. Exosomatic input is underused while endosomatic energy is overused and the combination of inputs is inefficient. Where combination of inputs is beyond D-F, it is stage 3 for exosomatic inputs where X_1 is overused, X_2 is underused and the combination of inputs is inefficient. Stage 2 represents the range of technically efficient use of variable inputs.

Ecological sustainability would occur closer to point A than point G on isoquant Y_1. Organic farmers would operate at point A. Conventional farmers would decide where to operate depending on the cost of exosomatic inputs. If these inputs were cheap, he might operate at point B instead of G. These are considerations taken by coffee farmers as to whether to grow shade-grown coffee or sun-grown coffee (Chap. 22). Sun-grown (point B, Fig. 24.1) means short-term profit. Shade grown (point A) means long term sustainability.

24.4 Case Studies

Figure 24.1 indicates that a ratio of endo- to exosomatic energy of 1.0 is the optimum compromise.

Table 23.2 in Chap. 23 gives endo/exosomatic ratios for a series of agricultural systems. The closest to the optimum of 1.0 are the organic legume system with a ratio of .91 and the no-till system with a ratio of .78. Energy input in the conventional systems is all exosomatic. Lack of pollution control is sacrificed for high yield.

24.4.1 *Pest Control by Services of Nature*

Pest control in agricultural systems also requires a balance between endosomatic and exosomatic controls. Ecologically based pest management helps cut down on poisons in the environment by taking advantage of interactions between pests and naturally occurring pest-controlling organisms. Population outbreaks of pests often occur because there are no predator populations to control them. Natural enemies such as ladybird beetles and parasitic wasps can sometimes be used in place of pesticides against insect pests (Luck 1986). Bats have been found to control insects in tropical Agroforestry systems (Williams-Guillen et al. 2008). Natural insecticides such as pyrethrum that is extracted from the flower of *Chrysanthemum coccineum* can be used when high levels of chemicals are not acceptable (Silcox and Roth 1995). Use of rotational schemes so that two crops from the same botanical family are not planted in the same field in successive years can be a big step toward reducing crop enemies such as parasitic nematodes. Close monitoring of pest populations and spraying only when a build-up is noticed can reduce pesticide use as compared to spraying regularly throughout the season.

Other innovations include the sterile insect technique and the use of pheromone traps. The sterile insect technique is a method involving the mass-rearing and sterilization, using radiation, of a target pest, followed by the systematic area-wide release of the sterile males by air over defined areas, where they mate with wild females resulting in no offspring and a declining pest population. Pheromones are chemicals used by insects and other animals to communicate with each other. Insects send these chemical signals to help attract mates, warn others of predators, or find food. Pheromone traps can be used to capture pests. In Georgia, pheromone traps are used to lure boll weevils that can destroy a cotton crop. Traps are inspected weekly. Pheromone traps are also used to capture Japanese beetles that can devastate blackberries and raspberries. These techniques increase the ratio of endosomatic to exosomatic control, where exosomatic control is strictly by industrial chemical pesticides.

24.5 Trends

During the Industrial Revolution, innovations emphasized increasing output by investing exosomatic energy (coal, petroleum) in production processes. Only towards the end of the twentieth Century did public awareness increase regarding the environmental consequence of high energy use. In agriculture, this awareness stimulated a movement towards farming with less polluting forms of energy. This type of farming substituted services of nature for exosomatic energy as factors of production.

References

R. Costanza, R. d'Arge, R. de Groot, et al., The value of the world's ecosystem services and natural capital. Nature **387**(6630), 252–260 (1997)

S.E. Jørgensen, Ecosystem services, sustainability and thermodynamic indicators. Ecol. Complex. **7**, 311–313 (2010)

S.E. Jørgensen, J.C. Marques, S.N. Nielsen, *Integrated Environmental Management: A Transdisciplinary Approach* (CRC Press, Taylor and Francis Group, Boca Raton, London, N.Y, 2016)

C.F. Jordan, *An Ecosystem Approach to Sustainable Agriculture: Energy Use Efficiency in the American South* (Springer Verlag, Heidelberg, 2013)

A. Kahan, *EIA Projects Nearly 50% Increase in World Energy Usage by 2050, Led by Growth in Asia* (U.S. Energy Information Administration, 2019). https://www.eia.gov/todayinenergy/detail.php?id=41433. Accessed 1 March 2021

R.F. Luck, Biological control of California red-scale, in *Ecological Knowledge and Environmental Problem Solving*, (National Academy Press, Washington, D.C., 1986), pp. 165–189

Macrotrends, *U.S. GDP Growth Rate 1961–2021* (2021). https://www.macrotrends.net/countries/USA/united-states/gdp-growth-rate. Accessed 1 Mar 2021

A. Ness, The shallow and the deep, long-range ecology movement. A summary. Inquiry **16**(1–4), 95–100 (1973). https://doi.org/10.1080/00201747308601682

C.A. Silcox, E.S. Roth, Pyrethrum for control of pests on agricultural and stored products, in *Pyrethrum Flowers*, ed. by J. E. Casida, G. B. Quistad, (Oxford, 1995), Oxford University Press, pp. 287–310

W.M. Smathers, C.F. Jordan, E.G. Farnworth, T.H. Tidrick, An economic production function approach to ecosystem management. Bioscience **33**, 642–646 (1983)

K. Williams-Guillen, I. Perfecto, J. Vandermeer, Bats limit insects in a neotropical agroforestry system. Science **320**, 70 (2008)

Chapter 25
Agriculture that Incorporates Services of Nature

25.1 Environmentally Benign Agriculture

The most environmentally desirable types of agriculture have a high rate of return on services of nature (endosomatic inputs). They produce less pollution, and are less dependent on exosomatic energy (fossil fuels). In that sense they are more sustainable systems. Chapter 23 showed that systems with an endo-exosomatic proportion of energy input close to 1.0 were the most sustainable. Systems included no-till agriculture and some types of organic agriculture. Other benign systems include permaculture and regenerative agriculture. Permaculture adopts patterns of species arrangements that occur in flourishing natural ecosystems. Regenerative agriculture is similar except it includes annual crops.

25.2 Intercropping

Intercropping is a type of permaculture. It is the cultivation of two or more crops simultaneously on the same field. Maintaining species and genetic crop diversity through intercropping can on average, lead to greater productivity in plant communities, less nutrient leaching and volatilization, and greater ecosystem stability (Tilman 2000). Greater productivity and nutrient retention come about through resource partitioning (Finke and Snyder 2008). When different species have different resource requirements, complimentary interactions between individuals become more important than competition (Francis 1989; De Kroon 2007).

Greater stability also results from greater diversity because diversity helps slow the spread of disease. Genetically uniform monocultures are susceptible to rapid outbreaks of disease. For example, a mutation in black stem rust of wheat resulted in a new outbreak of the disease in Northeast Africa (Stokstad 2007).

C. F. Jordan, *Evolution from a Thermodynamic Perspective*,
https://doi.org/10.1007/978-3-030-85186-6_25

The "Yellow shoot disease", spread by the Asian citrus psyllid, is devastating citrus groves in Florida (Karp 2008). In contrast to monocultures, a diverse cropland does not present an unlimited opportunity for growth and reproduction of pest species. In China, the level of rice blast caused by the fungus *Magnaporthe grisea* was greatly decreased by planting several varieties of rice together instead of just a single variety (Wolfe 2000). Managing rangelands for a diversity of grass and forb species can reduce grasshopper populations (Branson et al. 2006). Because of their more complex structure, agroforestry systems help conserve biodiversity of fauna. In southwest India, 90 percent of the birds species of the native forest were found in an arecanut palm production system (Ranganathan et al. 2008).

25.3 Regenerative Agriculture

Regenerative agriculture is based on the concept of mimicking natural succession. An interesting example exists in the Tome-Assú colony near Belèm Brazil (Jordan 1987). A cycle may start with the clearing of secondary forest growing on a site that is chosen for cultivation. Some trees may be used for lumber, and others converted to charcoal. After the plot is cleared, remaining slash is burned and seedlings of rubber trees tolerant of the low fertility soils may be planted at intervals of several meters. For several years before the canopies of the rubber trees close, other crops are planted to take advantage of the freshly fallowed soil and to maintain soil cover. First corn is planted between the rubber seedlings. When the corn is half a meter tall, ginger is planted. When the corn us harvested, the stalks and leaves are collected and this mulch is spread around the base of fruit tree seedlings to conserve moisture and improve soil quality in neighboring plantations. When the ginger is harvested, the husks, which the farmers claim repel nematodes, are spread on nearby beds of onions which are susceptible to nematodes. Following corn harvest, cotton, winged beans and peanuts are planted in the same plot. The crops do well because of the fertilizing effect of the decaying corn roots. After 2 years when the canopy of the rubber trees begins to close, a high yield graft may be added. In other plantations, palm species valuable for heart of palm and for fruit used for juice and pig feed are the tree species used. Cacao is another commercially valuable crop, and the pulp separated from the bean is used to make compost for other crops. Pepper is an important crop, and vines are supported by nitrogen fixing trees. Several years before a pepper plantation is abandoned, trees with high commercial value such as mahogany are planted. Successional agroforestry systems such as that at Tome-Assú combine restoration and biodiversity objectives as well as agricultural sustainability objectives.

25.4 Agroforestry

Agriculture that incorporates trees is another environmentally benign type of agriculture. The integration of trees and shrubs with crops in agroforestry has many beneficial effects (USDA 2020). Perennial plants in agroforestry systems bind the soil and prevent erosion. Trees through leaf litter fall and root sloughing replace soil organic matter, thereby maintaining the nutrient cycling feedback loop. Agroforestry systems are becoming increasingly relevant worldwide as society has come to recognize their multiple roles and services: biodiversity conservation, carbon sequestration, adaptation and mitigation of climate change, restoration of degraded ecosystems, and tools for rural development (Montagnini 2017).

Agroforestry has much in common with other approaches to sustainable agriculture such as organic agriculture, permaculture, and regenerative agriculture, where endosomatic energy inputs are equal or greater than exosomatic energy subsidies. Agroforestry systems resemble natural ecosystems in that they conserve feedback systems so system stability is maintained, and they store entropy, that is they keep the system organized. In conventional agricultural systems where bare soil is exposed for part or all of the year, the entropic gradient between solar radiation and heat flowing to the atmosphere is scarcely reduced because the soil is a heat radiator. In contrast, agroforestry systems like natural ecosystems reduce the energy gradient by storing entropy instead of allowing the energy to heat the soil. Agroforestry systems are especially beneficial in tropical regions where weed growth is continuous and high year-round temperatures accelerate organic matter decomposition.

25.4.1 Agroforestry in Tropical Regions

25.4.1.1 Taungya

Taungya agriculture combines production of staple crops and valuable timber species. In Thailand, teak plantations are established with wide spacing between the seedlings, so that thinning of the stand is not necessary as the trees mature. In the spaces between the seedlings, farmers plant upland rice. Weeding of the rice prevents establishment of woody species that would compete with the teak. After 3 years, rice cultivation ceases and litter fall from the trees reestablishes soil organic matter in the site (Jordan et al. 1992).

25.4.1.2 Alley Cropping

In Cameroon, alley cropping combines maize with *Inga sp.*, a tree with nitrogen fixing symbionts. *Inga* is planted in rows with maize planted in the alleys between the rows (Rainforest Saver 2020). The trees are periodically pruned back to allow light and water to reach the corn in the alleys. *Inga* prunings add organic matter and nitrogen to the soil.

25.4.1.3 Sloping Agricultural Land Technology

In the Philippines where mountainous terrain makes agriculture difficult, the sloping agricultural land technology (SALT) has proven advantageous. In that system, fast growing woody species are planted along contours several meters apart on the mountainside. While the trees are becoming established, rainfall causes the area between contours to become flattened, resulting in a series of terraces in which corn, upland rice, beans and pineapple are planted. Sometimes crops such as coffee, banana and fruit trees are planted in every third strip (Jordan 1995).

25.4.1.4 Shade Grown Coffee

The organic coffee production system in Costa Rica illustrated by Fig. 22.3, Chap. 22, is a type of agroforestry system. The tree component is a nitrogen fixing species. Prunings add organic matter to the soil that feeds the decomposers which recycle nutrients to the coffee bushes. It is this recycling service that provides sustainability and distinguishes agroforestry systems from fossil fuel powered systems such as the sun coffee system in Fig. 22.4.

25.4.1.5 Biodiversity Islands

Agroforestry Systems can represent "Biodiversity Islands" in human-dominated landscapes (Montagnini 2020). They are islands of biodiversity surrounded by a sea of partially degraded landscapes. Multistrata agroforestry systems of home gardens common in rural tropical regions represent a tremendous reservoir of biodiversity. Home gardens are rich in genetic resources as people domesticate preferred native species, and they are also sites for conservation of species that have been extirpated in the wild. They serve as nuclei from which species can radiate into degraded areas.

25.4.1.6 Tropical Agroforestry Products

Tropical agroforestry systems produce specialized food products such as coffee, cacao, passion fruit, papaya, clove, peach palm, patchouli, cardamom and vanilla (Young 2017). Table 25.1 lists other economically and culturally important products of tropical agroforestry systems. Production systems for these crops are better adapted to small economies that occur in tropical regions. Many of these crops are based on indigenous traditions.

Table 25.1 Examples of non-timber forest products, extracted and collected by local people from tropical forests (Montagnini and Jordan 2005)

Product	Examples
Fuel and fodder biomass	Leaves, branches, roots of trees and shrubs
Construction materials	Canes, rattan, bamboo, palms
Fiber	Palms, lianas, herbs
Flowers	Orchids, anthurium, passion flower
Ornamental plants	*Zamia, Chameadorea*
Fruits	Zapote, durian, Brazil nut, açai
Tubers	Yams, taro
Other edible plant parts	Heart of palm
Mushrooms	Variety of edible mushrooms
Seeds	Colorful seeds for crafts
Oils	*Dipteryx odorata*, palm oil
Medicinal plants	*Quassia amara, Smilax, Cinchona*
Gums and resins	Rubber, chicle
Tannins and dyes	Brazil tree (*Caesalpina echinita*)
Wildlife products, food and ceremony	Honey, eggs, feathers, birds, mammals, fish, insects

25.4.1.7 Importance of Rule of Law

Agroforestry is a feasible system only where there is rule of law over property ownership. In the Amazonian frontier where indigenous property rights are ignored by unfettered logging and ranching operations, agroforestry is not possible. Only along old established river bank communities where it is possible for local governments to enforce boundaries around forest reserves can sustainable agroforestry systems be established (Castellanet and Jordan 2002). Some of these reserves contain economically desirable Brazil nut trees that are hundreds of years old.

25.4.2 Agroforestry in the Temperate Zone

In contrast to tropical agroforestry, agroforestry in the temperate zone is a recent innovation. Workman and Allen (2004) have described temperate alley cropping systems where cash crops such as hay, corn, cotton and vegetables are planted between rows of pecans, persimmons, or walnuts, or species valued for timber such as oak. Silvopasture is a technique that combines trees with livestock and their forages on one piece of land. The trees provide timber, fruit, or nuts as well as shade and shelter for livestock and their forages, reducing stress on the animals from the hot summer sun, cold winter winds, or a downpour.

Fig. 25.1 Shade grown blueberry plantation with *Paulownia fortunei* as the overstory species at Spring Valley Ecofarms. (Photograph by C. F. Jordan)

25.4.2.1 Shade Grown Blueberries

On my farm in Athens Georgia, I took inspiration from a shade-grown coffee system (Fig. 22.3) to establish shade grown blueberries (Fig. 25.1). I used *Paulownia fortunei* as an overstory species. It is fast growing, so that every year it can be pruned to add organic matter to the base of the blueberries. After several years, the tree must be cut to prevent too much shading, but sprouts readily emerge from the stump so replanting is unnecessary. The irregular shape of the trees is because it acts as a trap crop for deer, a major pest species. They prefer to eat the *Paulownia* bark instead of blueberry shoots.

25.5 Disadvantages of Agroforestry

Like all sustainable agricultural systems, agroforestry systems take several years to become established. They often are labor intensive, disadvantageous where labor is expensive and fossil fuel energy is cheap, but advantageous where the opposite is true. They can be less profitable for the farmer than agriculture highly dependent on energy subsidies, but more profitable for society that does not have to deal with agricultural pollution.

25.6 The Governmental Perspective

Governments often support industrial agriculture because the large scale can help a country's balance of trade. Development agencies that encourage integration of developing countries into the world economy have traditionally supported industrial agriculture rather than organic or traditional methods. In some cases, traditional agriculture that sustained populations for many generations was displaced when government incentives favored export agriculture (Moseley and Gray 2008). Since industrial agriculture often requires less labor (Stockdale et al. 2001), one result has been a rise in unemployment. However, the World Bank is no longer relying on Industrial Agriculture for its development projects, but rather is funding projects that employ a wide variety of approaches that are more environmentally and culturally friendly. Kareiva et al. (2008) found that World Bank projects with biodiversity goals were as successful in all development objectives, including poverty reduction and private sector development, as those that focused solely on development. A return to sustainable agriculture in developing regions will require substantial policy, institutional and professional reform. For example, tariffs are a key issue. They can either help or hinder development of sustainable food production (Pinstrup-Andersen 2003, 2005). High tariffs imposed by the Organization for Economic Cooperation and Development member countries on agricultural products limit the ability of farmers in developing countries to compete in world markets. On the other hand, tariffs by developing countries on commodities from industrialized nations protect local farmers and enable them to produce in a more environmentally friendly manner.

References

D.H. Branson, A. Joern, G.A. Sword, Sustainable management of insect herbivores in grassland ecosystems: New perspectives in grasshopper control. Bioscience **56**, 743–755 (2006)

C. Castellanet, C.F. Jordan, *Participatory Action Research in Natural Resource Management* (Taylor and Francis, UK, 2002)

H. de Kroon, How do roots interact? Science **318**, 1562–1563 (2007)

D.L. Finke, W.E. Snyder, Niche partitioning increases resource exploitation by diverse communities. Science **321**, 1488–1490 (2008)

C.A. Francis, Biological efficiencies in multiple cropping systems. Adv. Agric. **42**, 1–43 (1989)

C.F. Jordan, *Amazonian Rain Forests: Ecosystem Disturbance and Recovery* (Springer Verlag, New York, 1987)

C.F. Jordan, *Conservation: Replacing Quantity with Quality as a Goal for Global Management* (Wiley, New York, 1995)

C. F. Jordan, J. Gajaseni, H. Watanabe (eds.), *Taungya: Forest Plantations with Agriculture in Southeast Asia* (C.A.B. International, Wallingford, 1992)

D. Karp, *Deadly Pathogen Harms Florida Citrus Groves* (2008). https://www.nytimes.com/2008/08/26/science/26citrus.html. Accessed 12 Apr 2021

P. Kareiva, A. Chang, M. Marvier, Development and conservation goals in world bank projects. Science **321**, 1638–1639 (2008)

F. Montagnini (ed.), Integrating landscapes: Agroforestry for biodiversity conservation and food sovereignty, in *Advances in Agroforestry 12*, Cham, Switzerland (Springer Nature, 2017)

F. Montagnini, The contribution of agroforestry to restoration and conservation: Biodiversity islands in degraded landscapes, in *Agroforestry for Degraded Landscapes*, ed. by J. C. Dagar, S. R. Gupta, D. Teketay, (Springer, Singapore, 2020), pp. 445–479

F. Montagnini, C.F. Jordan, *Tropical Forest Ecology: The Basis for Management and Conservation* (Springer Verlag, Berlin, 2005)

W. G. Moseley, L. C. Gray (eds.), *Hanging by a Thread: Cotton, Globalization and Poverty in Africa* (Ohio University Press, Athens, OH, 2008)

P. Pinstrup-Andersen, Food security in developing countries: Why government action is needed. UN Chron. **40**(3), 3 (2003)

P. Pinstrup-Andersen, Ethics and economic policy for the food system. Am. J. Agric. Econ. **87**, 1097–1112 (2005)

Rainforest Saver, *The Cameroon Inga Project* (2020). https://www.rainforestsaver.org/cameroon-inga-project Accessed 1 Jan 2021

J. Ranganathan, R.J. Ranjit Daniels, M.D. Subash Chandran, P.R. Erlich, G.C. Daily, Sustaining biodiversity in ancient tropical countryside. PNAS **105**, 17852–17854 (2008)

E.A. Stockdale, N.H. Lampkin, M. Hovi, R. Keatinge, E.K.M. Lennartsson, D.W. Mcdonald, S. Padel, F.H. Tattersall, M.S. Wolfe, C.A. Watson, Agronomic and environmental implications of organic farming systems. Adv. Agric. **70**, 261–327 (2001)

E. Stokstad, Deadly wheat fungus threatens world's bread basket. Science **315**, 1786–1787 (2007)

USDA, *Agroforestry* (2020). https://www.usda.gov/topics/forestry/agroforestry. Accessed 5 Jan 2021

D. Tilman, Causes, consequences and ethics of biodiversity. Nature **405**, 208–211 (2000)

M.S. Wolfe, Crop strength through diversity. Nature **406**, 681–682 (2000)

S.A. Workman, S. Allen, *The Practice and Potential of Agroforestry in the Southeastern United States* (University of Florida IFAS Extension CIR 1446, 2004). https://www.researchgate.net/publication/238671026_The_Practice_and_Potential_of_Agroforestry_in_the_Southeastern_United_States. Accessed 6 Jan 2021

K.J. Young, Mimicking nature: A review of successional agroforestry systems as an analogue to natural regeneration of secondary forest, in *Integrating Landscapes: Agroforestry for Biodiversity Conservation and Food Sovereignty*, ed. by F. Montagnini, (Springer Verlag, New York, 2017), pp. 179–209

Chapter 26
Rebuilding Natural Capital: A Case Study

26.1 The Nature of Capital

Capital consists of assets that can enhance one's power to perform economically useful work. In economic systems, it can be in the form of factories and trucks. It can also be in the form of information such as potential markets for a product. On a farm, there is economic capital – the machinery to cultivate the crops and the equipment to store and transport the harvest. There is also natural capital, the services of nature that contribute to sustainability. Natural capital can be in the form of stored energy such as that in soil organic matter, or in the form of information about interactions between predator and prey insects.

26.1.1 Depletion of Natural Capital

Soil organic matter supplies the energy for decomposers that enable recycling of nutrients, a service of nature. The deep layer of organic matter in soils of midwestern United States is natural capital. It has enriched the nation, but it is rapidly being depleted. Ground water reservoirs are also natural capital, and they also are being depleted. Most farmers recognize these problems, but they are helpless to do anything about them. They must continue exploiting the capital to keep their expenses low so they can sell their harvest at the lowest possible prices to win the competition for market share.

C. F. Jordan, *Evolution from a Thermodynamic Perspective*,
https://doi.org/10.1007/978-3-030-85186-6_26

26.2 Rebuilding Natural Capital

In the first part of my professional career, I studied how nutrient cycling facilitated by soil organic matter stabilized natural systems. After many years of basic research, I wanted to study a practical application. In 1993, I had an opportunity to do that on a 100-acre farm near Athens Georgia. It had been part of a cotton plantation before the Civil War. Cotton continued to be the major crop under the share-cropper system after the war until the 1930s when the boll weevil and low soil fertility ended cotton production. Most of the farms in the region were abandoned and pine quickly invaded. On the farm that I acquired, agriculture continued. Wheat and sorghum had been planted using deep plowing and heavy applications of fertilizer. Soil consisted of dense red clay with little or no organic matter content. The red clay is a sub-soil derived from the ancient rocks of the Appalachian Mountains and purged of nutrients for millions of years by a hot, wet climate.

I knew that the most important factor in restoring degraded soil was re-establishing nutrient cycling through building of soil organic matter. My question in 1993 was how to do that while still producing an economic product. I thought Georgia Organics could help. Georgia Organics is an organization to promote organic agriculture in Georgia. Every year they sponsor tours of organic farms throughout the state. I decided to join and take some of the tours to see what I could learn. My greatest surprise was that most organic farmers were not conserving soil organic matter. To prepare the soil for planting, they would plow or rototill the soil with the result that oxidation decreased the soil organic matter. Weed control was often with plastic weed barriers which are energetically expensive and an environmental nuisance. Degradable barriers were environmentally better but did not last long. I began by trying something else.

I had two students that wanted to try alley cropping, where open "alleys" between hedgerows of leguminous shrubs provided nitrogen for the crops, and mulch to suppress weeds. One student had been a Peace Corps Volunteer who had discovered the use of alley cropping in Cameroon. The other was a Brazilian who thought it might help agriculture in Brazil. Our hedgerow was the nitrogen fixing false indigo (*Amorpha fruticosa*). The system did not work as well in Georgia as it did in Cameroon. In Cameroon, the growing season is long enough whereby the hedgerow can be pruned, the corn can grow in the residue, and the hedgerow can regrow after the harvest. In Georgia, the growing season ends shortly after harvest, and there is not enough time for the hedgerow to regrow before planting season the next spring. After 4 years, our hedgerows still did not produce sufficient mulch to increase soil organic matter while vegetables were growing in the alley. The hedges did provide habitat for spiders that were good for pest control, and also served as barriers to deer, a major pest.

We also experimented with cover crops and no-till cultivation (Fig. 26.1). Cover crops such as clover (top left) or winter wheat (top right background) are planted in the fall. They germinate but remain stagnant over winter until spring when they put on most of their growth. They are then knocked flat with a roller-crimper (center

Fig. 26.1 Top: Crimson clover, winter wheat. Center: roller crimper, no-till vegetable planter. Bottom: Soil profile under cultivation with the plow, under no-till cultivation. (Photographs by C.F. Jordan)

left). Organic matter is added to the soil when the cover crop is rolled flat, crimped, and killed. Roots slough off and provide an energy source for the decomposer community. The nutrients are released in the spring when conditions are warm and wet, the optimal conditions for plant growth. A no-till planter (center right) opens a furrow into which vegetable seedlings are transplanted, and then the furrow is closed with two tamping wheels below the white bucket that holds weights for down-pressure on the wheels.

We did not use herbicides to control weeds. The mulch evident on the soil surface in the center right photo suppressed weeds for only about a month, after which they began to grow through the partially decomposed mulch. Usually by that time, the crop was well-enough developed that it could withstand the competition. When we combined no-till with alley cropping after the hedgerow was mature, a deep layer of organic matter developed that remained even while being farmed for vegetables.

Figure 26.1 bottom left shows what the soil looked like in 1993 when I acquired the farm, and bottom right was the soil between the hedges in 2019, after 20 years of cropping with cover crops, conservation tillage and periodic compost addition. The soil in the lower right photo is rich in organic matter, similar to the A horizon in mature hardwood ecosystems of the Piedmont (Fig. 11.2a, Chap. 11). The dark color is due mostly to recalcitrant carbon in humus. Organic matter containing recalcitrant carbon is slow to decompose because of high concentrations of humic acids that inhibit decomposing organisms. Undecomposed particulate matter increases aggregate stability. Aggregate stability means the tendency of fine particles to be bound together thereby improving the physical structure of the soil, enabling roots to grow easily and water to percolate more readily. Recalcitrant carbon compounds remain in the soil for long periods of time in contrast to labile carbon which is rapidly oxidized by decomposers.

26.3 An Obstacle to Rebuilding Natural Capital

A major obstacle to establishing sustainable farms is the length of time necessary to develop the soil organic matter that makes fertilizers unnecessary. At the Rodale experimental farm, 5 years passed before yields in the organic fields matched those in the conventional fields (Pimentel et al. 2005). A feasible approach for transitioning to organic is to convert only a small percent of the farm every year. That way, yearly decreases in total farm income are minor.

Reference

D. Pimentel, P. Hepperly, J. Hanson, D. Douds, T. Seidel, Environmental, energetic, and economic comparisons of organic and conventional farming systems. Bioscience **55**(7), 573–582 (2005)

Chapter 27
Can Organic Agriculture Feed the World?

27.1 Organic Agriculture vs. Low- Energy-Input Agriculture

Energy use efficiency in agriculture has been of interest since the 1970s when scientists became aware of an incipient world energy crisis. Organic agriculture was proposed as a possible solution. A contentious question among resource scientists then was "Can Organic Agriculture Feed the World?" (Badgley et al. 2007a, b). That question was raised because it was assumed that organic agriculture is low-energy-input agriculture. However, organic agriculture is not always low-energy-input agriculture. Sometimes it uses high amounts of exogenous energy. Compost generation can be energy intensive. Fossil fuel energy is commonly used to power tractors, irrigation pumps, and greenhouses. In a case study of a diversified, organic vegetable production system, Schramski et al. (2013) found that a typical mechanized, intensive organic agricultural system in the U.S. operates with an energy efficiency of only 0.025. Manure must be collected, composted, and spread. Cover crops must be planted, then crimped and rolled flat. Nitrogen fixing plants must be cultivated. Nevertheless, energy consumption by U.S. farms accounts for only about 2% of the nation's energy use (Canning et al. 2010). The bulk is from transportation, industrial, residential and commercial sectors of the economy. More important than energy use by agriculture is pollution caused by chemical intensive, agriculture. Organic farming does not use fertilizers and pesticides while industrial agricultural does (Pimentel and Patzek 2005). Equally important is efficient use of water. Organic agriculture uses efficient drip irrigation, while industrial agriculture uses inefficient center pivot systems. The question to be answered then does not involve energy use, but rather concerns pollution and wastage of fresh water. Can systems that use no chemical fertilizers and pesticides and that use water efficiently be as productive as those that do?

C. F. Jordan, *Evolution from a Thermodynamic Perspective*,
https://doi.org/10.1007/978-3-030-85186-6_27

27.2 Organic Agriculture vs. Conventional Agriculture

There have been a number of studies that compared high energy use agricultural systems with organic systems in which energy use actually *was* low. In these cases, organic agriculture can be used as a proxy for low energy input systems. Crop yields in organically farmed systems were 20–30% lower in organic plots, but energy inputs were much lower (Mäder et al. 2002; Refsgaard et al. 1998; Dalgaard et al. 2001). Ziesmer (2007) found that excluding labor, organic agriculture uses 30 to 50 percent less energy in production than comparable non-organic agriculture, and Hill (2009) concluded that organic agriculture is more energy efficient than conventional agriculture in some but not all cases.

Badgley et al's. (2007a) comments shed light on the reason for lower organic yield in many cases. Lower yields in organically farmed plots in the U.S. and western Europe occurred because the benefits of nature's services were still accumulating in the experimental plots, while in nations where organic agriculture and agroforestry are traditional, services of nature are already well developed. For comparing yield and energy efficiency of organic and conventional agriculture, the Rodale studies showed that trials have to extend over a 5 years or more or more to allow the services of nature to show their full potential.

In an analysis of 118 studies conducted over 50 years across 51 countries to assess how farm size affects outcomes other than food production, Ricciardi et al. (2021) found that farms less than 2 hectares in size tend to be more productive and biodiverse per unit area than large ones, and are roughly as profitable and resource-efficient. Their higher biodiversity suggests that services of nature were important and they were therefore low-energy-input systems.

27.3 Can Agriculture Dependent on Low Energy Input Feed the World?

There is yet no clear-cut yield advantage for organic systems. The crop yields and economics of organic systems, compared with conventional systems appeared to vary based on the crops, and technologies employed. We cannot state unequivocally that agriculture dependent on low energy, low chemical input can feed the world. In the Rodale plots however, soil erosion was less, and water conservation, soil organic matter and biodiversity were consistently greater in the organic systems than in the conventional systems. (Pimentel et al. 2005).

27.4 Yield Is Not the Problem

The relative yields of industrial vs. organic agriculture may not be the most important issue regarding hunger. There is evidence that hunger in some parts of the world is not caused by low agricultural yields, but rather because in some regions, people have insufficient income to purchase food (Alexandratos 1999). A 2007 conference entitled “Africa can Feed Itself” (Naerstad 2007) concluded that organic agriculture could improve food security, but important obstacles are unfair trade regimes, market access, food dumping, border wars, unpredictable weather, dissembling government policies, access to land and land use, expensive credits, delays in receiving inputs and human disease. Other problems include the tendency for agricultural development projects often ignore or exclude women, yet they are often the primary agricultural producers. Food aid itself is a problem. It has diminished Africa’s capacity to feed itself by undermining African farm production and local food markets.

According to the Food and Agriculture Organization of the United Nations (FAO 2009a, b) the world produces more than 1½ times enough food to feed everyone on the planet. Increasing food production is not a critical problem in today’s world (although distributing it to world populations may be). Can yield from organic agriculture feed the world is not the critical question. Can the world be fed without causing chronic pollution, or without depleting ground water reservoirs is the critical question. Systems to accomplish that are highly dependent on endosomatic energy. Studies to date have not shown that low energy input systems can always match yield of high exosomatic energy inputs. But industrial systems produce more than enough yield to eliminate world hunger, so matching the yield of industrial systems is not the problem. The question is, what type of systems can survive without causing pollution and water shortages that would definitely influence yield in the long term under industrial agriculture. The answer may be low-energy-input organic systems.

References

N. Alexandratos, World food and agriculture: Outlook for the medium and longer term. Proc. Natl. Acad. Sci. **96**, 5908–5914 (1999)

C. Badgley, J. Moghtader, E. Quintero, E. Zakem, M.J. Chappell, K. Avilés-Vázquez, A. Samulon, I. Perfecto, Organic agriculture and the global food supply. Renew. Agric. Food Syst. **22**, 86–108 (2007a)

C. Badgley, I. Perfecto, K. Cassman, J. Hendrix, Can organic agriculture feed the world? Renew. Agric. Food Syst. **22**(2), 80–85 (2007b)

P. Canning, C. Ainsley, S. Huang, K. Polenske, A. Waters, *USDA Economic Research Service Report No. ERR-94* (United States Department of Agriculture, 2010). Archived from the original on 18 September 2010 by Wikipedia Agriculture, https://en.wikipedia.org/wiki/Agriculture#Energy_dependence. Accessed 31 May 2021

T. Dalgaard, N. Halberg, J.R. Porter, A model for fossil energy use in Danish agriculture used to compare organic and conventional farming. Agric. Ecosyst. Environ. **87**, 51–65 (2001)

FAO, *1.02 Billion Hungry* (2009a). http://www.fao.org/news/story/en/item/20568/icode/. Accessed 17 Dec 2020

FAO, *The State of Food Insecurity in the World* (Economic and Social Development Department Food and Agriculture Organization of the United Nations, Rome, 2009b). www.fao.org/publications/sofi/2020/en/. Accessed 17 Dec 2020

Food and Agriculture Organization of the United Nations, *The State of Food Insecurity in the World* (Economic and Social Development, Department Food and Agriculture Organization of the United Nations, Rome, 2009)

H. Hill, *Comparing Energy Use in Conventional and Organic Cropping Systems* (2009). https://attra.ncat.org/attra-pub/viewhtml.php?id=301. Accessed 19 Dec 2020

P. Mäder, A. Fließbach, D. Dubois, L. Gunst, P. Fried, U. Niggli, Soil fertility and biodiversity in organic farming. Science **296**(5573), 1694–1697 (2002)

A. Naerstad, *Africa Can Feed Itself* (The Development Fund, Grensen 9B, N-0159 Oslo, Norway, 2007). https://www.utviklingsfondet.no/files/uf/documents/Africa_Can_Feed_Itself.pdf. Accessed 17 June 2021

D. Pimentel, T.W. Patzek, Ethanol production using corn, switchgrass, and wood: Biodiesel production using soybean and sunflower. Nat. Resour. Res. **14**(1), 65–76 (2005). https://doi.org/10.1186/1471-2156-15-23

D. Pimentel, P. Hepperly, J. Hanson, D. Douds, T. Seidel, Environmental, energetic, and economic comparisons of organic and conventional farming systems. Bioscience **55**(7), 573–582 (2005)

K. Refsgaard, N. Halberg, E.S. Kristensen, Energy utilization in crop and dairy production in organic and conventional livestock production systems. Agric. Syst. **57**, 599–630 (1998)

V. Ricciardi, Z. Mehrabi, H. Wittman, D. James, N. Ramankutty, Higher yields and more biodiversity on smaller farms. Nat. Sustain. (2021). https://doi.org/10.1038/s41893-021-00699-2. Accessed 17 Apr 2021

J.R. Schramski, K.L. Jacobsen, T.W. Smith, M. Williams, T.M. Thompson, Energy as a potential systems-level indicator of sustainability in organic agriculture: Case study model of a diversified, organic vegetable production system. Ecol. Model. **267**, 102–114 (2013)

J. Ziesemer, *Energy Use in Organic Food Systems* (FAO, Rome, 2007). http://www.fao.org/docs/eims/upload/233069/energy-use-oa.pdf. Accessed 6 Jan 2021

Chapter 28
What Has Thermodynamics Taught Us About Sustainability?

28.1 What Have We Learned by Viewing Resource Management from a Thermodynamic Perspective?

28.1.1 Services of Nature Are Not Free

Thermodynamic theory says sustainable resource systems depend on a high ratio of endosomatic to exosomatic energy inputs. Endosomatic inputs fuel services of nature, but incorporating these services into agricultural systems comes at a cost. There is resistance to paying for the cost of preserving nature's services, or for restoring services that have been damaged, because in the past these services were considered "free" (Worldwatch Institute 1998). "Free" implies no cost, but services of nature do have a cost as becomes apparent when these services are lost.

28.2 Valuing Nature's Services and Natural Capital

The biggest obstacle to conservation of natural resources is the failure of the economic system to recognize the value of natural capital and the services of nature. Costanza et al. (1997) and others reviewed by Gatto and De Leo (2000) proposed that putting a dollar value on natural capital would help conserve services of nature through recognition of their importance to the economy. Since then there have been arguments both for (Gómez-Baggethun et al. 2010) and against (Díaz et al. 2018) monetizing nature's services. Demonstrating their cost would bring attention to the importance of conserving them, but a problem in implementing policies for conservation is the "paradox of value" (Farnworth et al. 1981), the apparent discrepancy between value in use and value in exchange. Services of nature have high value in

C. F. Jordan, *Evolution from a Thermodynamic Perspective*,
https://doi.org/10.1007/978-3-030-85186-6_28

use but since they are thought of as free, their value as regulators of ecosystem function is ignored until they become scarce.

28.2.1 Natural Capital Is Not Recognized

The value of economic capital such as factories and machinery that contribute to the gross national product is recognized by the Internal Revenue Service. As capital depreciates, the value lost can be deducted from the income tax that owners have to pay. The rules also apply to agriculture. As a tractor wears out, the farmer can deduct its depreciation on his income tax. Services of nature are equally important as tractors in food production. They include pollination of crops by beneficial insects and recycling of nutrients by decomposers in soil organic matter. The value of this capital is not recognized by the Internal Revenue Service. The value of pollinators that have died due to pesticides used to kill pest insects cannot be deducted as depreciation. The degraded value of soil organic matter cannot be deducted when it disappears due to erosion or plowing.

28.2.2 National Capital

On a national level, natural capital consists of forests, fisheries, clean air and water, and other natural resources that support the wellbeing of a nation. Unfortunately, most economists and many governments do not recognize natural capital. It is taken for granted, and it is not valued until it is almost gone. Because it is not recognized, countries that once were rich in natural resources didn't realize how quickly they were becoming impoverished until the resource on which they depended were almost gone.

Between 1970 and 1989, the accumulated depreciation in the value of Costa Rica's forests, soils and fisheries exceeded $4.1 billion in 1984 prices. Relative to the size of the economy, the annual loss was huge. In the Philippines, annual losses from deforestation averaged 3.3% of the gross national product between 1970 and 1987. As a result of this factor alone, the Gross Domestic Product was 3.3% lower than the government said it was (Repetto 1992). The result was the same as if half the factories in the country disappeared and no one noticed they were gone until someone began to investigate why the economy was shrinking. If countries knew what they were losing, they might be more inclined toward conservation.

28.3 Energetic Value of Nature's Services

Another approach to safeguarding services of nature and natural capital has been to quantify the services of nature in terms of energy (Odum 1988) However there has been little agreement on the underlying assumptions necessary for calculations (Huettner 1976). Estimates have varied wildly. Jørgensen (2010) used the energetic value of work done by ecosystems to quantify their value. A comparison of these values with the values found by Costanza et al. (1997) shows that the value based upon the total work capacity is much higher.

28.4 Taxes, Fees, and Reimbursements

A traditional approach for discouraging economic activities that are deleterious for society is to put a tax on the activities or a fee for use of nature's services. Taxes on cigarettes and alcohol are an example. If a small tax fails to adequately control their use, taxes can be raised. A similar approach can be used for controlling economic expansion that threatens environmental services. When economic development results in the expansion of cities, nature's services are being diminished. To help decrease the exploitation of nature's services and to repair the damages caused by expansion, we could impose a fee for economic activities that diminish them. The fee can be used to pay for ecosystem restoration. The ability for soil to absorb runoff water and prevent flooding during storms is a service of nature, but that service is diminished by pavement and buildings. A fee to remedy that problem already exists in some cities. It is called a stormwater management fee, and is based on the amount of impermeable surface on the land that you own. The fee is used to build marshes or retention ponds to absorb surface runoff.

28.5 Case Studies

There are many examples of where a fee for the use of nature's services or a tax on exosomatic energy that replaces services could reduce the loss of services. Taxes and fees make users of nature's services pay for them, thereby charging their cost to those who actually deplete the services. The users of course pass these costs on to consumers, but that is how an economic system is supposed to work. To have a nice environment, you have to pay for it.

28.5.1 Agriculture (The Plowman's Folly)

Edward H. Faulkner was an agricultural agent in Ohio in the early twentieth Century. In 1943, he wrote a book titled Plowman's Folly in which he criticized the use of plows in farming. *"An objectionable feature of plowing is the merciless troweling administered by the moldboard to that portion of the furrow slice which is brought from the plowsole and exposed to wind and sunshine."* The reason that farmers cultivate the soil is to eliminate weeds, and to loosen the soil so that roots of germinating seeds can penetrate the soil. But "that troweling" Faulkner pointed out, was the cause of erosion that so quickly wore out soils. Loosening increases metabolic burning of soil organic matter with concomitant leaching of nutrients as subsurface bacteria are suddenly exposed to a pulse of oxygen. He was puzzled by the resistance of farmers to treat the soil more gently. Even today when no-till planters are available, many farmers still turn over the soil before planting. They can make more money in the short-term by plowing and fertilizing than they can by preserving soil structure because society pays the cost of fertilizer pollution. A carbon tax on fertilizers based on the amount of carbon produced in the manufacture of nitrogen fertilizer would raise the price of fertilizers and encourage farmers to use conservation tillage. Initially, yield might be lower than could be obtained with fertilizers, so farmers would increase the prices for their produce. A tax on herbicides necessary for no-till would also raise the cost of food, but would encourage development of sensors that recognize and apply herbicides only to the most competitive weeds. We would pay for a clean environment through higher prices for food.

28.5.2 Fisheries

Rivers are natural capital for fisheries. Migration of fish back to the spawning ground where they were hatched is guided by the chemical composition of the river water (Johnsen and Hasler 1980). As a result of dam construction, populations of many fish species have declined sharply. Of special concern have been populations of salmon that have commercial value, ceremonial value to indigenous tribes, and recreation for sportsmen and sportswomen. Coalitions of environmentalists have lobbied to remove these dams, and more than 1700 dams have been removed in the U.S. since 2012 (Flaccus 2020), but it has been a difficult battle. Opposition has come from power companies who use the dams for power generation, farmers and ranchers who use water from behind the dams for their crops and livestock, and homeowners who live on the banks of the reservoirs that disappear. Notable success stories include the removal of the Edwards dam on the Kennebec river in Maine (Taylor 2020), and the hope is that salmon migration will recover. Efforts also are underway to demolish four hydroelectric dams on the Klamath river in California. That project, if it goes forward, would be the largest dam demolition project in U.S. history and reopen 400 stream miles of potential salmon habitat that's been blocked off for more than a century. "We are saving salmon country, and we're doing it through reclaiming the West," said a Yurok tribal attorney fighting for dam removal (U.S. News and World Report 2020). Power companies that use dams to produce hydroelectric power should pay a usage fee for the use of the river. That would discourage the use of dams. The higher cost of electricity would enable preservation of valuable species of fish.

28.5.3 Forestry

There are a number of programs and organizations that certify wood from forests that are managed without clear-cutting or other unsustainable practices. Certification requires consumers to pay more for lumber so the higher price can cover the cost of sustainable management. Clear-cutting could be a sustainable management approach if seedlings from a nursery were replanted in a clear-cut, but that is not enough. More important is conservation of soil organic matter that is destroyed by clear-cutting. A thermodynamic definition is harvesting without damaging the soil organic matter, but that is often overlooked. Also overlooked are the trophic chains that are broken by clear cutting. Selective harvesting requires leaving the basic structure of the forest intact. It is sustainable, but it increases the cost of management. That increased cost is the fee for a guarantee of future harvests.

Prescribed burns are another way to ensure sustainability in some forests, especially conifer forests. Fire is a natural phenomenon that when used judiciously, can maintain the productivity of an ecosystem. Judiciously here means strong enough to kill back underbrush that under dry conditions could send flames to the canopy, but light enough so that it does not destroy the soil organic matter essential for nutrient recycling. For more than 50 years, Tall Timbers research station in Florida has been showing how prescribed burns are essential in preserving the longleaf pine ecosystem and bringing the longleaf pine back from extermination. Schiffman (2020) has described how indigenous tribes in the Northwestern United States used fire to reduce the tangle of small trees and underbrush that today create tinderboxes. Occasional light fires created a biodiverse mosaic of forests and woodlands interspersed with open meadows that were a far safer landscape than the uniform conifer stands that exist today, and had higher biodiversity (Kelly et al. 2020). Yet the reluctance to use controlled burns in California and Oregon has exacerbated damage from wildfires that have raged through more than 6.6 million acres in 2020, destroying homes as well as small towns. Adding to the problem is salvage logging of trees killed by a fire. Salvage logging carries away organic matter necessary to stimulate regeneration of forests. A fee for salvage logging would discourage the practice.

28.5.4 Species Conservation

Prairie potholes are depressional wetlands found most often in the Upper Midwest of the U.S. This formerly glaciated landscape is pockmarked with an immense number of potholes, which fill with snowmelt and rain in the spring. Prairie potholes are natural capital. Filled with water, they are a summer home to more than 50% of North American migratory birds. Potholes are destroyed by agricultural expansion.

The Agricultural Adjustment Administration in U.S. history was a New Deal program to restore agricultural prosperity during the Great Depression by curtailing farm production, reducing export surpluses, and raising prices. It paid producers of basic commodities for cutting their output. Its goal was the restoration of prices paid to farmers for their goods to a level equal in purchasing power to that of 1909–14, which was a period of comparative stability. Today, the U.S. farm program pays subsidies to farmers not to grow crops in environmentally sensitive areas and makes payments to farmers based on what they have grown historically, even though they may no longer grow that crop.

28.5.5 Wildlife Management

Wolves and big cats are killed because they sometimes prey on livestock. With no control, cattle and sheep overgraze vegetation sometimes to the point where they endanger their own existence. Predators that control herbivores in ecosystems close to ranches sometimes kill livestock. As part of a wolf conservation program, ranchers are paid for stock losses to wolves. In addition to receiving money for confirmed losses, ranchers are paid based on the number of wolves on their private land and public leases, the number of livestock exposed to wolves, and the rancher's efforts to avoid conflicts.

28.5.6 Flood Control

River meanders slow flow water velocity and in so doing, decrease erosion and decrease downstream flooding during periods of heavy rainfall. Meanders in the Mississippi River were eliminated for the convenience of barge traffic. Under the principle of "user pays", an increase in fees paid by barge operators to use locks on the river could be used to mitigate flood damage. Until 1978, US taxpayers covered the entire cost of maintaining the navigation channel. Since that time, barge operators have paid a diesel fuel tax. Money from that tax goes into the Inland Waterways Trust Fund, which pays for about half of the cost of new construction on all inland waters. The Trust Fund does not pay for routine maintenance and operating costs. Taxpayers cover the full cost of those expenses. They are indirectly paying for goods carried by barges, whether they want them or not. With higher fees, the cost of goods would be paid by those that buy them.

28.5.7 Landscape Management

Before development of South Florida, the everglades region provided a buffer against water level fluctuations that resulted in drought or floods. It was also the home to many endangered species including the wood stork. It was partially drained in the mid-twentieth century to provide more land for sugar plantations and vegetable farms. Efforts at restoration began in the late 1990s because of recurring floods and drought. Since restoration efforts began, the state of Florida has purchased about 100,000 of those acres and converted the land into environmental conservation projects. Florida residents have paid for the service of flood control through taxes.

28.5.8 Climate Change

Perhaps the most serious environmental problem of today is the increase in atmospheric carbon dioxide that causes global warming. There is growing awareness that "nature-based solutions" could help to slow further warming (Seddon et al. 2020). However, some purported solutions have the opposite effect. For example, one "solution" is to clear-cut a forest, and then replant, the idea being that the harvested wood sequesters carbon while the growing seedlings take carbon from the atmosphere (HSBC 2020). The problem is that clear-cutting a forest releases the carbon stored in soil organic matter. In mature forests, soil carbon can comprise the bulk of stored carbon (Spawn and Gibbs 2020, Meng et al. 2021), while in plantations, very little is stored below ground. Thus clear-cutting of immature forests or plantations results only in no net change, because carbon stored in wood eventually is released as the wood decomposes. Clear cutting of mature forests results in a net transfer of carbon from the soil to the atmosphere, thereby exacerbating global warming.

Another approach to global warming is the implementation of a carbon tax. A carbon tax is a tax levied on activities that emit carbon dioxide according to the amount of carbon emitted to produce goods and services. That encourages industries to search for energy sources that are less polluting, or to look for ways that reduce carbon dioxide output from current energy sources. That increases costs for producers, but they compensate by passing along a price increase to consumers. This ensures a more sustainable system. Higher prices force consumers to be more efficient in the way they use energy. Feedback, the invisible hand of the marketplace would be working in the way it is supposed to work.

28.6 Natural Resources and the Free Market System

Imposing taxes and regulations interfere with the working of the invisible hand of the marketplace, or so goes the arguments against them. They impede the capitalistic model of free market enterprise. As a result, conflicts are increasing between economic growth and the biophysical limits of the environment that supports growth. The invisible hand is not working because there is no marketplace to provide feedback that tells the economic system when the services of nature are stressed. Conflicts can only be moderated if there is efficient feedback between a resource base and the economic systems that depend upon them. As much as we all hate taxes, they are necessary to compensate for a lack of marketplace that supplies feedback. We also hate fees for something that we thought of as being free, but taxes and fees remind us that nature's services are not free. If you want a healthy environment, you must pay for it.

References

R. Costanza, R. d'Arge, R. de Groot, et al., The value of the world's ecosystem services and natural capital. Nature **387**(6630), 252–260 (1997)

S. Díaz, U. Pascual, M. Stenseke, et al., Assessing nature's contributions to people. Science **359**(6373), 270–272 (2018)

E.G. Farnworth, T. Tidrick, C. Jordan, W. Smathers, The value of natural ecosystems: An ecological and economic framework. Environ. Conserv. **8**, 275–282 (1981)

G. Flaccus, *Largest US Dam Removal Stirs Debate Over Coveted West Water* (2020). https://www.usnews.com/news/politics/articles/2020-03-29/largest-us-dam-removal-stirs-debate-over-coveted-west-water. Accessed 28 Dec 2020

M. Gatto, G.A. DeLeo, Pricing biodiversity and ecosystem services: The never-ending story. Bioscience **50**(4), 347–355 (2000). https://doi.org/10.1641/0006-3568(2000)050[0347:PBAEST]2.3.CO;2

E. Gómez-Baggethun, R. deGroot, P.L. Lomas, C. Montes, The history of ecosystem services in economic theory and practice: From early notions to markets and payment schemes. Ecol. Econ. **69**(6), 1209–1218 (2010)

HSBC, *Financing the Earth's Assets. The Case for Mangroves as a Nature-Based Climate Solution* (2020). https://www.sustainablefinance.hsbc.com/carbon-transition/financing-the-earths-assets. Accessed 16 June 2021

D.C. Huettner, Net energy analysis: An economic assessment. Science **192**(4235), 102–104 (1976)

P.B. Johnsen, A.D. Hasler, The use of chemical cues in the upstream migration of coho salmon, Oncorhynchus kisutch Walbaum. J. Fish Biol. **17**(1), 67–73 (1980)

S.E. Jørgensen, Ecosystem services, sustainability and thermodynamic indicators. Ecol. Complex. **7**, 311–313 (2010)

L.K. Kelly et al., Fire and biodiversity in the Anthropocene. Science **370**(6519), 929 (2020)

Y. Meng, J. Bai, R. Gou, et al., Relationships between above- and below-ground carbon stocks in mangrove forests facilitate better estimation of total mangrove blue carbon. Carbon Balance Manag. **16**(8) (2021). https://doi.org/10.1186/s13021-021-00172-9. Accessed 10 July 2021

H.T. Odum, Self organization, transformity and information. Science **242**(4882), 1132–1139 (1988)

R. Repetto, Accounting for environmental assets. Sci. Am. 94–100 (1992)

R. Schiffman, Fighting wildfires with fire. Wall St. J. **3–4**, C4 (2020)

N. Seddon, A. Chausson, P. Berry, et al., Understanding the value and limits of nature-based solutions to climate change and other global challenges. Philos. Trans. R. Soc. Lond. B Biol. Sci. (2020). https://doi.org/10.1098/rstb.2019.0120

S.A. Spawn, H.K. Gibbs, *Global Aboveground and Belowground Biomass Carbon Density Maps for the Year 2010* (ORNL DAAC, Oak Ridge, Tennessee, USA, 2020). https://doi.org/10.3334/ORNLDAAC/1763. Accessed 10 July 2021

P. Taylor, *From the Mountains to the Sea* (Islandport Publishers, Yarmouth Maine, 2020)

U.S. News and World Report, 29 March, 2020 (2020). https://www.usnews.com/news/politics/articles/2020-03-29/largest-us-dam-removal-stirs-debate-over-coveted-west-water. Accessed 28 Dec 2020

Worldwatch Institute, *Putting a Value on Nature's "Free" Services* (1998). http://www.worldwatch.org/node/447. Accessed 25 Feb 2021

Part III
Conservation

Conservation of Resource Systems Means Preserving the Services of Nature

A Philosophy of Conservation

In my 1955 course in conservation of natural resources, I was taught that conservation meant "wise use". The problem with that definition is that wise use is in the eye of the beholder. It has been used to justify everything from clear-cutting to filling in wetlands. Framing conservation in terms of thermodynamic laws is more objective. Thermodynamics concerns energy transformations. In ecosystems, energy transformations occur through species interactions. Species interactions are services of nature that underlie evolution and sustainability. *Conservation of resource systems means preserving the services of nature.*

Chapter 29
In Wilderness Is the Preservation of the World

29.1 Sacred Groves

"*A legacy of prehistoric traditions of nature conservation, sacred groves are patches of forest that rural communities in the developing world protect and revere as sacrosanct. Deeply held spiritual beliefs ensure that not a tree is felled nor a creature harmed within its boundaries*" (Gadgil 2018). English colonial forestry in India and Burma, while providing wood for construction of ships that at one time ruled the waves, devastated the ancient legacy, but despite centuries of colonial extraction, many sacred groves have survived.

In ancient Greece, a sacred grove was a grove of trees dedicated to a god or gods, and the use of its trees was restricted (Bowe 2010). A current example of sacred sites is from Thailand, where all young men are required to serve for a period of time as monks in Buddhist temples. One young Thai became a graduate student of mine, and when we were not researching taungya agriculture, he took me to visit a number of Buddhist temples. The grounds of each are dedicated to preservation of a particular species or group of species. One temple is dedicated to macaques. Another is for birds. A temple dedicated to fish is popular with local villagers that come on religious holidays to feed the fish. A temple that impressed me greatly was for fruit-eating bats. They are endangered because farmers kill them. The bats sleep during the daytime, and the trees around the ornate temple were loaded with thousands of the furred creatures hanging upside down, waiting like vampires for the sun to set.

C. F. Jordan, *Evolution from a Thermodynamic Perspective*,
https://doi.org/10.1007/978-3-030-85186-6_29

29.2 Information Is Stored in Sacred Groves

The interactions between species that control the stability of natural systems are not random. They are highly specific, and are fine-tuned to expedite the flow of energy and recycling of nutrients in particular environments. This organization of species consists of information about who interacts with whom and how they interact. It is information that has been built up through feedback over millions of years of trial and error. That information is coded in the feedback systems of natural ecosystems. When ecosystems are simplified for economic expediency, information that took millions of years to code is lost in the twinkling of an eye. To ensure information is available when needed for ecosystem restoration, storage banks are needed where services of nature are preserved and can be borrowed.

29.3 Wilderness as a Resource Bank for Nature's Services

In 1962, The Sierra Club published a book of quotations from the writings of Henry David Thoreau along with photographs by Eliot Porter of the landscapes that Thoreau visited in the seasons that he visited them. The title of the book is, "In Wilderness is the Preservation of the World". What does that mean? How can the World be preserved by wilderness?

Wilderness is sustainable because its flows of energy and cycles of energy are unbroken, its species interactions are intact, and the natural capital incorporated in these cycles and interactions is preserved. Wilderness connotes places undisturbed by human activities. Humans, wherever they migrate and settle, interrupt these flows and cycles and deplete the natural capital. Conquering the frontier wilderness was the goal of pioneers that settled the American West. It is the goal of migrants from the parched northeast of Brazil moving to the Amazon rain forest. It means redirecting flows of energy toward sustaining human populations and away from sustaining ecosystems. How much redirection can take place before the Amazon turns to desert? We don't know. But what we do know is redirecting energy flow away from a feedback system will cause the spotted owl to become extinct. Is that important? For some people yes, for others, no. But what is important to everyone is that the intact mature forests where the spotted owl lives or the rain forests of Brazil where the golden lion tamarin survives hold the secret keys to the sustainability of managed systems. The secret keys are the services of nature and the natural capital unique to that local environment. Wilderness is the bank from which the services can be borrowed to maintain endangered species or to restore the productivity of the land.

References

P. Bowe, The sacred groves of ancient Greece. Stud. Hist. Gard. Des. Landsc. **29**(4), 235–245 (2010)
M. Gadgil, Sacred groves: An ancient tradition of nature conservation. Sci. Am. **319**(6), 48–57 (2018)

Appendix 1: Cycling of Strontium-90 in a Tropical Rain Forest

Trophic-dynamic models proposed in the mid-20th century gave the first suggestion of the connectiveness of all living things in nature. Later studies that followed the cycles of radioisotopes in the environment were the first direct evidence of ecosystems, that is of the interconnectedness between all living things in a forest, a prairie, or a lake. This evidence was an inadvertent result of studies that had a health-safety purpose.

1.1 Why the Study Was Done

My first professional position was with one of the U.S. Atomic Energy Agency projects to determine the fate of radioisotopes in the environment. Our site was in the rain forest of Puerto Rico, and the aspect that we studied was the half-residence time of Strontium-90 in that ecosystem. Half-residence time is the time that it takes for half of the isotope in the ecosystem to disappear. Strontium-90 is a radioactive isotope that causes genetic mutations and its cycling is analogous to cycling of its stable strontium isotope in the environment.

1.2 The Model

We started with a model of stable strontium dynamics in an undisturbed system (Jordan Kline and Sasscer 1973). To quantify the model, we measured stable strontium in the soil, wood, canopy and humus, and its flow in rain, dry fall, throughfall, litterfall, and saturated flow in soil (Fig. 1.1). It was assumed that biomass of each compartment was neither increasing nor decreasing, and therefore input equaled

C. F. Jordan, *Evolution from a Thermodynamic Perspective*,
https://doi.org/10.1007/978-3-030-85186-6

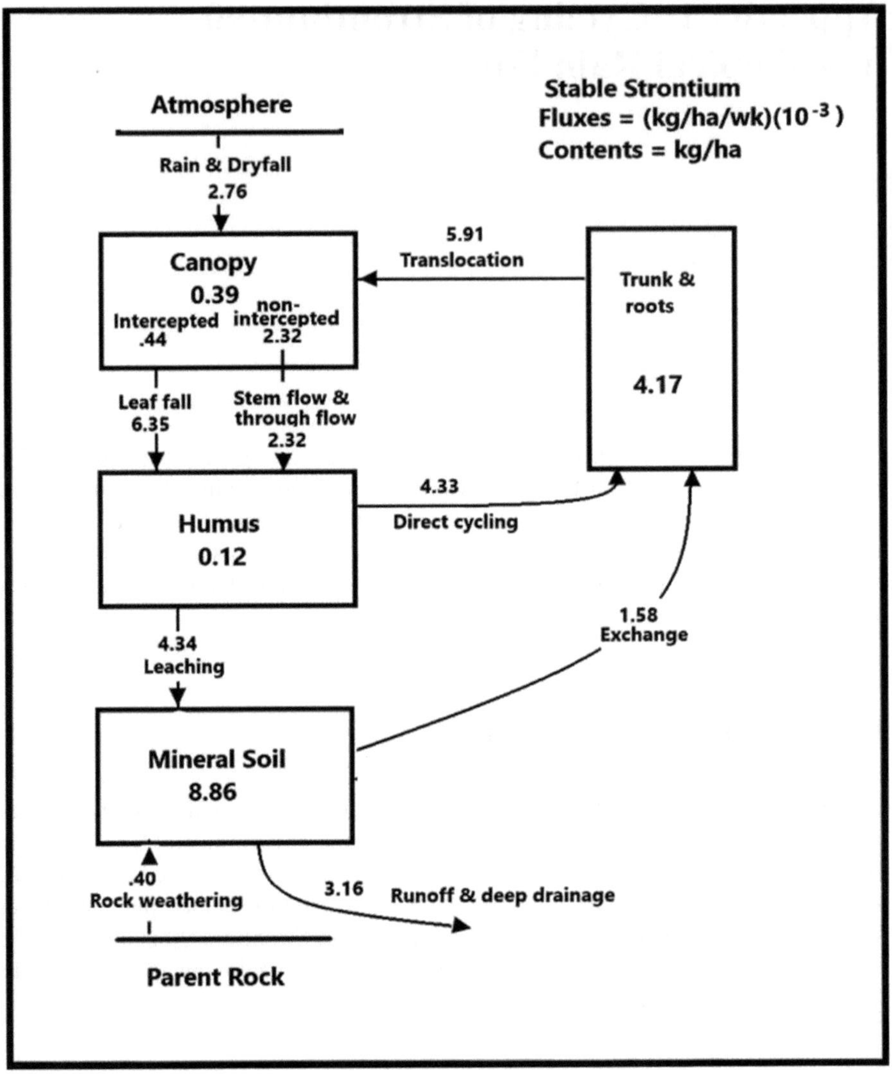

Fig. 1.1 Stable strontium cycle in the Puerto Rican rainforest. (Modified from Jordan Kline and Sasscer 1973)

output. With these values, we calculated transfer coefficients that were used for a predictive model of Sr^{90} flow.

1.3 Model Validation

To test the model, we used input equivalent to fallout events in Puerto Rico from 1962 to 1967 represented by circles in the canopy compartment (Fig. 1.2). The

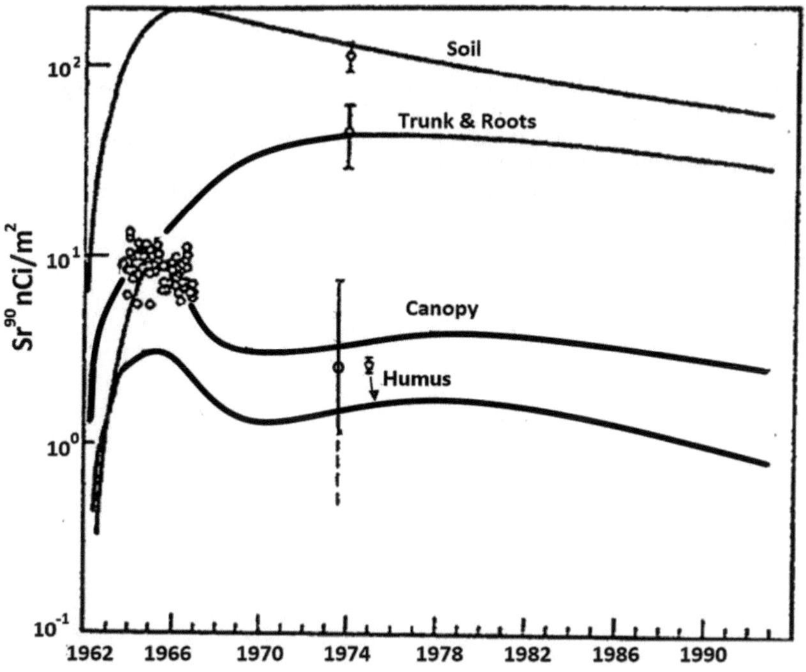

Fig. 1.2 Predictive model of radioactive strontium dynamics in the Puerto Rican rainforest and validation points. (Modified from Jordan and Kline 1976). Sr^{90} in each compartment is radioactivity per square meter of forest floor

model predicted that after a decrease in activity around 1970, amount of Sr^{90} in the canopy and humus should rise toward a secondary peak.

To validate the model, a series of samples were taken from soil, wood, canopy and litter and analyzed for Sr^{90} in 1974. Observed levels are indicated by the circles in four compartments. The secondary pulse was detected in the humus and in the canopy, albeit with high variation in the canopy due to two outliers. The high amounts of activity in the wood of the trunk also is evidence of recycling. Most of the chemical elements entering tropical forest ecosystems through fallout are scavenged by mosses and lichens growing on the leaves (Jordan et al. 1980). While there may have been some translocation of Sr^{90} from leaves to tree trunk, the much higher radioactivity in tree trunks per square meter of forest floor than in the leaves could only have happened through uptake of the isotope after moving from the leaves to the humus to the soil.

The model predicted a half- residence half time of Sr^{90} in the rain forest environment of 20.04 years. Since the half-life of Sr^{90} is 28.8 years, most of the loss from the system was due to radioactive decay of the isotope exchanged on the mineral soil. Relatively small amounts were leached out of the ecosystem.

References

C.F. Jordan, J.R. Kline, D.S. Sasscer, A simple model of strontium and manganese dynamics in a tropical rain forest. Health Phys. **24**, 477–489 (1973)

C.F. Jordan, J.R. Kline, Strontium-90 in a tropical rain forest: 12th year validation of a 32-year prediction. Health Phys. **30**, 199–201 (1976)

C.F. Jordan, F. Golley, J. Hall, J. Hall, Nutrient scavenging of rainfall by the canopy of an Amazonian rain forest. Biotropica **12**, 61–66 (1980)

Appendix 2: Ecosystem Boundaries

2.1 Ecosystem Boundaries

An ecosystem can be defined by a boundary line in which there are more species interactions within the perimeter than across the line. This works for a pond, but in other systems, interactions can be hard to detect. An ecosystem could also be defined by the dominant vegetation, but sometimes the vegetation changes gradually along a transect so that the location of a boundary is arbitrary.

Ecosystem boundaries are often difficult to discern in the highly disturbed landscape of temperate zones. They are clearer in a landscape such as the Amazon region where effects of human disturbances in past centuries have disappeared as indigenous populations have declined (Sanford et al. 1985). Figure 2.1 illustrates seven ecosystem types that exist along a 1000-meter environmental gradient, extending Eastward from the Rio Negro near San Carlos Venezuela. The gradients between the

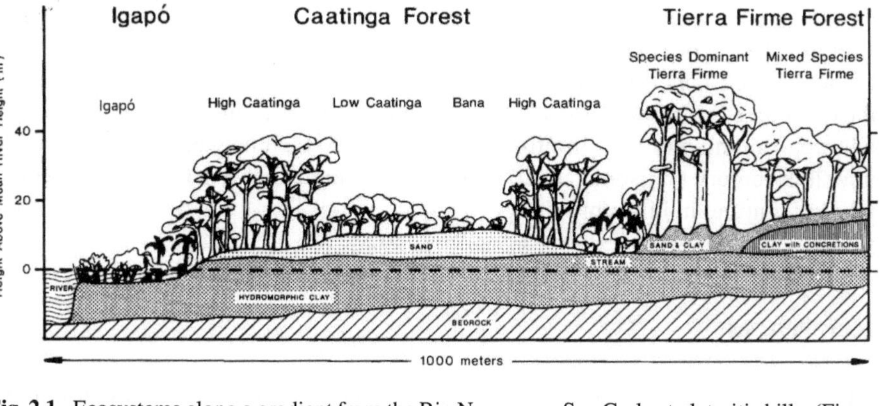

Fig. 2.1 Ecosystems along a gradient from the Rio Negro near San Carlos to lateritic hills. (Figure is from Jordan (1989), courtesy of United Nations)

C. F. Jordan, *Evolution from a Thermodynamic Perspective*,
https://doi.org/10.1007/978-3-030-85186-6

ecosystems reflect the underlying continuum of hydromorphic clay and sand. The gradient results from changes in both moisture and nutrient content of the soil. Along the continuum, the boundaries of each vegetation type can be identified within a few meters by the depth of sand in the Caatinga forest and by the appearance of clay and concretions in the Tierra Firme forest. The stream ecosystem is enriched in nutrients from leaves that fall into the water (Meyer et al. 1998).

The top predator in all of these ecosystems is the jaguar. The system that defines the jaguar's territory is the landscape. The caatinga and tierra firme forest types exist in a hierarchical relationship with the landscape. Within each forest type there are several ecosystem types.

Ecosystems also are defined by vertical gradients. Alexander von Humboldt, during his stay in Ecuador in 1802, observed stratification of vegetation types on the undisturbed slopes of volcanos Pichincha, Cotopaxi, Antisana, and Illinica. "He saw the Andes as a vertical exhibit of the earth's climates, plants, and animals, from hot tropical lowlands to the eternal snow and ice of the towering summits. The slopes disclosed distinct life zones, each zone consisting of a definite grouping of soils, land forms, rainfall, temperature, plant species, and animal species. These environments – today we would call them ecosystems – had not assembled themselves accidentally or arbitrarily but had grown as communities with specific characteristics determined by interactions. All component forces continuously acted on each other and created a balanced unit" (E.L. Jordan 1981).

Boundaries in marine ecosystems are much less distinct except for corals and mangroves. Nevertheless, Sonnewald et al.(2020) identified over 100 "ecoprovinces" in oceans of the world based on seven plankton functional group biomasses and nutrient fluxes.

2.2 Cultural Boundaries

Cultures, like ecosystems have indistinct boundaries, but no one would argue that French culture is not different from German culture. Just because one grades into the other across the boundary does not mean that separate cultures do not exist. The "discrete community" vs. "continuum" debate that raged between ecologists in the mid 20th century (Wilson et al. 2004) seems as pointless as a debate about community vs. continuum in Anthropology.

References

C.F. Jordan, An Amazonian rain forest. The structure and function of a stressed ecosystem and the impact of slash and burn agriculture. Man and the biosphere series, volume 2. UNESCO, Paris and Parthenon, Carnforth, England (1989)

E.L. Jordan, The birth of ecology: An account of Alexander von Humboldt's voyage to the equatorial regions of the new continent, Jordan CF (ed) Tropical ecology. Benchmark papers in ecology/10. Hutchinson Ross Publishing Co. Stroudsburg Pennsylvania (1981), p 4-15

J.L. Meyer, J.B. Wallace, S.L. Eggert, Leaf litter as a source of dissolved organic carbon in streams. Ecosystems **1**, 240–249 (1998)

R. Sanford, J. Saldarriaga, K.E. Clark, C. Uhl, R. Herrera, Amazon rain-forest fires. Science **227**(4682), 53–55 (1985)

M. Sonnewald, S. Dutkiewicz, C. Hill, G. Gorget, Elucidating ecological complexity: Unsupervised learning determines global marine eco-provinces. Sci. Adv. **6**(22), eaay4740 (2020). https://doi.org/10.1126/sciadv.aay4740

J.B. Wilson, A.D.Q Agnew,M.T. Sykes, Ecology or mythology? Are Whittaker's "gradient analysis" curves reliable evidence of continuity in vegetation? Preslia **76**(3), 245–253 (2004)

Appendix 3: How to Study Something when You Are Standing in the Middle of It

3.1 The Need for a Macroscope

In contrast to the laboratory scientist, the ecosystem scientist stands in the midst of the object being studied. He or she cannot directly observe the whole ecosystem. To determine ecosystem characteristics and the effect of disturbance on the system, the scientist measures the parts of the system and the interactions between them, and then synthesizes the system in a mathematical, graphical, or verbal model. Before beginning the study, the scientist is faced with two decisions. Which of the thousands of organisms, organism interactions, flows and cycles surrounding her should she measure, and how accurately should she measure them?

3.2 What Should Be Measured?

Network analysts have found that there is a trade-off between the usefulness of a model and its complexity. Often modelers are seduced by the idea of adding complexity in an attempt to capture reality more accurately. It does not. As modelers incorporate more phenomena, the model's predictions typically become less accurate. The error is unknown. For example, one of the inputs of a grassland ecosystem model would be the rate at which herbivores consume plant biomass. This would be represented by a transfer coefficient from autotrophs to herbivores. This transfer coefficient would represent the aggregation of hundreds of different plant-consumer interactions, each of which has a certain probability of occurring on any particular day. In the ecosystem level model, the coefficient is represented as a single value, but it is really a probability that depends on the averaged probability of each of the hundreds of plant-consumer interactions during a season or year. The decomposer unit of an ecosystem contains the aggregations of dozens of units of arthropods

C. F. Jordan, *Evolution from a Thermodynamic Perspective*,
https://doi.org/10.1007/978-3-030-85186-6

which in turn contains the aggregations of many units of insects. If each of these parameters were part of the ecosystem model, total uncertainty would build up, and the error would increase to the point where predictions are meaningless (Saltelli et al. 2020).

3.2.1 Why Ecosystem Models Can't Include Everyone's Favorite Species

At one time, there were workshops that brought together a diverse group of ecologists to pull together information from all hierarchical levels and all taxonomic groups, the objective being to construct an ecosystem model that would give insights above and beyond those gained from individual population and process level studies. It was thought that such workshops would produce new understanding of basic ecosystem properties including productivity (energy flow), total biomass, nutrient cycling efficiency, stoichiometry, stability, resistance, resilience and the influence of species diversity and disturbance on these characteristics. Typically, a modeler led the workshop by soliciting input from each of the specialists. It never worked very well. All that an ecosystem modeler needed from each participant was a single value for biomass or energy flow, but each scientist despaired at having to give up the complexity of her own specialty to condense it all into a single value.

Modelers now heed Albert Einstein's advice that "Everything should be made as simple as possible, but no simpler." That means there should be only the number of compartments in a model needed to answer the question being asked. A compartmental model with four trophic levels including a decomposer level can answer questions about ecosystem stability, as long as the compartments are connected in a feedback system. The output value for each compartment is the aggregation of inputs from dozens of species interactions within each compartment, but aggregation is not done by summing individual sub-system models of ecosystem processes. It is done by going to the field and sampling compartments of the top-down model, the actual ecosystem itself (Jordan et al. 1972).

The model in Fig. 1.1, Appendix 1 does not even contain animals. Their activity would be an extra step of the transfer from canopy to humus but it does not significantly change the transfer value as determined by a sensitivity analysis. The model does not contain decomposers either. They are taken to be part of the humus. An effort to tease apart living from non-living humus would not change the transfer from humus to soil. Including those transfers would complicate the model without helping to answer a question about the environmental half-time of strontium-90. A model of stability for the Serengeti ecosystem however would include animals, and ecosystem scientists would aggregate all herbivores and all predators because changes in their populations would affect system ecosystem stability. So when should animals be included in an ecosystem model? The only way to tell is to construct preliminary models of the ecosystem of interest with and without animals. If their presence affects the outcome, they should be measured and included. Location of biomass in the system makes the difference. In forests, the most significant energy storage is in autotrophs. In grasslands it is in herbivores.

Ecologists interested in particular species and their stability would not aggregate, but focus on particular predator-prey interactions. Population ecology is based on a different paradigm than ecosystem ecology.

3.3 How Accurately Should Each Parameter Be Measured?

Students are taught that all measurements should be as accurate as possible, but this cannot be the guiding rule when time and money allotted for the study are limited. This is often the case for evaluation of environmental impacts of economic development. An iterative approach can be used to determine the sensitivity of system output to variations in each of the input parameters. This will indicate how accurately each parameter needs to be measured. In the first round, the effort deployed on each parameter is set by intuitive guesswork. After completion of the field work, the value of each parameter is put in a mathematical model. Then a sensitivity analysis is performed to determine the degree that model output changes as a result of changes in input values for each parameter and combination of parameters. In a second approximation, those parameters to which output is most sensitive are measured again, this time with more accuracy. Often a second iteration is not possible, but a sensitivity analysis of the first iteration can determine how much the system outcome is affected by variation in parameter estimates.

3.3.1 How Much Time and Effort Must Be Given to Quantifying each Ecosystem Parameter When Time and Money Are Limited?

Appendix 1 describes a study where this question played an important part in the design of the study. The objective was to predict the half-residence time in the forest of 90-strontium resulting from radioactive fallout. The predictive model was based on the stocks and flows of stable strontium in and between four compartments of the rain forest. We replicated samples as much as time and money allowed. The model predicted and environmental half time of Sr^{90} in the rain forest environment of 20.04 years (Jordan et al. 1976). Several years after I left the project, I ran into the scientist who replaced me. He told me that he had spent several years with an expanded array of throughfall collectors and found that my estimate of throughfall was off by as much as 10 percent. I answered that we had done a sensitivity analysis and that an error of that size could change the estimated environmental half time by about two years, and that kind of an error would not change the conclusion that strontium-90 is strongly bound in the ecosystem. His answer was, "Still, you were off by 10 percent".

A second example came from my experience as part of the San Carlos project (Herrera et al. 1981) one objective of which was to determine the effect of shifting cultivation on the energy flow and nutrient cycling in the Amazon rain forest. One of the necessary parameters was to determine the standing biomass of the undisturbed forest in order to determine nutrient input to the soil when the forest was cut and burned. To do that, I had a crew fell 42 trees ranging in size from 10 cm to 1.4 m in diameter, cut them into manageable size, separate the leaves, weigh them, and take sub-samples for dry-weight conversion and nutrient concentration in the biomass. Results were used to produce an allometric curve that predicted biomass of a tree as a function of diameter. Then we measured the diameter of trees greater than 10 cm in diameter along two transects, 10 meters wide by 10 kilometers long and used the data along with the allometric curve to estimate the biomass of the forest (Jordan and Uhl 1978). Nutrient content of the biomass was then used to determine the quantity of nutrients released into the soil. Our results showed that availability of most nutrients in the soil remained high throughout the three years of cultivation. Phosphorus was an exception. While total phosphorus remained high, soluble phosphorus available to plants declined rapidly. Laboratory studies showed that that immobilization of phosphorus by iron and aluminum oxides in the soil was the cause of decreased yield (Jordan 1989). Further research showed that in the intact forest, organic acids leached from living biomass chelated the iron and aluminum oxides in the soil and maintained the phosphorus in soluble condition, but in the slash and burn plot where organic matter was depleted, phosphorus was fixed as iron and aluminum phosphate and rendered unavailable to plants (Lee et al. 1990). Therefore productivity declined.

Another scientist on the project took a different approach to determining forest biomass. He laid out one single plot 10 meters on a side, cut down everything, and weighed every bit of above ground living plant material on the plot, including mosses and lichens. He was proud of his study that resulted from very precise measurements of above ground biomass in that one 100 m^2 plot. I asked him if that plot was representative of the entire forest. He answered that that was not the objective of his study. He was very critical of my estimate of forest biomass. Our allometric equations did not include height of trees, because it is difficult to determine the height of an individual tree in a closed canopy, and the transect that we ran passed through an ecosystem that was not completely homogeneous. However, if we had made an error of 10 or 20 percent in the estimate of forest biomass, it would have made no difference in our ecosystem-level conclusion regarding why productivity in Amazonian soils rapidly decreases following and cutting and burning.

3.3.2 The Answer

The answer regarding the time and effort needed for each ecosystem parameter is this: enough to answer the ecosystem question with a deviation that will not change the conclusion. In the case of the Sr^{90} environmental half-time calculation, the deviation of two years from the half-time estimate of 20 years would not change the conclusion that the isotope is strongly retained in the soil. If on the other hand the deviation were 20 years or more, the conclusion would be the first approximation model was not satisfactory and a second iteration would be necessary.

3.4 The Tyranny of Small Decisions

Asking the right question is important when evaluating environmental impacts. The problem of impacts arises when many single decisions are made without considering their cumulative impacts. Cutting down one acre of old growth forest, or spraying one field with pesticide can reasonably be said to have negligible environmental impact. The problem arises when there are many acres or many fields. In 1970, I was involved in developing an impact statement for Palisades Nuclear Generating Station, a nuclear power plant on the shores of Lake Michigan. Population ecologists predicted only a local increase in plankton and the fish populations that feed upon them due to discharge of warm water from the cooling towers. The important question not asked was, what would be the impact on Lake Michigan if a dozen or more plants were built along the shore, and if the impact statements for each of them were the same as for the Palisades. Basing a decision on separate analysis of many projects results in what W. E. Odum (1998) called the tyranny of small decisions. The impact of one plant is negligible, but the cumulative impact of many small impacts can alter the stability of populations within a system. The cumulative impact of small decisions grows faster than the increase in number of decisions.

References

R. Herrera, C.F. Jordan, E. Medina, H. Klinge H, How human activities disturb the nutrient cycles of a tropical rainforest in Amazonia. Ambio **10**(2/3, MAB: A Special Issue), 109–114 (1981)

C.F. Jordan, J.R. Kline, D.S. Sasscer Relative stability of mineral cycles in forest ecosystems. Am. Nat. **106**, 237–253 (1972)

C.F. Jordan, J.R. Kline Strontium-90 in a tropical rain forest: 12th year validation of a 32-year prediction. Health Phys. **30**, 199–201 (1976)

C.F. Jordan, C. Uhl, Biomass of a “tierra firme” forest of the Amazon Basin. Oecolog Plantar **13**, 387–400 (1978)

C.F. Jordan, An Amazonian rain Forest. The structure and function of a stressed ecosystem and the impact of slash and burn agriculture. Man and the biosphere series, volume 2. UNESCO, Paris and Parthenon, Carnforth, England (1989)

D. Lee, X.G. Han, C.F. Jordan Soil phosphorus fractions, aluminum, and water retention as affected by microbial activity in an Ultisol. Plant Soil **121**, 125–136 (1990)

W.E. Odum, Environmental degradation and the tyranny of small decisions, E.P. Odum (ed) Ecological Vignettes. Harwood Academic Publishers, Australia (1998), pp. 221–224

A. Saltelli, and 22 others, Five ways to ensure that models serve society: A manifesto. Comment. Nature **582**(7813), 482–484 (2020)

Appendix 4: Problems of Industrial Agriculture

Agricultural systems are nested within economic systems. As a result, there are two levels of problems associated with food systems: Ecological problems and economic problems

4.1 Ecological Problems

- Prevalence of monocultures in industrial food production systems leads to loss of genetic diversity (Soule and Piper 1992). Low genetic diversity increases the risk of disease or insect outbreak (Real 1996).
- Inorganic nitrogen leached from fertilizers spread on agricultural fields enters waterways and causes hypoxia. "Within and near oxygen-depleted waters, finfish and mobile macroinvertebrates experience negative effects that range from mortality to altered trophic interactions" (Breitburg et al. 2009, Tomasetti and Gobler 2020).
- Nitrate leached into water supplies can interfere with the ability of blood to carry oxygen to vital tissues of the body in infants 6 months old or younger. The resulting illness is called methemoglobinemia or "blue baby syndrome" (Department of Environmental Quality 2016).
- Nitrogen volatilized from fertilizers enters the troposphere and poses direct health threats to humans and causes substantial losses of agricultural production (Galloway et al. 2008).
- Animal waste lagoons and spray fields near aquatic environments can significantly degrade water quality and endanger health (Mallin 2000).
- Overuse of antibiotics in the livestock industry has resulted in increasing resistance of pathogens (Mlot 2000).

C. F. Jordan, *Evolution from a Thermodynamic Perspective*,
https://doi.org/10.1007/978-3-030-85186-6

- Increasing resistance of weeds to a single type of herbicide is resulting in the need for an expensive series of herbicides (Service 2007).
- Use of insecticides kills beneficial insects that can help control pest species (Soule and Piper 1992).
- Plowing and other methods of tillage that disrupt the structure of the soil result in erosion that destroys croplands (McNeill and Winiwarter 2004).

4.2 Economic, and Social Problems

- Industrial agriculture is highly dependent upon petroleum to synthesize fertilizers, pesticides, and herbicides, and to fuel the airplanes, trucks, and tractors that deliver and spread these compounds. Because petroleum supply is erratic (Kerr 2008), prices for commodities are uncertain and profits for farmers fluctuate broadly depending on economic cycles.
- Use of genetically modified crops can put the farmer under control of international corporations that own patents on the crops. As use of these crops spreads, the world's food supply becomes increasingly dependent on the economic goals of a handful of corporations and not on the needs and desires of consumers (Then 2000).
- The simple, vertically integrated economic food chains common in industrial agriculture can be highly susceptible to economic and political disturbances (Striffler 2005).
- Factory farms have adverse effects on rural communities, such as deteriorating socioeconomic conditions and loosening of community social fabric (Lobao and Stofferhan 2007).
- Industrial agriculture is leading to a depletion of water resources. Groundwater level in Ogallala Aquifer that supplies irrigation water to the High Plains of the USA has dropped 150 feet or more, forcing many farmers to abandon their wells (Little 2009)

References

D.L. Breitburg, D.W. Hondorp, L.A. Davias, R.J. Diaz, Hypoxia, nitrogen, and fisheries: Integrating effects across local and global landscapes. Annu. Rev. Mar. Sci. **1**, 329–349 (2009)

Department of Environmental Quality, Nitrate in drinking water. http://www.deq.state.or.us/wq/pubs/factsheets/ground water/nitratedw.pdf. Accessed 7 Dec 2016 (2016)

J.N. Galloway, A.R. Townsend, J.W. Erisman, M. Bekunda, C. Zucong, J.R. Freney, L.A Martinelli, S.P. Seitzinger, M.A. Sutton, Transformation of the nitrogen cycle: Recent trends, questions, and potential solutions. Science **320**, 889–892 (2008)

R.A. Kerr, World oil crunch looming? Science **322**, 1178–1179 (2008)

J.B. Little, The Ogallala aquifer: Saving a vital US water source. Scientific American. http://saullosdhs.weebly.com/uploads/5/7/9/4/57946963/182-the-ogallala-aquifer-saving-a-vital-us-water-source.pdf. Accessed 26 Dec 2020 (2009)

L. Lobao, C. Stofferhan, The community effects of industrialized farming: Social science research and challenges to corporate farming laws. Agric Human Values **25**, 219–240 (2007). https://doi.org/10.1007/s10460-007-9107-8

M.A. Mallin, Impacts of industrial animal production on rivers and estuaries. Am. Sci. **88**, 26–37 (2000)

J.R. McNeill, V. Winiwarter, Breaking the sod: Humankind, history, and soil. Science **304**, 1627–1629 (2004)

C. Mlot, Antidotes for antibiotic use on the farm. Bioscience **50**, 955–960 (2000)

L.A. Real, Sustainability and the ecology of infectious disease. Bioscience **46**, 88–97 (1996)

R.F. Service, A growing threat down on the farm. Science **316**, 1114–1117 (2007)

J.D. Soule, J.K. Piper, Ecological crises of modern agriculture. J.D. Soule, J.K. Piper (eds) Farming in Nature's Image, Island Press, Washington (1992), p. 11–50

S. Striffler, Chicken: The Dangerous Transformation of America's Favorite Food. Yale Agrarian Studies Series, New Haven (2005)

C. Then A danger to the world's food; genetic engineering and the economic interests of the life-science industry. M. Qaim, A. Krittiger, J. Von Braun (eds). Agricultural Biotechnology in Developing Countries: Towards Optimizing the Benefits for the Poor. Kluwer, Boston (2000), p 229–236

S.J. Tomasetti, C.G. Gobler, Dissolved oxygen and pH criteria leave fisheries at risk. Science **368**(6489), 372 (2020)

Appendix 5: Thermodynamic Niches

An ecological niche was defined by Hutchenson (1965) as something implicit in the environment (an *n* dimensional hypervolume). That is like saying that there are many different sizes and styles of houses in a newly built neighborhood all waiting to be occupied by different types of people. At first, only one or two are inhabited, but eventually many but not necessarily all will be filled. Thermodynamic theory has a different view of the niche concept. It says that there are not a set number of ecological niches with fixed dimensions waiting to be filled, but that species create their own custom niches as they adapt to a new energy environment. The dimensions of each niche is not pre-determined but depends on the innate characteristics and skills of each newcomer. Cuthill et al. (2020) presented an evolutionary model that indicated the largest radiation of species occurred not after ecological niches were vacated by extinction, but when life exploited new realms of opportunity.

5.1 Energetic Niches in Forests

In a monoculture tree plantation, there is only one thermodynamic niche for trees. Leaves of all individuals occupy the same location in the canopy. All the leaves have the same characteristics so they all are competing for the same quantum of sunlight. Because the canopy is not entirely filled with leaves, there is some sunlight they do not capture and that penetrates to the soil. In diverse forests, the system is more efficient in capturing all the photosynthetically available solar radiation (Wu 2007). Tree species have complimentary thermodynamic niches. Understory species capture light that diffused through the canopy. Ground level species use light not captured by understory species (Fig. 6.5, Chap. 6). Root distribution also contributes to the thermodynamic niche. In monocultures, roots tend to be concentrated in the same soil horizon. They have no access to nutrients and water in other horizons

C. F. Jordan, *Evolution from a Thermodynamic Perspective*,
https://doi.org/10.1007/978-3-030-85186-6

(Kunert and Cárdenas 2015). In natural forests, different species develop different root distribution patterns. Some reach deep soil horizons thereby accessing water and nutrients not available to shallow rooted species (Sanford 1989). Deep rooting allows some trees to photosynthesize when others cannot because of drought. In a study from Panama, trees planted in mixtures were able to maintain higher transpiration rates during the dry season (Kunert et al. 2012) due to complementary soil water use between different species (Schwendenmann et al. 2015). Differential combinations of elemental compositions in trees have also been used to identify ecological niches (Sardans et al. 2021).

If energy flow to ecosystems and through ecosystems were unvarying in space and time, there would develop during the course of evolution fewer species at each trophic level – the best competitors for that static condition. But energy flow varies with time, day by day, season by season, decade by decade, and century by century. This results in a distribution of energy over time and space in a way that it can be partitioned among many species. Energy availability depends on photosynthetic activity, which depends on soil moisture. Some species reduce power output more during drought than do others. Some species are short-lived and produce seeds quickly, thus taking a chance that seeds will encounter favorable conditions for germination. Other species are long -lived and produce seeds over a long period, thus ensuring that at least some seeds will encounter a year favorable for growth. Over a period of 10–20 years, there are opportunities for several different strategies. Energy flow also varies according to space. Available energy for photosynthesis is less on a mountain top than in a valley because the lower temperatures on mountain tops reduces photosynthetic activity. Trees at high altitudes have a different thermodynamic niche than species at low altitudes.

This separation of niches avoids competition between species and spreads out resource use more evenly over a longer period of time and a wider expanse of space. Because of the variability in energy flow not only through the ecosystem but also with time, there has been an opportunity for many different species to evolve within each trophic level (mega-niche), each developing their own thermodynamic micro-niche.

5.2 Macroecology

Macrosystems science strives to integrate patterns and processes that span regional to continental scales. A major goal of macrosystems research is to advance understanding of interconnected ecological patterns and processes that operate from regional to continental scales, which are difficult to identify through individual studies conducted over short time periods. Macroecology is a field that looks for laws that govern the abundance, distribution and diversity of species. It is a nonexperimental statistical investigation of the relationships between the dynamics and interactions of species populations and the processes of speciation, extinction, and expansion and contraction of ranges on regional to global scales (Brown 1995). A particular avenue of enquiry relevant to thermodynamic theory is how a mega-niche is divided into many micro-niches. High temporal variability in environmental conditions is one mechanism (Patrick et al. 2021). Range size of animals and the topographic relief of their ranges make energy sources differentially available to organisms of different sizes is another. Interactions between apex predators and landscape features such as availability of water that vary seasonally result in different thermodynamic niches for different species (Kittle et al. 2016). Behavior is also important in niche differentiation. Lions and hyenas compete for prey in the Serengeti, but hyenas hunt in packs while lions hunt solo.

5.3 Energetic Niches on the Farm

Temperature response is an important component of the energetic niche of autotrophs. On my farm there are several tree species that leaf out early in the spring, as soon as temperatures remain above freezing for a few weeks. Among them are princess tree, persimmon, wild plum, and red maple. In contrast, oaks, hickories, ash, and beech are photoperiod sensitive, and only respond to temperature when day length is sufficiently long. The difference represents alternative strategies. Speaking teleologically, oaks are conservative, and leaf out only when they are pretty sure that no more cold snaps will occur and kill their leaves. Princess tree (*Paulownia tomentosa*) leafs out during a February warm spell and gambles that there will be no further spring frosts. If they lose, they are unable to muster a full set of leaves until July. If they win, they enjoy a competitive advantage throughout the spring. Temporal differences in energy availability results in increased species diversity, when different species are favored by different windows of thermodynamic opportunity.

Spatial heterogeneities also provide opportunities for development of separate niches. On my farm close to a pond where soil organic matter is thickest and soil remains damp, the trees that have developed a niche are fast-growing red maples and tulip poplars. On drier soils with less organic matter, slower growing white oak, post oak, southern red oak and scarlet oak predominate.

Some grass species in my pastures also have distinct thermodynamic niches determined in part by how fast the soil heats up during spring and summer. When I first got the farm in 1993, annual rye grass was the first to flower in May. Then as the summer arrived there came switch grass, orchard grass, and Indian grass. Later came purple top, and in the fall as day length decreased, broom sedge and little blue stem. Life cycle characteristics (flowering and fruiting) of specialist species are important. A specialist species develops its own temporal thermodynamic niche. In contrast, fescue in my farm seems to be a generalist in that it grows and reproduces by tillers throughout the year. Because it competes with other grasses for light energy, its thermodynamic niche overlaps the niches of the more specialized species.

Old fields in the Georgia Piedmont have recently been overwhelmed by Johnson grass (*Sorghum halepense*), a tropical C4 species whose vigorous growth in a hot climate combined with energy storage in underground rhizomes allow it to outcompete other grass species and suppress perennial herbaceous and woody plants. (University of Georgia 2016). Johnson grass is a tropical species adapted to hot climates and is an invasive pest with characteristics similar to other tropical pest grass species such as cogon grass in Asia and elephant grass in Africa. It photosynthesizes more efficiently than non-tropical C3 species, and consequently has a high-power output that gives it an advantage over C3 species. The appearance of Johnson grass in Georgia simultaneously with the 2 degree increase there in average yearly temperature since 1980 (Climate Reality, 2018) suggests (but does not prove) that the species is responding to an increase in temperature resulting from global warming. This is a niche that previously did not exist in Georgia but was developed by the invader.

Herbivores also have differential thermodynamic niches. In 2011, caterpillars that transform into black-dotted brown moths (*Cissusa spadix*) completely defoliated several post oaks on my farm but did not damage the other species of oaks. The brown moth is a type of herbivore that restricts its foraging to a particular species. It has a narrow energetic niche. Other species are omnivorous. Omnivorous species such as the gypsy moth have a wide energetic niche. They take energy from many different tree species.

References

J.H. Brown, Macroecology. Univ. of Chicago Press. Chicago (1995)

Climate Reality, How does climate change affect Georgia? https://www.climaterealityproject.org/blog/how-does-climate-change-affect-georgia. Accessed 12 Dec 2020 (2018)

J.F.H. Cuthill, N. Guttenberg, G.E. Budd, Impacts of speciation and extinction measured by an evolutionary decay clock. Nature **588**, 634–641 (2020)

G.E. Hutchinson, The Ecological Theater and the Evolutionary Play. Yale Univ. Press, New Haven (1965)

A.M. Kittle, J.K. Bukombe, A.R.E Sinclair, S.A.R. Mduma, J.M. Fryxell, Landscape-level movement patterns by lions in western Serengeti: comparing the influence of inter-specific (2016). https://doi.org/10.1186/s40462-016-0082-9

N. Kunert, L. Schwendenmann, C. Potvin, D. Hölscher, Tree diversity enhances tree transpiration in a Panamanian forest plantation. J. Appl. Ecol. **49**, 135–144 (2012)

N. Kunert, A.M. Cárdenas, Are mixed tropical tree plantations more resistant to drought than monocultures? Forests **6**(6), 2029–2046 (2015)

C.J. Patrick, K.E. McCluney, A. Ruhi, A. Gregory, J. Sabo, J.H. Thorp, Multi-scale biodiversity drives temporal variability in macrosystems. Front Ecol Environ (2021). Accessed 10 Feb 2021. https://doi.org/10.1002/fee.2297

R. Sanford, Root systems of three adjacent, old growth Amazon forests and associated transition zones. J. Trop. For. Sci. **1**(3), 268–279 (1989)

J. Sardans, H. Vallicrosa, Zuccaarini and 9 others, Empirical support for the biogeochemical niche hypothesis in forest trees. Nature Ecol Evol **5**, 184–194 (2021). https://doi.org/10.1038/s41559-020-01348-1

L. Schwendenmann, E. Pendall, R. Sanchez-Bragado, N. Kunert, D. Hölscher, Tree water uptake in a tropical plantation varying in tree diversity: Interspecific differences, seasonal shifts and complementarity. Ecohydrology **8**, 1–12 (2015)

University of Georgia, UGA scientists using $5 million grant to combat invasive weed Johnsongrass. https://newswire.caes.uga.edu/story.html?storyid=5718&story=Combatting-Johnsongrass. Accessed 5 Jan 2021 (2016)

J.D. Wu, S. Han, T. Shi, C. Jin, T. Pei, G. Yu, Energy budget above a temperate mixed forest in northeastern T China. Hydrol. Process. **21**, 2424–2434 (2007). https://doi.org/https://doi.org/10.1002/hyp.6395

Appendix 6: The Keystone Concept

A keystone species is one that has a disproportionately large effect on its natural environment relative to its abundance. Mills et al. (1993) pointed out that the term is impractical because keystone-ness depends on scale. A beetle might be keystone species in the water inside the bract of a heliconia plant, but in the soil, earthworms could be a keystone species. In a lake, a beaver might play the role, while in the Serengeti plains of Africa, lions are a keystone species (Carroll 2016). In a pond, bass may be keystone, but what predators are keystone in the open ocean – sharks? whales? Are predators always keystone species? Mycorrhizae interact with almost all terrestrial plant species. Shouldn't they be keystone species? How about pollinators? Don't hundreds of angiosperms depend on pollinators. Although ecosystems have grown in complexity since autotrophic eukaryotes and bacteria interacted in the primordial soup, even today the only two trophic levels that are essential for ecosystems to function are the producers (autotrophs) and the decomposers. Animals in ecosystems merely aid in the flow of energy and recycling of nutrients between autotrophs and decomposers.

References

S.G. Carroll, Serengeti Rules: The Quest to Discover how Life Works and Why it Matters. Princeton Univ. Press, Princeton (2016)

L.S. Mills, M.E. Soulé, D.F. Doak, The keystone-species concept in ecology and conservation. Bioscience **43**(4), 219–224 (1993)

C. F. Jordan, *Evolution from a Thermodynamic Perspective*,
https://doi.org/10.1007/978-3-030-85186-6

Appendix 7: The Maximum Power Principle

The Maximum Power Principle Biological says that systems operate at the optimum efficiency for maximum power (Odum and Pinkerton 1955). That means that because of friction, systems do not operate at a theoretical maximum power output but at a power output that is the optimum compromise between efficiency and power. The relation between efficiency and power output can be demonstrated by a system consisting of a pulley, rope, and two weights. Figure 7.1 shows a pulley with

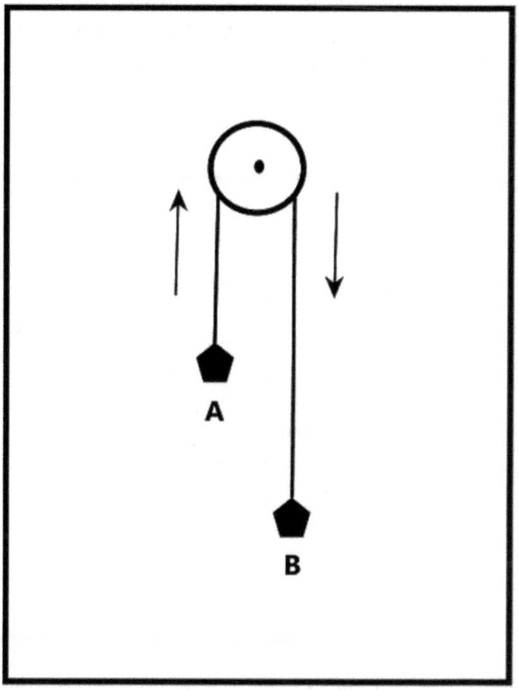

Fig. 7.1 Weights suspended from a pulley. The power input due to the falling weight B drives the rise of weight A, which stores potential energy

C. F. Jordan, *Evolution from a Thermodynamic Perspective*,
https://doi.org/10.1007/978-3-030-85186-6

a rope threaded through it and a weight on each end of the rope. Weight B moving downward is the power going into the system, and weight A being raised is the work being done, or power output. The relative mass of the two weights determines the efficiency that weight A is raised. If B is much heavier than A, say an A/B ratio of 1 kg to 10 kg, the force ratio is low. A will be raised quickly, but little work (power output) will have been accomplished. The falling weight B converts potential energy into kinetic energy, but that kinetic energy is converted into wasted heat as the B strikes the floor. Both power output and efficiency are low. If A is almost as heavy as B, say in a ratio of 9/10, the force ratio is high. A will be raised very slowly (low power output), so efficiency will be low since efficiency is work done per unit time. The best compromise in a system with no friction is where efficiency is 50% (Fig.7.2). There, A would be 5 kg and B would be 10 kg for a force ratio of .5.

Force ratio is a measure of power output. The relationship between efficiency and force ratio is shown in Fig. 7.3. The relationship is the core of the maximum power principle.

Line 1 is the function relating efficiency to power output (force ratio) for a theoretical system with no friction or other constraints. If the force ratio is low (.1), efficiency is low because much energy is lost as heat when **B** strikes the ground. Power output is low because not much work has been accomplished. At a high ratio (.9), both efficiency and power output are low because the reaction is slow. For any system operating at maximum power (force ratio of .5), line 2 indicates its efficiency. In the theoretical system of line 1, both maximum power and maximum efficiency occur at point **A**. (Note that Fig. 7.2 says that efficiency can never be higher than 50% for a real-world system).

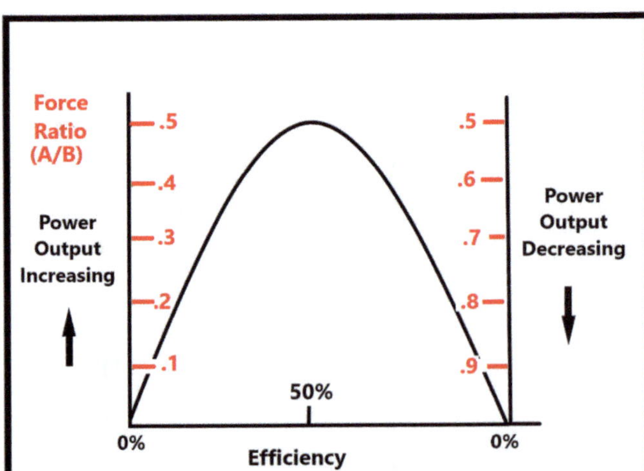

Fig. 7.2 Power output in systems as a function of efficiency. In this theoretically optimum case, the power output of the system is maximum at an efficiency of 50% and the force ratio (power out/power in) ratio is also 50%. At either end of the curve, the efficiency is close to zero

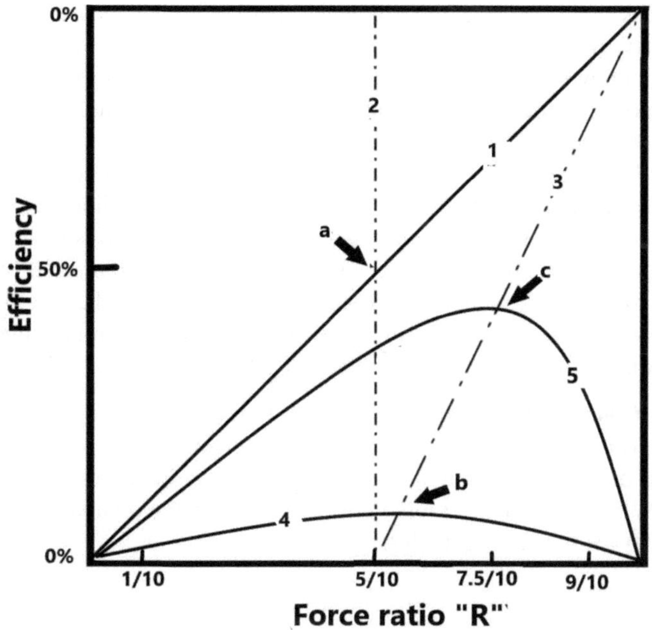

Fig. 7.3 Efficiency as a function of power output for various systems, and the optimum efficiency for maximum power output (where the curves intersect line 3). (Figure modified from Odum and Pinkerton 1955)

In a real world where friction and other constraints are present, line 1 bends at point a to form an arc. As the arc moves toward the right (decreasing efficiency, y axis), power output (x axis) decreases faster than efficiency decreases but only up to a point. Line 3 indicates where that point would occur for any system. Beyond that point, efficiency drops sharply with each decrease in power output, so line 3 represents the optimum efficiency for maximum power output for any system. It shows where the optimum compromise between efficiency and power output would occur. For the system indicated by line 5, point **c** shows that point. Efficiency is about 40%. The force ratio below point **c** is .75, the force ratio (power output) for the optimum compromise between efficiency and power output. If the system were to operate at a force ratio greater than .75, efficiency would drop faster than power output, so c is the optimum compromise.

Line 4 represents an early successional system with high power output and low efficiency. Point **b** is the optimum compromise between efficiency and power output for that system. Power output is high but efficiency is low. Early in succession, systems sacrifice efficiency for high power output. As ecosystems change from early to late succession, point **b** moves upward along line 3 (but not beyond 50% efficiency). Efficiency increases but power output decreases. Line 5 represents a late successional system with lower power output but higher efficiency (point **c**) than the early successional system (point b).

Regardless of the system, the efficiency that maximizes power for that system will always occur along line 3. Odum called that efficiency the optimum efficiency for maximum power output. Since the point represents a compromise between efficiency and power output, a description that clarifies the point might be *The Efficiency for Optimum Compromise Between Energy Efficiency and Power Output.*

Here we have been talking about time efficiency, the efficiency where the speed of the transformation is optimal. In Chap. 20, we are talking about resource conversion efficiency. There we call it energy returned on energy invested (EROI). EROI does not have a time component. A system with a high EROI would have a high resource conversion efficiency, but its time efficiency would be low if it operated slowly. A system with a low EROI could have a low or a high time efficiency, depending on the time interval over which EROI is measured.

Imaginary Case Study Imagine an organic farming system operating at point **c**. The farmer wants to make more money. He adds fertilizer to the system until it operates at point **b**. The power output (yield) of the system increases, but the efficiency decreases. Whether he makes more money depends on the relative cost of the fertilizer and the price for his vegetables.

Reference

H.T. Odum, R.C. Pinkerton, Time's speed regulator: The optimum efficiency for maximum power output in physical and biological systems. Am. Sci. **43**, 331–343 (1955)

Appendix 8: Gaia

The Gaia theory proposes that living organisms interact with their inorganic surroundings on Earth to form a synergistic and self-regulating, complex system that helps to maintain and perpetuate the conditions for life on the planet (Lovelock 1979). Critics have interpreted Gaia as being teleological because it seems to say that evolution deliberately shaped the planet in a way that made it an optimal environment for life as a whole (Kirchner 1989). Evolution they say, is a random process and has no "goal", except for survival of species. The thermodynamic theory of ecosystem evolution challenges that view. Evolution is not a random process but moves toward increased power output due to competition between species for energy. Increased power output of species results in increased energy flow through ecosystems to which species must subsequently adapt. There is reciprocal adjustment of energy flow between species and ecosystems. Between the great extinction events, adjustment has been towards increased energy flow, while during the events, adjustment has been toward a decrease.

Before the great oxidation event some 2.4 billion years, the Earth's atmosphere was weakly reducing, consisting of predominantly N_2 and CO_2 produced by volcanism. Early forms of life utilized an anaerobic metabolic pathway, a very inefficient path of energy conversion that produced methane. Methanogenic organisms produced low power output. Cyanobacteria, which arose during the oxidation event, were able to use energy from the sun and in combination with CO_2, produced oxygen, a much more efficient mode of energy conversion. Because of their greater power output, organisms with an oxidizing pathway were able to outcompete those with an anaerobic pathway. This resulted in a tremendous "jump start" for species that metabolized oxygen. From there on, evolution towards increasing power output was more rapid.

When energy flow was disrupted during the Earth's great extinction events, power output decreased sometimes drastically, but later as the skies cleared and energy availability again increased, simultaneous and reciprocal evolution between species and the Earth's energy environment again propelled species and ecosystems

C. F. Jordan, *Evolution from a Thermodynamic Perspective*,
https://doi.org/10.1007/978-3-030-85186-6

toward maximum power. However there was no deliberate attempt on the part of species to change energy flow to make the environment "optimal" for their existence. It was simply that species interact with the Earth's energy flow to form a synergistic and self-regulating, complex system that helps to maintain and perpetuate the conditions for life in ecosystems. The principles underlying the thermodynamic vision of evolution are the same as those on which the Gaia theory rests. Evolution shaped the planet in a way that seems today to make it an optimal environment. But the environment was always optimal. It's just that optimality changed as evolution progressed. The optimum environment for the first eukaryotes was different from the optimum environment for species today.

References

J.W. Kirchner, The Gaia hypothesis: Can it be tested? RevGeophys **27**(2), 223 (1989)
J. Lovelock, Gaia: A New Look at Life on Earth. Oxford Univ Press, Oxford (1979)

Appendix 9: Thermodynamic Principles in Ecosystem Studies

A principle is something whose prediction has a high probability of being correct but there is not enough evidence to assure the probability is 100%. Thermodynamic theory of ecosystem evolution is based partially on the first and second laws of thermodynamics and partially on thermodynamic principles. Nielsen et al. (2020) in their book, *A New Ecology Systems Perspective* present "a workable ecosystem theory" that encapsulates the thermodynamic view of ecosystems and their evolution. "We, the nine authors, conclude that we do have a workable ecosystem theory that can be presented in different ways but under all circumstances can be used to explain and understand ecological observation, properties, and processes, and even be applied in environmental management. The theory should be considered an early attempt to present working hypotheses about how ecosystems function from a thermodynamic and complex systems perspective" (Nielsen et al. 2020).

Their nine "working hypotheses" that resulted from decades of collaboration are based on thermodynamic principles. Following each "working hypothesis" there is in italics its interpretation as an ecological principle or thermodynamic law.

1. Mass and energy are conserved. *Biomass (potential energy) in one trophic level of an ecosystem becomes kinetic energy in the next.* (First law of thermodynamics).
2. All ecosystem processes are dissipative, and therefore irreversible. *During growth and metabolism of organisms, in an ecosystem energy used to build structure and keep it alive "dissipates" from high quality energy (glucose from photosynthesis) to low quality energy in the form of heat. The process cannot be reversed. Heat released by decomposers when metabolizing a carcass cannot be used to assemble a new organism.* (Second law of thermodynamics).
3. All life uses largely the same biochemical constituents and processes. *All living organisms have approximately the same relative proportion of chemical elements. The more closely this proportion resembles the proportion in the environment, the more readily a species can adapt.* (Principles of stoichiometry).

C. F. Jordan, *Evolution from a Thermodynamic Perspective*,
https://doi.org/10.1007/978-3-030-85186-6

4. An ecosystem uses surplus energy to move further from thermodynamic equilibrium. *In thermodynamic equilibrium, all interactions within a system are random. As an ecosystem evolves and power output increases, interactions become more ordered and less random.* (Law of Maximum Entropy Production).
5. Ecosystems coevolve and adapt to prevailing conditions. *In the thermodynamic sense, prevailing conditions means the global energy gradient (entropic gradient)) that varies according to geographic location and time. Ecosystems and their embedded species adapt reciprocally and simultaneously through feedback to this gradient* (Cybernetic Theory).
6. Ecosystems have diversity of structure and function. *Ecosystems and their embedded species adapt to resource gradients by regulating their energy flow in many diverse ways.* (Thermodynamic niches).
7. Ecosystems have many levels of organization and operate hierarchically. *Control is both top-down and bottom-up. Energy flow through a forest controls energy flow through the soil ecosystem by supplying decomposers with energy to recycle nutrients. Nutrient flow out of soil organic matter controls solar energy capture by the ecosystem.* (Hierarchy theory).
8. Ecosystems work together in networks that improve the resource flow utilization. *Across a landscape, diverse ecosystems are complementary in energy use to increase energy flow through the landscape.* (Network theory). The interactions between ecosystems of the Serengeti landscape result from the migration of wildebeests and other herbivores.
9. Ecosystems have an enormous amount of genetic, biochemical and process information. *As diversity increased during evolution, there was a concomitant increase in genetic, biochemical and process information.* (Information theory).

Reference

N.N. Nielsen, B.D. Fath, J.C. Marques, F. Müller, B. Patten, R.E. Ulanowicz, S.E. Jørgensen, E. Tiezzi, A New Ecology Systems Perspective. Elsevier, Amsterdam (2020)

Appendix 10: Facts in Ecology

Is there such a thing as a fact in ecology? One way to define scientific "facts" is to consider them as hypothesis with a high probability of occurring. A good place to look for facts is in the "hard" sciences. Within science, there is a series of hierarchical levels from "hard" to "soft". Hard science is often considered to be more rigorous, that is, it has better predictive ability than soft science. The sequence from hard to soft would be: Newtonian physics, Chemistry, Biology, Ecology, Economics, Social sciences (sociology, anthropology).

Newtonian physics is the most rigorous of the sciences simply because it has the good fortune to have a very good statistical base. Each physical object used in an experiment consists of billions and billions of experimental objects (atoms), all of which are moving randomly. To test Newton's hypothesis that every physical body continues in its state of rest unless it is compelled to change that state by forces impressed upon it, we lay a pencil on the desk and watch it for a while to see if it moves. If it doesn't, we have not proven the hypothesis, but merely failed to falsify it. Modis (1992) calculated that if someone started watching the pencil at the instant that the universe was created, he would not yet have witnessed a levitation. To have a chance for such a sighting, the observer would have to wait another period of that length, but this time with every second itself stretched to have an age of the universe. Yet the possibility is there, and it is only a question of time.

In the field of chemistry, the statistics, while not as good as in Newtonian physics, still are good enough that many hypotheses are considered facts. Take for example the periodic table of elements, considered by most scientists to be confirmed. In reality, the periodic table is a probability table. In an ordinary mixture of hydrogen gas, 0.0156% of the atoms are deuterium and 10^{-18}% are tritium. As a result, the atomic weight of hydrogen is not really 1.00000, but 1.00784. Deuterium and tritium have different properties than hydrogen, but since there are so few in relation to stable hydrogen, their effect is so small that it is not detectable without the use of highly sensitive analytical instruments. Predictions in biochemistry are also pretty good, because biochemical pathways vary little between organisms of the same

C. F. Jordan, *Evolution from a Thermodynamic Perspective*,
https://doi.org/10.1007/978-3-030-85186-6

species, or between species in the same trophic level. When we consider quantum physics, predictions have a lower probability of being correct. In contrast to Newtonian Physics, quantum physics considers interactions between only two atoms. Predictions have been only as good as predictions of the outcomes of a coin toss (Reynolds et al. 2020).

Biology is taught as though it consists of facts that students must memorize (Bohndiek 2020). Are there really any facts in biology? A few years ago, a student of mine became curious about established "facts" in ecology. We searched the literature but could not find one in ecology (Jordan and Miller 1996). We found one in physics that always has been valid when observations are restricted to time spans greater than one millionth of a second and to locations within our solar system. Under these conditions, there has never been any known observations of an exception to the first law of thermodynamics. Another law of physics is gravity, which is based on the fact that two bodies in the universe interact with each other. Other than those, there are few facts anywhere in science. Even the second "law" of thermodynamics is not a fact but a statistical statement. Boltzmann showed that because certain processes do not seem to occur – for example the spontaneous conversion of heat into mechanical energy – does not mean they are impossible, but extremely unlikely. In microscopic systems consisting of only a few molecules, the second law is violated regularly, but in macroscopic systems which consist of vast numbers of molecules the probability that entropy will increase becomes virtual certainty.

If anything in biology is a fact, "species" should be a fact. Ways of defining species include karyotype, DNA sequence, morphology, behavior or ecological niche. Between closely related species, these characteristics vary along a continuum. A species is not a fact but merely a conglomeration of probabilities. If in my field studies of birds, I capture an individual that I don't recognize, I can test its DNA and compare the result to a known data bank. Depending on the similarity of the best match, I can determine the probability that my catch belongs to a known species. The concept of species is based on probabilities (Nosil et al. 2021). Like "species", "ecosystem" is not a fact but merely a concept. Nevertheless, that does not diminish the usefulness of both of the concepts.

There is a distinction between scientific facts and what is called institutional facts that are simply a convention that enables humans to organize their existence. Institutional facts are used routinely in court. I had beer with Andy Capp at the Rose and Crown the night of a crime is an institutional fact that can be verified by the bartender. His testimony indicates a high probability that I did not commit the crime.

Some observations that are sometimes thought of as scientific facts but are not include: classifications (pine trees are conifers); descriptions (giraffes are animals with long necks); comparisons (rabbits run faster than turtles). Many people believe it is a fact that the sky is blue. Blue is not a *scientific* fact but merely a label. Our sensation upon looking at the sky on a clear day is called "blue", but that is an administrative designation. In Spain that sensation is called "azul". Changing the name does not change the sensation. A scientific fact is that when sunlight with a wave length of between 450–495 nm strikes gases and particles in the Earth's atmosphere, it is scattered more than radiation at other wave lengths. Light between those wave lengths produces the sensation we call "blue".

Reference

C.F. Jordan, C. Miller, Scientific uncertainty as a constraint to environmental problem solving: Large-scale ecosystems. Lemons J (ed) Scientific Uncertainty and Environmental Problem Solving. Blackwell Science. Cambridge Mass (1996). p 91–117

Appendix 11: Site Information and Calculations Used for Data in Table 23.2 of Chap. 23

Chapter 23 showed how the ratio between endosomatic and exosomatic energy input into ecosystems differed between different systems and how that affected sustainability. Here the calculations are given that led to those results.

11.1 Decomposition Constant

A decomposition constant for labile carbon molecules determines the rate at which the energy contained by agricultural stover (residue biomass) becomes available to the community of soil organisms that conserve and recycle nutrients. It is used to calculate the rate of endosomatic input given for the systems in summary Table 3 below. The rate of biomass decomposition depends on the proportion of original organic matter remaining after the cropping period. X_0 is the original amount of biomass and X is the biomass at the end of the cropping cycle. Cropping cycle is the length of the growing season. Percentage used (X/X_0) is the proportion of the original biomass that was consumed by decomposers during a cropping cycle. It represents the energy value of services of nature used for production. Values of λ and t for biomass substrates in various types of agriculture along with the percentage of X_0 used are given in Table 1. Decomposition constant λ and % used were calculated using Table 1 data and the equation

$$N(t) = N_0 e^{-\lambda t}$$

where

N_0 is the initial quantity

N_t is the remaining quantity after time, t

λ is the decomposition constant

C. F. Jordan, *Evolution from a Thermodynamic Perspective*,
https://doi.org/10.1007/978-3-030-85186-6

Table 1 Decomposition constant λ for organic substrates used to calculate the percentage of organic matter substrate remaining after time period t for labile carbon

Substrate	Management	Location	λ	Cropping cycle - time	% used per cycle	Reference
Coarse wood	Shifting cultivation	Tropics	.082	2 years	15	Zell et al. (2009)
Corn stover	No-till	Mid-west	0.10	4 months	34[a]	Al-Kaisi (2009)
Green manure	Organic	Wet North temperate	0.23	16 weeks	97[a]	Shi (2013)
Cow manure	Organic	Wet North temperate	.004	120 days	37	Rashid et al. (2013)
Grass roots	Conventional	Dry North temperate	1.17	.4 years	80	Vivanco (2006)

[a]X/X_0 given by authors

This is an alternative statement of the decay law in Chap. 5. It is solved using www.calculator.net/half-life-calculator.

11.2 Site Descriptions

11.2.1 The Amazon Region of Venezuela (Uhl and Murphy 1981)

This study was conducted near the village of San Carlos de Rio Negro in the Amazon region of Venezuela, close to the borders of Colombia and Brazil. The terrain is rolling hills. The low areas between the hills have soils composed of quartz sand (Spodosols). The higher areas, called "tierra firme", contain Oxisols of highly leached clay. A typical shifting cultivation cycle begins with the cutting of a 0.5–2.0 hectare patch of mature tierra firme forest. Cut vegetation is left to dry for several months before burning. After burning, yuca, the local name for cassava or manioc (*Manihot esculenta*) is planted throughout the clearing, locally called a conuco. Often a conuco has crops such as pineapple, plantain, banana, cashew, sugar cane, yams, and beans planted in combination with yuca, but yuca's tuberous roots generally account for more than 90% of the edible crop production. Conucos are usually farmed actively for two to three years, and then abandoned due to declining yield of yuca. Yield at the study site the first year was 4.31 mg/ha, and 2.81 mg/ha the second year. The decline is caused by a decrease in soluble phosphorus in the soil, caused by its fixation by iron and aluminum oxides in the clay (Jordan 1989).

Table 2 shows the energy balance for the study over two cropping cycles. All inputs are exosomatic energy (labor), since they arise from outside the forest's biological cycle. The output (food yield) is endosomatic energy since it arises from the potential energy resulting from photosynthesis.

Table 2 Energy outputs of Manihot esculenta and energy inputs from shifting cultivation during a two- crop cycle in the Amazon region of Venezuela. (Adapted from Uhl and Murphy 1981)

Inputs	kj/ha
Cutting forest	202,907
Burning	9660
Gathering stem stocks	42,664
Planting	177,066
Cleaning and weeding	425,048
Harvesting	341,322
Peeling	395,133
Grating and soaking[a]	227,547
Squeezing and sifting[a]	66,078
Fire making and baking	182,958
Traveling	467,060
Sum of inputs	**2,537,443**
Output	**35,172,147**
Output/Input	**13.9:1**

[a]Energy to remove the hydrocyanic acid from tubers

Soil at the site was extremely low in nutrients. When the trees are felled, they are allowed to dry, and then the site is burned. The fire consumes only the smaller branches, which is turned to ash that supplies some nutrients and raises the soil pH. Increasing soil pH solubilizes some of the phosphorus that is bound in the clay by iron oxides (Brady 1974). This is a very fleeting phenomenon. The main contribution of the decomposing trunks and larger branches is to provide energy to decomposers that release nutrients bound in wood

11.2.1.1 Calculation of Energy Used to Release Nutrients Bound in Biomass

(Results are summarized in Table 3, column 1).

Biomass of trees including roots: 320 mg/ha (Table 2.2 in Jordan1989).

16.305 kj/g of rain forest trees (Golley1969)

16.305 kj/g × 320 (10^6) g/ha = 5,217.600(10^6) kj/ha (row a)

X = 5217 Gj of energy contained in one hectare of forest trees.

Services of nature (cost of energy used by microbes) is % from Table 1 times 5217 Gj/ha:

.15 × 5217 = 782 Gj/ha (row b)

Exosomatic energy input (row c) is sum of inputs in Table 2.

Total energy input in kj/ha to produce food is energy subsidies in the form of labor from Table 2 plus energy used by microbes for recycling nutrients: 2.53 + 782 = 784.5 Gj/ha (row d). This is the total energy cost used to produce food energy.

Food energy yield from Table 2 = 35.1 Gj/ha (row e)

Output/exosomatic energy invested (EROI) = 35.172/2.53 = 13.9 (row f)

Table 3 Yields and energy ratios for nine agricultural systems used to produce food energy

	1	2	3	4	5	6	7	8	9
Site	Amazon Shifting Cult. (cassava)	New Guinea Shifting Cult. (yams)	Mich. No-till (corn)	Mich. Conven-tional (corn)	Penn. (Rodale) Organic Manure (corn)	Penn. (Rodale) Organic Legume (corn)	Penn. (Rodale) Conven-tional (corn)	Iowa, Conven-tional (corn)	Georgia Conven-tional (corn)
Soil order/ fertility	Oxisols, very low	Entisols moder-ate	Alfisols, medium high	Alfisols, medium high	Alfisols, medium high	Alfisols, medium high	Alfisols, medium high	Mollisols very high	Ultisols low
[a]Energy in biomass, Gj/ha	5217	1826	11.0	0	13.3	13.8	0	0	0
[b] Energy value of SoN* Gj/ha	782	274	3.8	0	4.9	13.4	0	2.5	0
[c]Exoso-matic energy Input, Gj/ha	2.53	5.8	4.9	7.1	16.9	14.8	21.7	13.7§§	13.7
[d]Total input Gj/ ha [b + c]	784.5	279.8	8.7	7.1	21.8	28.2	21.7	16.2	13.7
[e] Food energy Yield Gj/ha	35.1	95.3	78.5	72.7	93.9	93.0	95.7	197.9	176.3
[f] EROI (exo) e/c	13.8	16.4	16.0	10.2	5.6	6.3	4.4	14.5**	12.9
[g] % of input from SoN [b/d]	99.9	98	44	–	22	48	–	15.0	–
[h] EROI (endo) § (e)(g)/b	.02	.34	9.09	–	4.22	3.3	–	11.9	–
[i]EROI total [e/d]	.04	.34	9.0	10.2	4.3	3.3	4.4	12.2	12.9

	1	2	3	4	5	6	7	8	9
[j] endo/exo b/c	309	47	.78	–	.29	.91	–	.18	–

[a]Energy equivalent of total biomass input for each management system

[b]Energetic value of biomass energy used to recycle nutrients: The amount from [a] available to decomposers during a growing season = (a) (% available from Table 1)

[c]Energy input from fossil fuel subsidies such as fertilizers or from labor (exosomatic energy)

[d]Total energy input [b + c] for food production

[e]Food energy yield for each management system

[f]Food energy returned on exosomatic energy invested

[g]% food energy returned attributed to services of nature

[h]Food energy returned on total energy invested

*SoN = Services of Nature

**Includes input from deep storage organic matter, not part of current services of nature

§Calculated by % of yield due to endosomatic energy divided by endosomatic input

§§Assume exogenous input excluding deep storage organic matter was the same as for Georgia

Output/endosomatic input = 35.172 (10^6) kj/ha out/782 (10^6) kj/ha in = .0449 (row h)

Output/ total input = 35.172 (10^6) kj/ha out/784.5 (10^6) kj/ha in = .0448 (row i)

11.2.2 The Central Highlands of New Guinea (Rappaport 1971)

This was a study of swidden agriculture by the Tsembaga tribe in the Bismark Range of New Guinea. Their territory ranges in altitude from about 2200 feet to 7200 feet, but crops are cultivated in cleared areas of secondary forest vegetation at an altitude between 2200 and 5000 feet. In any year, 90 percent of the land lies fallow, slowly returning again to forest. The farmers plant taro, sweet potato, cassava, bananas and many other crops. Feral and domestic pigs are an important source of protein, and sweet potatoes and cassavas are used as pig feed as well as for human consumption. Considerable labor is used to construct fences to keep pigs out of the gardens. As in Amazonian shifting cultivation, trees are cut and burned. Since the layer of fertile soil under tropical forests is remarkably thin, (seldom more than two inches in the Tsembaga territory) and is easily depleted, the nutrients freed from the fallen trees by burning are beneficial if not crucial to the growth of garden plants (Rappaport 1971).

Table 4 shows the energy balance for a yearly cycle of agriculture by the Tsembaga. The energy input from the forest that resulted in nutrient availability for the crops was not accounted for. The nutrient recycling service of nature is fueled by a percentage of the energy in the reduced carbon compounds that comprise the input biomass.

Table 4 Endosomatic energy output and exosomatic inputs from swidden gardens of the Tsembaga during a yearly cycle in the highlands of New Guinea, using labor inputs only. (Adapted from Rappaport 1971)

Inputs	kj/ha
Clearing underbrush	585,477
Clearing trees	234,176
Fencing garden	353,222
Weeding and burning	196,110
Placing soil retainers	149,667
Planting and weeding until harvest	1,864,494
Other maintenance	475,594
Sweet potato harvest	463,549
Taro harvest	57,981
Cassava harvest	22,580
Yam harvest	162,322
Cartage	1,238,240
Sum of inputs	**5,803,415**
Output	**95,306,846**
Output/Input	**16.4:1**

Biomass increment of secondary tropical forest = 450 g/m^2/yr (Jordan 1971)

Land had been fallow for 25 years. 450 × 25 = 11,250 g/m^2 after 25 years, or 112 mg/ha at the time forest was felled for cultivation

Kilojoules per gram of rain forest trees: 16.305 kj/g (Golley1969)

Energy content of secondary forest trees per hectare = 16.305 kj/gram × 112 (10^6) grams/ha = 1826.16 (10^6) kj /ha (row a)

Services of nature (energy available to microbes) is % from Table 1 times energy content of trees: .15 × 1826.16 Gj/ha = 274 Gj/ha (row b)

Exosomatic energy input (row c) is sum of inputs from Table 3.

Total input into gardens = 5.80 Gj/ha from labor +274 Gj/ha from trees = 279.8 Gj/ha (row d)

Rows e-h were calculated as in column 1.

Calculations summarized in Table 3, *column 2*

11.2.3 Experimental Comparison of No-till and Conventional Agriculture in Michigan (Snapp et al. 2015)

The Kellog Biological Station in southern Michigan carried out research between 1989 and 2007 on the grain yield and energy balance of various agricultural methods. The studies cited here are of corn under no-till cultivation and conventional cultivation (columns 3 and 4 in Table 3).

Energy value for corn yield was 78.5 Gj/ha for the no-till system, and 72.7 Gj/ha for the conventional input system (row e). For exosomatic energy inputs, we took the published value of "farming energy inputs", which were 4.9 Gj/ha for the no-till and 7.1 Gj/ha for the conventional input (row c). We assumed that corn stover left on the soil surface represents energy available to decomposers to recycle nutrients. To determine this endosomatic input (services of nature) for no-till, we used the data of Paul et al. (2015) from the same research project. They estimated that rates of soil organic matter accumulation from stover in no-till was about 30 gC/m^2/yr or 600 kg biomass/ha. The energy content of corn stover varies from 16.7 to 20.0 Mj/kg. (Pordesimo et al. 2005). Using an average value of 18.4 Mj/kg, energy input due to corn stover added to soil would be 600 kg/ha × 18.4 Mj/kg, = 11.04 Gj/ha (row a). 34% of that becomes available to decomposers for nutrient recycling during the growing season (Table 1). 11.04 × .34 = 3.8 Gj/ha, the energy value of services of nature (row b). Other rows were calculated as in column 1.

Calculations summarized in Table 3, *columns 3 and 4*

11.2.4 Experimental Comparison of Organic and Conventional Agriculture at Rodale Farms in Pennsylvania (Pimentel et al. 2005)

Since 1981, the Rodale Institute Farming Systems Trial near Kutztown Pennsylvania has compared energetic efficiency in organic and conventional grain-based farming systems. Pimentel et al. (2005) reported the results of a 22-year study of these farming systems. The conventional cropping system, based on synthetic fertilizer and herbicide use, represented a typical row-crop farm. There were two organic systems, one which relied on cattle manure for nutrient input, and the other on nitrogen fixing green manure.

Corn grain yields were 6431 kg/ha for the aged manure, 6368 kg/ha for the organic legume, and 6553 kg/ha for the conventional systems. Estimates of the energy content of corn kernels vary from about 13.9 to 15.3 Mj/kg depending on the variety of corn and the method used for analysis. An average value of 14.6 Mj/kg, corn is used for food yield. Food energy yield = 93.9 Gj/ha for the organic manure system, 93.0 for the organic legume, and 95.7 Gj/ha for the conventional system (row e).

Exosomatic energy input for the conventional system was 21.76 Gj/ha, exosomatic energy input for the manure organic system was 16.9 Gj/ha, and 14.8 for the legume organic (row c).

The net annual above ground carbon input for the cow manure based organic system was 981 kg/ha carbon or 1962 kg/ha biomass. Energy content of cow manure is highly variable but 6.772 Mj/kg is an intermediate value.

Biomass energy input then is 1962 kg/ha × 6.772 Mj/kg = 13.28 Gj/ha. For the legume based system, input was 574 kg/ha carbon or 1148 kg/ha biomass. Energy value of legume is 12 Mj/kg (Ruthergien 2002). Biomass energy input then is 1148 kg/ha × 12 Mj/kg = 13.77 Gj/ha. (row a).

Using the column "percent used" from Table 1, total biomass energy used from cow manure was 13.28 × .37 = 4.91 Gj/ha. For the legume it was 13.77 × .97 = 13.36 Gj/ha (row b).

Calculations summarized in Table 3, *columns 5, 6 and 7*

11.2.5 Conventional Agriculture in Iowa (Cox et al. 2011)

Iowa has some of the most productive agricultural soils of the country. Yield for Iowa corn calculated from data of USDA (2016) averaged 215 bu./acre of corn, which translates to 197.9 Gj/ha/yr. of food energy, a value higher than for any of the other sites in Table 3 (row e). Fertilizer inputs are important for high yield. Their energy equivalent, calculated from data of Iowa State University (2013) and Sawyer (2015) and using energy values of fertilizer and lime (Pimentel and Pimentel 2008, Table 10.7), was 13.2 Gj/ha/yr. (row c). However, soil organic matter also is an important contributor to high yield. Many of Iowa's soils were formed under tall prairie grasses with a deep, fibrous root system that resulted in fertile soils with high organic matter content. This organic matter supplies a major proportion of the nutrients that provide high corn yields.

This stored soil organic matter is being continually depleted. Erosion is caused by mechanical tillage, which loosens the soil to where it is easily washed from the farmer's fields into nearby creeks and rivers. Soil erosion not only washes away soil, but eroded soil carries with it a cargo of fertilizers, pesticides and manure that flows into local creeks and streams and from Iowa into the Mississippi River. Ultimately it ends up in the Gulf of Mexico, generating a zone of depleted oxygen that suffocates marine life when it forms each year (Environmental Protection Agency 2017). Pollution has a huge energetic cost on the economy of fishing, shrimping, and sport fishing in the region (Jordan 2016). Pollution from agricultural chemicals is a cost paid for at the societal level. While sheet and rill erosion are primarily responsible for soil loss in Iowa, gully erosion can cause losses of up to 64 tons per acre.

Nutrient supply from decomposing organic matter comes from both deep stored organic matter and organic matter resulting from decomposition of recently produced organic matter that. Only the latter represents current services of nature. We take services of nature to represent energy currently moving through trophic levels. It was calculated as follows:

A U.S.D.A. definition of the "sustainable" rate of soil loss was 5 U.S. tons/acre (11.2 Mt/ha) according to the Environmental Working Group (Cox et al.2011). Since about 7% of the total weight of fertile top soils is organic matter (Overstreet et al. 2017), a loss of 11.2 Mt/ha amounts to .784 Mt/ha of organic matter per year. We take this to represent the rate of organic matter supplied by root decomposition of current year's crops, since that is what it would take to make the system "sustainable". Root production of fertilized corn in Iowa is .93 Mt/ha (Liebman et al. 2013). So only .930 – .784 = .146 tons of organic matter per hectare are left after erosion. At 17 kj/kg, the energy that remained and was taken up by corn equals 2.48 Mj. We take this to represent the current services of nature.

Calculations summarized in Table 3, *column 8*

11.2.6 Conventional Agriculture in Georgia (Lee 2019)

Native vegetation in the Southeastern U.S. consists of forests. Native forests survived on nutrients recycled from 10 or more inches of topsoil, but when this topsoil was eroded away following a century of cotton farming, all that was left was soil almost completely devoid of soil organic matter (Jordan 2013; Dobbs 2017). Corn is the most widely grown crop in Georgia. Yields were low until 1948 when hybrids better adapted to growing conditions in the southern U.S. were developed. Most of the corn production in Georgia occurs in the Upper Coastal Plain, and is primarily used for animal feed and not human consumption. According to the University of Georgia Experiment Station in Tifton Georgia, a typical yield of corn on these soils is 180 bushels/acre, or 12,105 kg/ha (Lee 2019). With an energy value of 14.6 Mj/kg corn, yield is 176.3 Gj/ha (row e). Using the data from Lee (2019) for the weight of nutrients removed by a typical corn crop, and energy values of fertilizers (Pimentel and Pimentel 2008), exosomatic energy inputs were calculated to be 13.7 Gj/ha (row c).

Calculations summarized in Table 3, *column 9*

11.3 Comparisons by Row

(a) Energy in the biomass that was a source of energy for the services of nature. The biomass of the forests cut for shifting cultivation was a huge source, but only a small proportion was used for the services of nature.
(b) Value of Services of Nature for producing yield. This is the energy content of the biomass consumed by decomposers. It is the energy cost used by the decomposers to recycle nutrients. The costs for shifting cultivation is very high compared to other systems because the biomass is not used efficiently to produce agricultural yield.
(c) Exosomatic energy inputs: For shifting cultivation, this is mainly labor. For conventional corn, it includes costs of fertilizers and pesticides. For the Rodale organic corn, the high costs include tractor fuel for transferring manure to the crops, and incorporating legumes into the soil. For the Iowa site, it includes the contribution of deep stored organic matter.
(d) Total endosomatic plus exosomatic energy inputs into the systems: Total energy costs for the organic systems is similar to or greater than that for industrial corn.
(e) Food energy yield is highly dependent on soil fertility. Yield from the shifting cultivation site in New Guinea site was higher than other systems, with the exception of the heavily fertilized sites.
(f) Energy returned on exosomatic energy invested is high for shifting cultivation. That energy though is in the form of labor in the hot sun. The EROI for no-till nearly matches that for Iowa corn. Both have a store of recalcitrant carbon built up over time that provides a superior environment for root growth.
(g) Organically based systems do not always have a high proportion of endosomatic energy.
(h) Yield returned on endosomatic energy invested shows that shifting cultivation is a very inefficient use of the services of nature. Legume-based organic corn stands out as low because leguminous biomass decomposes quickly. Rate of return for the Iowa site is high because of the large contribution of energy and nutrients stored in non-renewable deep layers of organic matter.
(i) Yield as a function of total energy input: The conventional systems (except for Pennsylvania) have the highest values because pollution is not counted as a cost. If pollution costs were counted, rate of return could be lower than that of the no-till system. The Rodale systems have low returns because of high input costs.
(j) Except for the shifting cultivation sites where endosomatic energy is wasted, the endosomatic/exosomatic ratio is an indicator of sustainability. Using that indicator, the no-till and the organic legume system are the most sustainable.

11.4 Summary

Agricultural systems are powered by both endo-and exosomatic sources of energy. The proportion of each type is way to evaluate the impact of a particular agricultural management system on the environment and the sustainability of the system.

References

M. Al-Kaisi, Myths and facts about residue breakdown. Iowa State University extension and outreach. https://crops.extension.iastate.edu/cropnews/2014/04/myths-and-facts-about-residue-breakdown. Accessed 8 Dec 2020 (2014)

N.C. Brady, The Nature and Properties of Soils. Macmillan, New York (1974)

C. Cox, A. Hug, N. Bruzelius, Environmental Working Group. Losing Ground. https://www.ewg.org/losingground/report/executive-summary.html. Accessed 14 Dec 2020 (2011)

C. Dobbs, Georgia encyclopedia, Geography and Environment, Soils. http://www.georgiaencyclopedia.org/articles/geography-environment/soils#Piedmont. Accessed 17 Dec 2020 (2017)

Environmental Protection Agency, Northern Gulf of Mexico hypoxic zone. https://www.epa.gov/ms-htf/northern-gulf-mexico-hypoxic-zone. Accessed 17 Dec 2020 (2017)

F.B. Golley, Caloric value of wet tropical forest vegetation. Ecology **50**, 517–19 (1969)

Iowa State University, A general guide for crop nutrient and limestone recommendations in Iowa. Iowa State University, Extension and Outreach.PM 1688, revised Oct. 2013. Accessed 21 Dec 2020 (2013)

C.F. Jordan, Productivity of a tropical forest and its relation to a world pattern of energy storage. J. Ecol. **59**, 127–142 (1971)

C.F. Jordan, An Amazonian rain Forest. The structure and function of a stressed ecosystem and the impact of slash and burn agriculture. Man and the biosphere series, volume 2. UNESCO, Paris and Parthenon, Carnforth, England (1989)

C.F. Jordan, An Ecosystem Approach to Sustainable Agriculture: Energy Use Efficiency in the American South. Springer Verlag, Heidelberg (2013)

C.F. Jordan, The farm as a thermodynamic system: Implications of the maximum power principle. Biophys Econ Res Qual **1**, 9 (2016). https://doi.org/https://doi.org/10.1007/s41247-016-0010-z

D. Lee, A Guide to Corn Production in Georgia. The University of Georgia College of Agriculture and Environmental Sciences (2019), p. 101. http://caes2.caes.uga.edu/commodities/fieldcrops/gagrains/documents/2017_CornProduction_Final.pdf. ACcessed 26 Dec 2020

M.Z. Liebman, M.E. Jarchow, R.N. Dietzal, D.N. Sundberg, Above- and below-ground biomass production in corn and prairie bioenergy cropping systems. Iowa state research farm Progress reports. Paper 2078. http://lib.dr.iastate.edu/farms_reports/2078. Accessed 26 Dec 2020 (2013)

L.F. Overstreet, J. De-Jong-Hughes, The importance of soil organic matter in cropping systems of the northern Great Plains. http://www.extension.umn.edu/agriculture/soils/soil-properties/importance-of-soil-organic-matter/. Accessed 30 Dec 2020 (2017)

E.A. Paul, A. Kravchenko, S. Grandy, S. Morris, Soil organic matter dynamics: Controls and management for sustainable ecosystem functioning. S.K. Hamilton, J.E. Doll, G.P. Robertson (eds) The Ecology of Agricultural Landscapes: Long Term Research on the Path to Sustainability. Oxford Univ. Press, New York (2015). p 104-134

D. Pimentel, P. Hepperly, J. Hanson, D. Douds, T. Seidel, Environmental, energetic, and economic comparisons of organic and conventional farming systems. Bioscience **55**(7), 573–582 (2005)

D. Pimentel, M.H. Pimentel, Food, Energy and Society, 3rd edn. CRC Press: Boca Raton, Florida (2008)

L.O. Pordesimo, B.R. Hames, S. Sokhansang, W.C. Edens, Variation in corn stover composition and energy content with crop maturity. Biomass Bioenergy **28**, 366–374 (2005). https://doi.org/10.1016/j.biombioe.2004.09.003

R.A. Rappaport, The flow of energy in an agricultural society. Sci Am, 1117–1132 (1971)

M.I. Rashid, R.G.M. deGoede, L. Brussard, E.A. Lantinga, Home field advantage of cattle manure deposition affects the apparent nitrogen recovery in production grasslands. Soil Biol. Biochem. **57**, 320–326 (2013)

D.C. Ruthergien, Feed value of selected foodstuffs. http://agriculture.vic.gov.au/agriculture/livestock/beef/feeding-and-nutrition/feed-value-of-selected-foodstuffs. Accessed 2 Jan 2021 (2002)

J.E. Sawyer, Nitrogen use in Iowa corn production. Iowa State University digital repository. https://lib.dr.iastate.edu/extension_pubs?utm_source=lib.dr.iastate.edu%2Fextension_pubs%2F107&utm_medium=PDF&utm_campaign=PDFCoverPages. Accessed 21 Dec 2020 (2015)

J. Shi, Decomposition and nutrient release of different cover crops in organic farm systems. Dissertation, Univ. of Nebraska, Lincoln http://digitalcommon.unl.edu/natresdiss/75. Accessed 4 Jan 2021 (2013)

S.S. Snapp, R.G. Smith, G.P. Robertson, Designing cropping systems for ecosystem services. S.K. Hamilton, J.E. Doll, G.P. Robertson (eds) The Ecology of Agricultural Landscapes: Long Term Research on the Path to Sustainability. Oxford Univ. Press, New York (2015). p. 378-408

C. Uhl, P. Murphy, A comparison of productivities and energy values between slash and burn agriculture and secondary succession in the upper Rio Negro region of the Amazon Basin. Agro-Ecosystems **7**, 63–83 (1981)

USDA, Iowa Ag News. 2016 Corn and Soybean County Estimates. https://www.bing.com/search?q=Iowa%20Ag%20News.%202016%20Corn%20County%20Estimates&qs=ds&form=QBRE. Accessed 5 Jan 2021 (2016)

L. Vivanco, A.T. Austin, Intrinsic Effects of Species on Leaf Litter and Root Decomposition: A Comparison of Temperate Grasses from North and South America Oecologia **150**(1), 97–107 (2006)

J. Zell, G. Kandler, M. Hanewinkel, Predicting constant decay rates of coarse woody debris—A meta-analysis approach with a mixed model. Ecol. Model. **220**(7), 904–912 (2009)